D0845588

STRANGERS IN A NEW LAND

WHAT ARCHAEOLOGY REVEALS ABOUT THE FIRST AMERICANS

STRANGERS IN A NEW LAND

WHAT ARCHAEOLOGY REVEALS ABOUT THE FIRST AMERICANS

A PETER N. NÉVRAUMONT BOOK

FIREFLY BOOKS

J. M. ADOVASIO

DAVID PEDLER

Published by Firefly Books Ltd. 2016

First printing

Publisher Cataloging-in-Publication Data (U.S.)
Names: Adovasio, J. M., author. | Pedler, David, author.
Title: Strangers in a new land : what archaeology reveals about the first Americans / J. M. Adovasio & David Pedler.
Description: Richmond Hill, Ontario, Canada : Firefly Books, 2016. | Includes bibliography and index. | Summary: "The history of North America, from Christopher Columbus to the present day, is a chaotic struggle for ownership of the land. This text will address questions such as the history of native communities that were displaced or conquered, and even further back to how Native North, Central and South Americans came to these continents in the first place. Historical facts are mainly supported through archaeological findings" — Provided by publisher.
Identifiers: ISBN 978-1-77085-363-8 (hardcover)
Subjects: LCSH: Indians — Antiquities. | Paleo-Indians – America — Origin. | Indians – Origin. | Indians – History.
Classification: LCC E61.A385 | DDC 970.00497 – dc23

Library and Archives Canada Cataloguing in Publication
Adovasio, J. M., author
Strangers in a new land : the first Americans / J.M. Adovasio, David Pedler.
Includes bibliographical references and index.
ISBN 978-1-77085-363-8 (bound)
1. Paleo-Indians. 2. Excavations (Archaeology)--America. 3. America—Antiquities. I. Pedler, D. R., author II. Title.
E61.A33 2016 970.01 C2015-903444-2

Published in the United States by
Firefly Books (U.S.) Inc.
P.O. Box 1338 Ellicott Station
Buffalo, New York 14205

Published in Canada by
Firefly Books Ltd.
50 Staples Avenue, Unit 1
Richmond Hill, Ontario L4B 0A7

Printed in China

Front cover image: Anatomical Origins
Back cover image: Top: courtesy of J.M. Adovasio;
bottom: photograph by Don Giles

Produced by Névraumont Publishing Company
Brooklyn, New York

Book cover and interior design by:
Cathleen Elliott, flyleaf design

PART ONE

QUESTIONS AND ANSWERS

PART TWO

THE EVIDENCE

PART ONE

QUESTIONS AND ANSWERS

1. Landfall at Guanahani

At about two o'clock in the morning of 12 October 1492, one Rodrigo de Triana, serving as lookout on the ship *La Pinta*, sighted an island landfall called *Guanahani* in the local aboriginal dialect. Later that morning, the soon to be styled Admiral of the Ocean Sea Christopher Columbus waded ashore on what was most likely San Salvador Island in the Bahamas, where he encountered *Guanahani's* soon to be extirpated aboriginal inhabitants. These Arawak-speaking people, called Taíno, were assumed by Columbus and his entourage to be Asians, perhaps from the semi-mythical Spice Islands in the East Indies. In any case, he called them *Indios* (Indians), and that name would remain with them to the present.

Columbus Discovers the Americas for Spain by John Vanderlyn.

The actual identity of the *Indios* would become even more of a mystery on 25 September 1513, when Vasco Núñez de Balboa crossed the Isthmus of Panama and became the first European to set eyes on the Pacific Ocean from the New World. Though it probably did not immediately occur to Balboa, this fresh perspective would have profound consequences for future European explorers and adventurers. It conclusively showed that the land mass and off-shore islands "discovered" by Columbus were instead new continents, immense new lands bounded by vast oceans on both coasts.

Among the first implications of this geographic epiphany was the recognition that the occupants of this *terra nova* were not Asians, or at least not Asians as was then understood. This revelation—the profoundness of which might be lost on modern readers—would ultimately lead to a series of questions which, in one form or another, have been asked about Native Americans ever since. From where did they come? How did they get here? When did they arrive? What were they doing? Despite half a millennium of mostly fanciful speculation, intermittent scholarship, and more-recent intensive research, the answers to these seemingly simple questions have remained tantalizingly elusive.

Once the full impact of Balboa's discovery sank into the educated heads of Europe, the question of who are the aboriginal inhabitants of the New World gained somewhat greater urgency. Columbus not only assumed that he had encountered Asians, he also most likely presumed that they had always been there. Based on the literal interpretation of the Book of Genesis that would have prevailed at the time, Earth and everything on it was believed to be little more than 6,000 years old. Hence, the "Asians" encountered by Columbus would have been in place at least since their arrival from the Garden of Eden.

Parenthetically, this shallow time perspective would persist until relatively recently, and in fact still does in certain circles. Bishop John Lightfoot formally established Earth's creation at 3929 BC in a series of works published between 1642 and 1644, and a bit later the Archbishop of Armagh, James Ussher, would refine the chronology and place that event before dawn on Sunday 23 October 4004 BC, following the Julian calendar. As will be discussed later, until such quaint notions were dispelled in the nineteenth century, cogent answers to the questions posed above could not be formulated.

Columbus and other chroniclers of his and somewhat later times also assumed that the *Indios* were a discrete people who varied little in physique, culture, or language. Disabused of the Asian origin assumed for the *Indios* before Balboa's discovery and soon confronted by the great diversity of the physical types, cultures, and languages among the New World's peoples, the newcomers were compelled to reformulate their answers to the "from where did they come?" question. European scholars responded with a series of fanciful fables. For some, the *Indios* were the long-lost tribes of Israel. For others, they were thought to be the descendants of any ancient society known to have built boats, such as the Greeks, Romans, Phoenicians, Egyptians, or Norsemen. For yet others, they were the fortunate survivors of the catastrophic inundation of Atlantis.

A singular early exception to this fevered speculation was the work of a Spanish Jesuit priest, José de Acosta, who traveled extensively in Central and South America and was among the first Europeans to repeatedly cross the American Cordillera. (Indeed, in doing so, he suffered from and described the consequences of altitude sickness, one variety of which is actually named after him.) In 1590, de Acosta published a genuinely remarkable document entitled *Historia Natural y Moral de las Indias* (*The Natural and Moral History of the Indies*) that was soon translated into several languages. Much of de Acosta's enduring fame derives from this seminal work, which provides vivid descriptions of the New World's peoples and places while also offering some truly insightful observations about Native American origins. He correctly intuited that Native Americans are of Asian descent and, even more astutely pinpointed northeastern Asia as their ancestral homeland, postulating that their forebears arrived on foot via some sort of dry land route which he believed connected northeastern Asia to the contiguous reaches of North America. Finally, he even attempted to answer the "when did they arrive?" question by speculating that the initial diaspora occurred some 2,000 years prior to Columbus's arrival in the Caribbean.

Perhaps the most salient point that modern readers can take away from de Acosta's *Historia*, a massive work that runs well over 500 pages in the most-recent modern English translation, is that his observations were made at a time when the geography of northeastern Asia was very poorly known. It would be almost 140 years before Vitus Bering would first sail through the strait that now bears his name. Moreover, it would be centuries before the episodic existence of the Bering Land Bridge connecting Siberia and Alaska would be known, let alone the attendant fluctuations in sea level during recurrent glacial cycles.

"They should be good servants and very intelligent, for I have observed that they soon repeat anything that is said to them, and I believe that they would easily be made Christian, for they appeared to me to have no religion, God willing, when I make my departure I will bring half a dozen of them back to their Majesties [King Ferdinand and Queen Isabella], so that they can learn to speak."

—Christopher Columbus, from his Log Book upon meeting the inhabitants of Guanahani during his first journey to New World

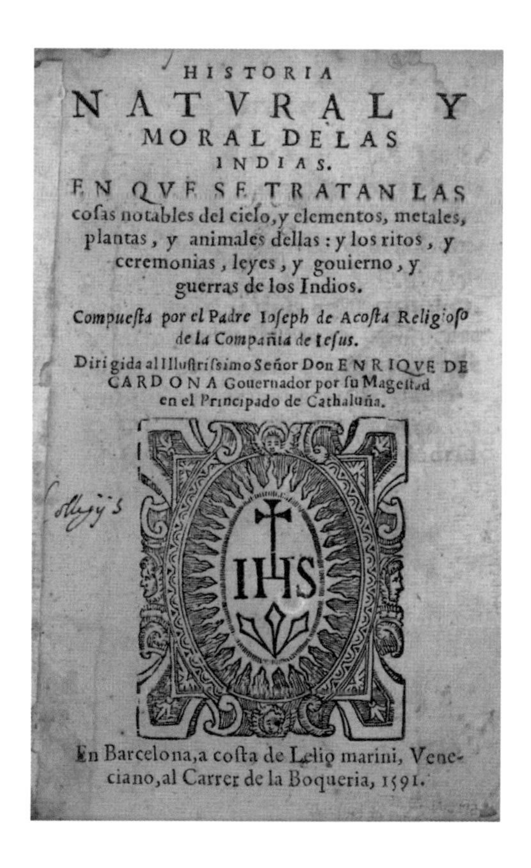
HISTORIA
NATVRAL Y
MORAL DE LAS
INDIAS.
EN QVE SE TRATAN LAS
cosas notables del cielo, y elementos, metales, plantas, y animales dellas: y los ritos, y ceremonias, leyes, y gouierno, y guerras de los Indios.
Compuesta por el Padre Ioseph de Acosta Religioso de la Compañia de Iesus.
Dirigida al Illustrissimo Señor Don ENRIQVE DE CARDONA Gouernador por su Magestad en el Principado de Cathaluña.
IHS
En Barcelona, a costa de Lelio marini, Veneciano, al Carrer de la Boqueria, 1591.

"Notwithstanding all that hath been said, it more likely that the first inhabitants of the Indies came by land."

—José de Acosta, Chapter XX, Book I, *The Natural and Moral History of the Indies*

Unfortunately, few read de Acosta's work when it was published or during the following three centuries. This was regrettable in a number of ways, not the least being the fact that he was among the first Europeans to champion the humanity, rights, and civil liberties of the New World's natives. More than three centuries after the appearance of de Acosta's *Historia*, and despite the fact that at least one other writer (Thomas Gage [1597–1656], an English clergyman) noted the similarity of Native Americans to Asians and identified their homeland as Asia, most commentators opted for more exotic origins for the *Indios*.

Toward the latter part of this long time span, a greater awareness and appreciation of the Asiatic origin of Native Americans might have been achieved were it not for the

Josiah Priest in his 1833 bestseller *American Antiquates and Discoveries in the West* concluded the mound at Marietta, Ohio, was part of a fort built by Roman soldiers who had made their way to the New World along with Polynesians, Egyptians, Greeks, Romans, Israelites, Scandinavians, Welsh, Scotts, and Chinese. From Ephrain Squter and Edwin Davis *Ancient Monuments of the Mississippi Valley*.

confounding effects of the Mound Builder controversy. Many North American authorities with an unabashedly racist perspective believed the great earthworks of the Mississippi and Ohio Valleys could not have been the work of uncouth "savages," and instead presumed that they must have been constructed by more-sophisticated builders whose origins lay in the circum-Mediterranean or Europe. Interestingly, one of the most eminent scholars researching the Mound Builder issue, William Bartram, ultimately recanted his own position that Danes had built the mounds and concluded that their architects were ultimately of Asian provenance. Bartram's position was echoed by United States President Thomas Jefferson, whose own research into the earthworks on his Virginia property (probably conducted in 1783, and recognized as the first stratigraphic archaeological excavation ever conducted in North America) also led him to conclude that their builders were indeed Indians and—like de Acosta and Gage long before him—that their ultimate origin was Asia.

The questions "from where did they come?" and "how did they get here?" would remain inconclusively resolved for some time thereafter. Indeed, after decades of debate these interrelated questions have recently been resurrected by Dennis Stanford of the Smithsonian Institution and Bruce Bradley of the University of Exeter. These two Paleoindian specialists have long argued for a European, not Asian, identity for at least some Native Americans. Whatever the validity of the Stanford and Bradley "Iberia not Siberia" scenario, by the late 1980s a vast majority of scholars concurred that the progenitors of most Native Americans were of Asian extraction. The present virtual consensus on the "from where did they come?" question has given rise in the more recent past to questions about precisely where the natal cradle of Native Americans lay in Asia.

2. From Where Did They Come?

While there is nearly universal consensus that the Arawak-speaking *Indios* encountered by Columbus—and indeed all other *Indios*—are of Asian origin, considerable disagreement remains concerning exactly where in Asia the contributing population(s) originated. A variety of data sets have been used to resolve this matter, including those pertaining to anthropometry (*i.e.*, the scientific study of the measurements and proportions of the human body[1]), dentition, blood groups and genetic marker frequencies, linguistics, molecular biology, and of course, archaeology. Though various combinations of these data sets and statistical analysis of them occasionally appear to be mutually supportive, the "goodness of fit" (*i.e.*, how well a statistical model "fits" a given set of observations) is more often illusory.

Anthropometry—Measurements and Proportions of the Human Body

As has been touched upon before, several early commentators who noted the resemblance of Native Americans to Asiatic peoples also naively and erroneously stressed an apparent lack of variation in observable characteristics (*e.g.*, outward appearance, development, behavior, *etc.*) between their numerous individual groups, which numbered at least 350 at the time of European contact. For example, Aleš Hrdlička, founder and first curator of the Department of Physical Anthropology at the Smithsonian Institution, was a central figure in the early scientific study of the initial peopling of the New World who amassed a very large collection of very carefully measured Native American skeletal remains. From this body of anthropometric data, Hrdlička attempted to isolate a constellation of skeletal features common to all Native Americans. In so doing, however, he was actually promoting his own personal views of the shallowness of the time depth for the arrival of the *Indios*.

Hrdlička reasoned that if humans had entered the New World only recently, specifically in post-glacial times (about 10,000 yr BP), extensive skeletal variation would not exist between Native American populations. By advocating an essential "sameness" in the form and structure of the human skeleton, Hrdlička simultaneously promoted his short chronology of New World human antiquity. Significantly, his research also confirmed that the skeletal characteristics of Native Americans closely corresponded to those of northeastern Asians, thereby supporting a northeastern Asian origin for the New World's first colonists.

Though its boundaries are well defined, northeastern Asia encompasses a vast area which in the recent past has been populated by groups with quite different anthropometric attributes (the size and proportions of the human body).[2] This variability is also geographically conditioned, with significant differences evident between the coastal east and interior west. Because of this variability, most scholars agree that a specific, circumscribed homeland for Native Americans cannot be established exclusively on anthropometric criteria.

Pinpointing a specific sub-region in northeastern Asia as the point of departure for the initial emigration is further compounded by the existence of only a very diminutive subset of aboriginal skeletal remains older than 7000 yr BP—at best, composed of the remains of fewer than 350 individuals. Of this subset, fewer than 5 percent are almost certainly older than 10,000 yr BP. Though such

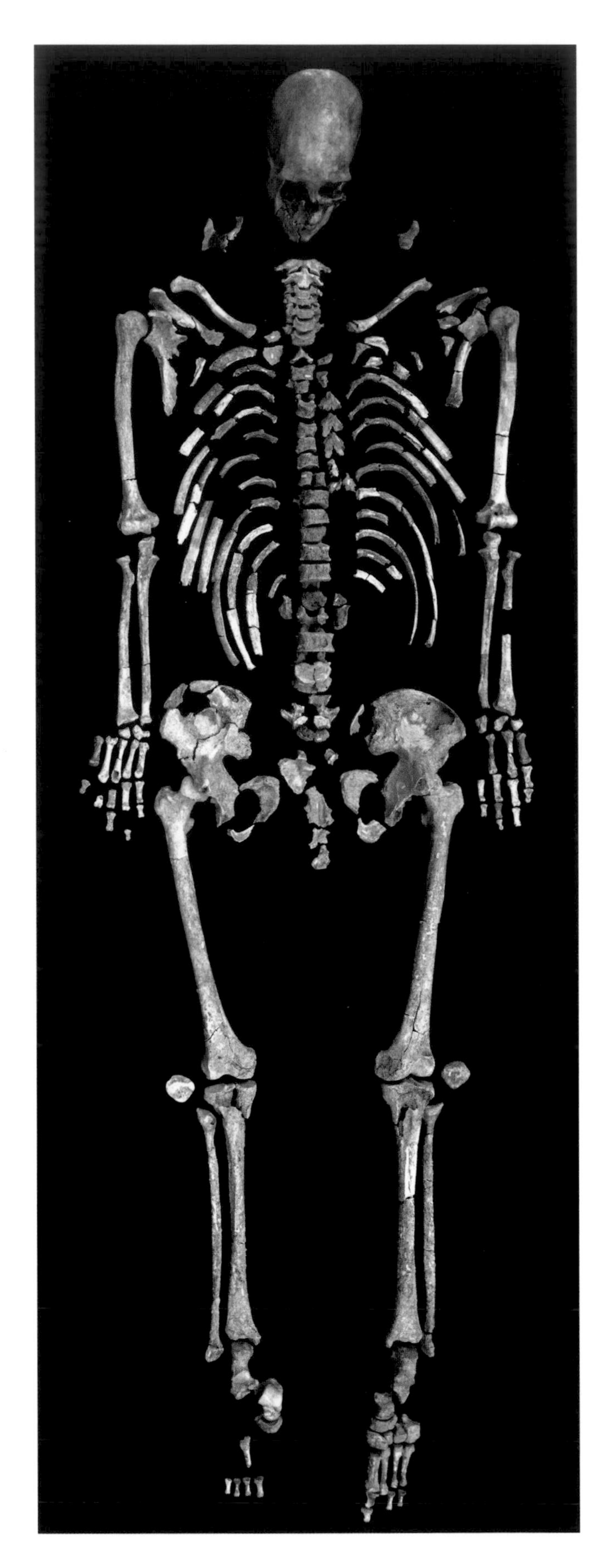

Kennewick Man skeleton. In the summer of 1996, two college students in Kennewick, Washington, stumbled on what turned out to be 9,400-year-old human bones eroding out of the banks of the Columbia River. The skeleton was brought to the attention of Pacific Northwest forensic anthropologist and archaeologist James Chatters. Based on the anatomy of the skull, Chatters and others saw affinities to European populations. In 2015, a team led by Natural History Museum of Denmark scientists was able to extract and sequence DNA from the skeleton. They found that "Kennewick Man is closer to modern Native Americans than to any other population worldwide." Courtesy of Douglas Owsley, National Museum of Natural History.

a small sample cannot possibly be statistically representative of the skeletal diversity of the earliest Americans, it has nonetheless been the focus of much recent research. Several scholars (*e.g.*, Max Blum, James Chatters, Glenn Doran, Walter Neves, Joseph Powell, and André Prous, among others[3]) have vigorously advocated that the earliest Native American skeletons are different from, and therefore distinct from, more-recent Native American populations. This rather strict interpretation, however, has recently softened.[4]

To this day, probably the most extensive study of Native American skeletal populations was published by the late physical anthropologist George Neumann in 1952.[5] Working with materials of mixed age, Neumann identified eight clusters or groups which spanned the entire post-glacial period. The earliest of these groups, called Otamid, was said to be osteometrically (*i.e.*, from osteometry, the study and measurement of human skeletons) similar to the *Homo sapiens* remains from the Upper Cave at Zhoukoudian, the renowned Chinese Middle and Upper Paleolithic archaeological site located near Beijing. Whatever the reality of the alleged north Chinese connection or, indeed, of any of Neumann's groups, his study does illustrate that much greater diversity existed among Native Americans than was acknowledged by Hrdlička. It also demonstrated that the "from where did they come?" question cannot be answered on anthropometric terms alone—even with larger sample sizes.

Dentition

In order to circumvent some of the limitations of anthropometric analyses that focus on the human skeleton, some scholars have elected to work with teeth instead of bones. There are several compelling reasons for doing this. First, the tyranny of preservation tends to favor the survival of teeth over bone, so there are many more of them available for study. Second, though dental attributes are inherited, teeth are generally much less subject to alteration or transformation by environmental influences, differential dietary preferences, use parameters, or vicissitudes in the health of the individuals from whom the teeth derive.

The analysis of teeth also provides another critical advantage over that of bone. Though all humans have the same basic 32-tooth pattern, considerable diversity is evidenced in so-called secondary dental attributes such as the presence or absence of an extra cusp on the upper first molar (known as Carabelli's Cusp) or the expression of lateral winging or "shoveling" of the upper incisors, to name but two such features. Because these secondary dental attributes are unevenly expressed around the world and their rate of expression changes through time, studying such features theoretically affords the possibility of establishing population relationships in space and through time.

Though the study of dentition to address the "from where did they come?" question has a pedigree in American archaeology which extends at least to Hrdlička, without question the most extensive study of this kind is the often-cited work of the late Arizona State University anthropologist Christy Turner, who analyzed dental features in a 200,000-tooth sample drawn from a population of over 9,000 native peoples of various ages from both North and South America.[6] By comparing New World teeth to each other, as well as to extensive tooth samples from the Old World (specifically including Europe, Africa, and southern and northeastern Asia), Turner demonstrated that it was possible to segregate the various tooth populations from each other and that two distinct subsets existed within the Asian sample. The earliest of these Asian subsets, called Sundadont, derives from southeastern Asia and the later subset, called Sinodont, derives from northern Asia and is thought to have evolved from the earlier Sundadont group. In Turner's scheme, Sinodont allegedly characterizes virtually all northern Asians, and literally all of the native peoples of the New World.

Within the New World Sinodont sample, Turner distinguished three sub-groups or varieties that were different not only from each other, but also from their supposed northern Asian progenitors. Originally broadly and somewhat vaguely defined, Turner later reflagged these three sub-groups with names corresponding to those employed in Joseph Greenberg's monumental tripartite New World linguistic classificatory schema: Eskimo-Aleut, Na-Dene, and Amerind,[7] described below. Turner also went so far as to propose that these groups emerged from their northern China ancestors around 20,000 yr BP, with each one following a different route into the New World. While the details of this speculative scenario make for fascinating reading, the specific homelands of Native Americans cannot presently be defined with greater precision using dental attributes than they can with anthropometric data.

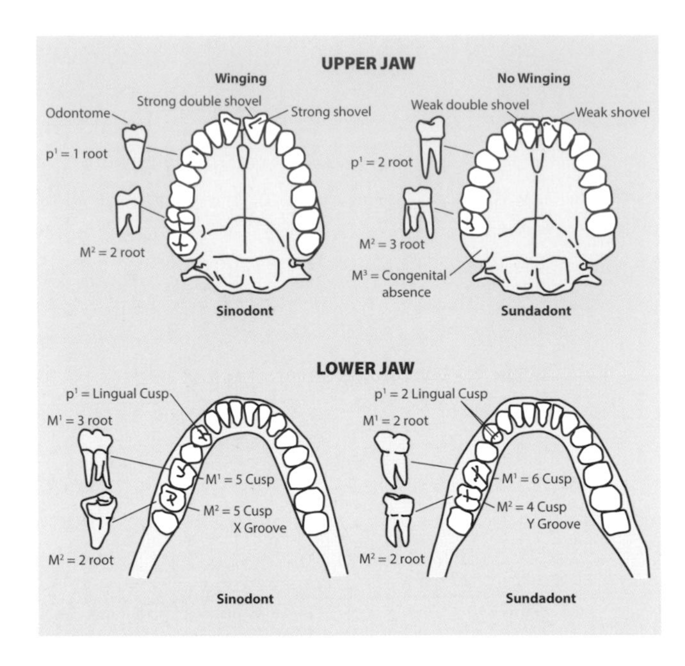

Arizona State University anthropologist Christy Turner described a pattern of dental features including shovel-shaped incisor, single-rooted upper first premolars and, tripe rooted first molars. These features which he called Sinodont only occur in northern Asian and Native American populations. He hypothesized people that bearing Sinodont tratits migrated into Mongolia about 20,000 years ago and across the Bering Strait about 14,000 years ago.

Blood Groups and Genetic Marker Frequency

The human genome is currently believed to contain some 20,000 genes, a number which incidentally is far lower than previous estimates. Of this total, 25 percent are contained within genes (coding DNA) and most of the remainder are non-coding DNA or "filler" between genes. The non-coding DNA, in turn, is composed of repeated sequences or "letters" which can be used, for example, in DNA fingerprinting associated with forensic applications and to establish degrees of relationship or distance between populations.

Though genetically less diverse than African populations, Native Americans nonetheless display considerable genetic diversity.[8] This variability has been studied for over fifty years as a means to examine the internal "relatedness" of various American Indian populations and their degree of affinity to specific populations in Asia. One of the most extensive studies of this kind was conducted by University of Michigan population geneticist and anthropologist James Spuhler, who was not particularly interested in the "from where did they come?" issue or genetic homelands, but instead examined the relationships between genetic frequencies of red blood cellular antigens and language distributions in macro-culture areas.[9] Using a population of fifty-three different socio-political/ethnic units (*i.e.*, tribes), Spuhler concluded that the correlation between genetic markers and languages is imperfect at best.

Although Spuhler eschewed the use of serologic (*i.e.*, the scientific study of serum and other bodily fluids) markers to establish genetic homelands, others such as physical anthropologist Emöke Szathmáry have addressed this issue,[10] often with mixed and controversial results. Specifically, Szathmáry contends that genetic marker data do not support Greenberg's tripartite linguistic classification (see below), and that some allegedly disparate populations such as Eskimo and Na-Dene are much more closely related than others have suggested. Szathmáry also believes that though affinity can be shown to exist between some North American groups and their contemporaries in Siberia (such as the St. Lawrence Island Eskimo and the Nenets of northern Arctic Russia), it is not possible to answer the "from where did they come?" question with the limited suite of genealogical attributes currently being used or the limited sampled populations from North America or northeastern Asia.

Linguistics

At the time of European contact, the New World was a place of considerable linguistic diversity. It is cstimated that some 1,000 mutually unintelligible languages, or a total of 2,500–3,000 if one includes dialects, were spoken in 1492. Of those, some 600 languages (or 1,060 dialects) survive to the present day. Many of these languages are on the verge of extinction.

Since the beginning of the nineteenth century, a series of attempts have been made to classify and phylogenetically arrange Native American languages based on a variety of similarity/dissimilarity attributes. The first serious attempt to order Native American languages was made in 1848 by ethnologist Albert Gallatin,[11] whose scheme defined some thirty-two language families. This number was multiplied nearly six-fold in 1891 by archaeologist and ethnologist Daniel Brinton,[12] who argued for eighty language families in North America alone, with an equal number in South America.

Shortly thereafter, American geologist and explorer John Wesley Powell reduced the number of North American language families to fifty-eight,[13] a number that subsequently was further reduced to fifty-five by anthropologist Franz Boas in 1911[14] and to only six macro-families or superstocks by anthropological linguist Edward Sapir, who studied under Boas at Columbia University, in 1921.[15] Though Sapir's scheme was viewed as hyper-reductionist by some of his peers, it has had a lasting influence to this day. Sapir's six superstocks were renamed phyla and formally subdivided into sixteen language families (and numerous additional language isolates) via the consensus reached at a 1964 conference on the mapping of Native American languages organized by two of the leading authorities on indigenous languages of North America, Charles Voegelin and Florence Voegelin.[16]

This pronouncement was scarcely the last word in North American language classification, and the tweaking of the 1964 consensus continued through the 1970s and into the 1980s. Perhaps the most extreme, and certainly the most influential, rendering of this thorny issue was forwarded by Joseph Greenberg, who as noted earlier in respect to Turner, further reduced Sapir's six superstocks or phyla to three: Eskimo-Aleut, Na-Dene, and Amerind. Greenberg also suggested that Amerind was the ancestor or parent of the other two and was among the language(s) spoken by the First Americans.

Greenberg's classificatory scheme was vituperatively assailed by other linguists who attacked both his methodology—called multilateral comparison—and his results. Refuting Greenberg's work, Smithsonian Institution linguist Ives Goddard and University of Hawaii historical linguist Lyle Campbell,[17] for example, charged that languages cannot be tracked beyond six millennia in the past, especially with the allegedly "specious," "superficial," and "fundamentally flawed" methodology used by Greenberg. Another vocal critic, historical linguist Johanna Nichols—though she believes that it is possible to trace languages beyond 6,000 years—employs a very different methodology which is based on approximately two dozen grammatical regularities in languages and their geographic distribution.[18] Nichols's research concludes that there are fifty-one language stocks and sixty-nine language families within those stocks in North America that have resulted from multiple migrations by different linguistic parent populations, which is radically different from the scenario proposed by Greenberg.

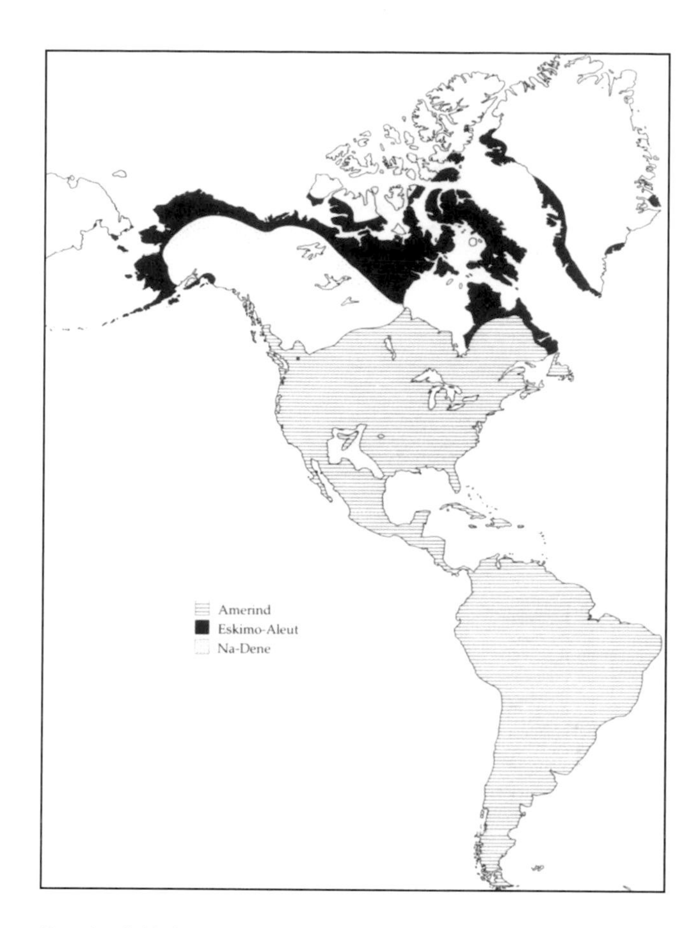

Stanford University anthropologist Joseph Greenberg became convinced that many of the language groups considered unrelated could be classified into larger groupings. In his 1987 book *Language in the Americas*, while supporting the Eskimo-Aleut and Na-Dené groupings as distinct, he proposed that all the other Native American languages belong to a single language macro-family, which he termed Amerind. From Joseph Greenburg *Language in the Americas*.

Ultimately, subscription to any of the schemes forwarded by Nichols, Greenberg, and others matters little in the ultimate resolution of the "from where did they come?" question. Despite the relatively recent and generally well received attempt to link the Na-Dene and northeastern Asian Yeniseian languages,[19] the fact remains that the only group of languages that can be unambiguously and conclusively demonstrated to have been spoken on *both* sides of the Bering Strait, and which continue to be spoken to the present day, are the Yupik languages of the Eskimo-Aleut language family. Given this fact, the recent and current distributions of these languages—throughout coastal Alaska, Canada, Greenland, and from the Aleutian Islands to the far eastern coast of Siberia's Chukchi Peninsula—neither illuminate nor permit any meaningful delineation of a geographically specific homeland for Native Americans.

Molecular Biology

Since the mid-1980s, perhaps the best lens through which to view the possible resolution of the "from where did they come?" question has been molecular biology. Three lines of inquiry have been employed to focus on the directly related issues of the ultimate genetic relationships of the First Americans and their geographic homeland: autosomal DNA, mitochondrial DNA (mtDNA), and Y-chromosome DNA (Y-DNA). Deoxyribonucleic acid (DNA), of course, is a molecule that provides the genetic instructions for the development and overall functioning of all known living things. Within cells, DNA is arranged into linear structures called chromosomes. Autosomal DNA is passed through both the maternal and paternal lines. Mitochondrial DNA is passed only through the female line (*i.e.*, mother to daughter), and Y-DNA is transmitted only through the male line (*i.e.*, father to son).

Autosomal DNA, because it is inherited from both parents, contains the complete genetic record of an individual. It has been used extensively to establish possible connections between the genetic or ethnic origins of individuals and populations via the comparative identification of particular genetic sequences called ancestry-informative markers. The accuracy and reliability of autosomal testing alone, however, has been questioned for a variety of reasons—not the least of which is that autosomal DNA is recombinant. For our present purpose, it is only necessary to note that recombination may result in the generation of offspring with combinations of genetic attributes that are different from either parent. For this reason, and also because autosomal data bases have been proven to be too small for determining valid affiliation, it is usually preferable to employ mtDNA or Y-DNA similarities to establish ancestry and possible genetic origins.

Because mtDNA is deleted from sperm cells at fertilization, the mtDNA that survives in the egg is exclusively derived from the female parent and is, hence, non-recombinant. Mitochondrial DNA also accumulates mutations in a very linear fashion, and these mutations are often geographically correlated and, at least originally, geographically circumscribed. As a result, mtDNA profiles can yield very accurate records of genetic drift, population movements, and/or population isolation. In short, mtDNA can illuminate the "from where did they come?" question.

Several areas of the mtDNA genome have been extensively examined in human populations. Ninety-four percent of the genome subsumes all of its coding functions and the remainder controls replication. Several different techniques have been employed to analyze both the coding and central regions. One of these techniques, as discussed by University of Pennsylvania anthropologist Theodore Schurr,[20] has employed a technique known to molecular biologists as restriction fragment length polymorphisms analysis to define a series of haplogroups in anatomically modern human populations. Put simply, a haplotype is a particular configuration or cluster of genes inherited from one parent, and a haplogroup is composed of a number of similar haplotypes which share a common ancestor.

Haplogroups are labeled with letter suffixes, and living Native American populations minimally include five such constructs: A, B, C, D, and X. (A sixth haplogroup, M, has recently been reported as ancestral to Haplogroups C and D, but is not represented among living Native Americans). These five groups represent the living descendants of the founding colonizing populations—or at least the ones that appear to have survived extinction. The majority of all Native Americans, without regard to their home continent, belong to Haplogroups A–D in differential frequencies (*i.e.*, A at 40 percent, B at 32 percent, C at 15 percent, and D at 5 percent). Interestingly, there appear to be gradual trends in the current geographic distribution of Haplogroups A, C, and D. Haplogroup A decreases in frequency from north to south, and Haplogroups C and D generally increase in frequency in the same direction. Haplogroup B shows no such pattern, but is most common in the North American Southwest and absent in the extreme northern and southern reaches of the Americas.

This variation among mtDNA haplogroups has suggested to some researchers that the New World was colonized in at least two episodes (or perhaps as many as four), with the earliest one(s) bringing Haplogroups A, C, and D in the first wave and Haplogroup B deriving from a later event. Others have attributed the current distribution of all four haplogroups to a single colonization event followed by subsequent genetic differentiation and population movement. But Haplogroup X, which shows a strong association with Algonquian-speaking populations of northeastern North America and is absent elsewhere, introduces an even greater degree of untidiness to this picture. Some researchers have attributed Haplogroup X's presence to the somewhat later movements of ancestral Na-Dene peoples, while others have attributed the haplogroup's presence to ancient Europeans. This latter interpretation figures prominently in the "Iberia not Siberia" scenario, which is discussed below in Archaeology.

Y-Chromosome DNA (Y-DNA), which is transmitted paternally, is fundamentally non-recombinant and accumulates several kinds of mutations through time and successive generations. Investigations of Y-DNA since the mid-1990s have employed two kinds of mutations to establish Y-DNA haplogroups and a growing number of haplotypes. Currently, there are eighteen Y-DNA haplogroups which, like their mtDNA counterparts, are labeled with a sequence of letter suffixes (A–R). Eight of these Y-DNA haplogroups (*i.e.*, Haplogroups C, E, F, G, I, J, Q, and R) are represented in living Native Americans.

Many of these haplogroups (specifically, Haplogroups E, F, G, and I) are believed to be post-Columbian introductions which collectively account for only 5 percent of the total Native American Y-DNA sample. The remaining haplogroups, (*i.e.*, Haplogroups C, Q, and R) constitute 95 percent of the New World sample, though Haplogroup R may also be a post-Columbian intrusion. If Haplogroup R does represent a later admixture during historic times, then only Haplogroups C and Q represent the descendants of founding populations—though again, other early haplogroups may have become extinct. Haplogroup Q occurs across the length and breadth of the New World, but Haplogroup C is confined to North American populations. Perhaps unhelpfully, if one is wishing for a specific geographic provenance, both Haplogroups C and Q are also generally present in northeastern Asia.

Collectively, molecular genetic data from autosomal DNA, mtDNA, and Y-DNA analyses appear to conclusively validate de Acosta's prescient prediction made over 400 years ago. The genetic homeland of Native Americans is indeed Asia—but again, Asia is a vast place whose landmass is larger than that of the entire New World. The mtDNA and Y-DNA data suggest that the New World's founding populations originated in a region that extends from central Siberia to the Altai Mountains on the west and southeastern Siberia and northern China on the east. The Y-DNA data also support a south-central and eastern Siberian contribution, but only if the current distribution of haplogroups reflects aboriginal reality. As illuminating as current molecular data may be about the genetic cradle(s) of Native Americans, a more exact circumscription of their homeland(s) is not presently possible.

Archaeology

Even though osteometric, anthropometric, linguistic, and molecular biological data appear to point conclusively to central-eastern or northeastern Asia as the homeland of the first travelers to the New World, archaeologists—especially American ones—have demanded the evidentiary equivalent of "Show me the money!" Or, in other words, where is the excavation-derived evidence for this population movement, and to what particular place in Asia does this evidence point? As we have previously indicated[21] and continue to maintain, such data are not presently available. On one hand, this is not to say that no credible information exists for the initial percolation of humans into eastern or northeastern Asia, as currently there is a general consensus that ancestral hominids penetrated the temperate environments of eastern Asia by 1 million yr BP.[22] On the other hand, the timing of the first arrival of humans—however loosely one defines "human"—in the climatically forbidding, northern reaches of eastern Asia remains the subject of no little dispute.

Russian Academy of Sciences archaeologist Yuri Mochanov has long argued for a very early human presence in the Lena River basin of central Siberia based upon a 12,000 square meter excavation at the site of Diring-Yuriakh in the early 1980s.[23] These investigations produced some 3,166 artifacts, allegedly from sixteen circumscribed clusters. Based on morphological criteria, all of these artifacts could in very broad terms be subsumed into the eastern Asian Chopper-Chopping Tool or Pebble Tool traditions.[24] Early claims of very deep antiquity for

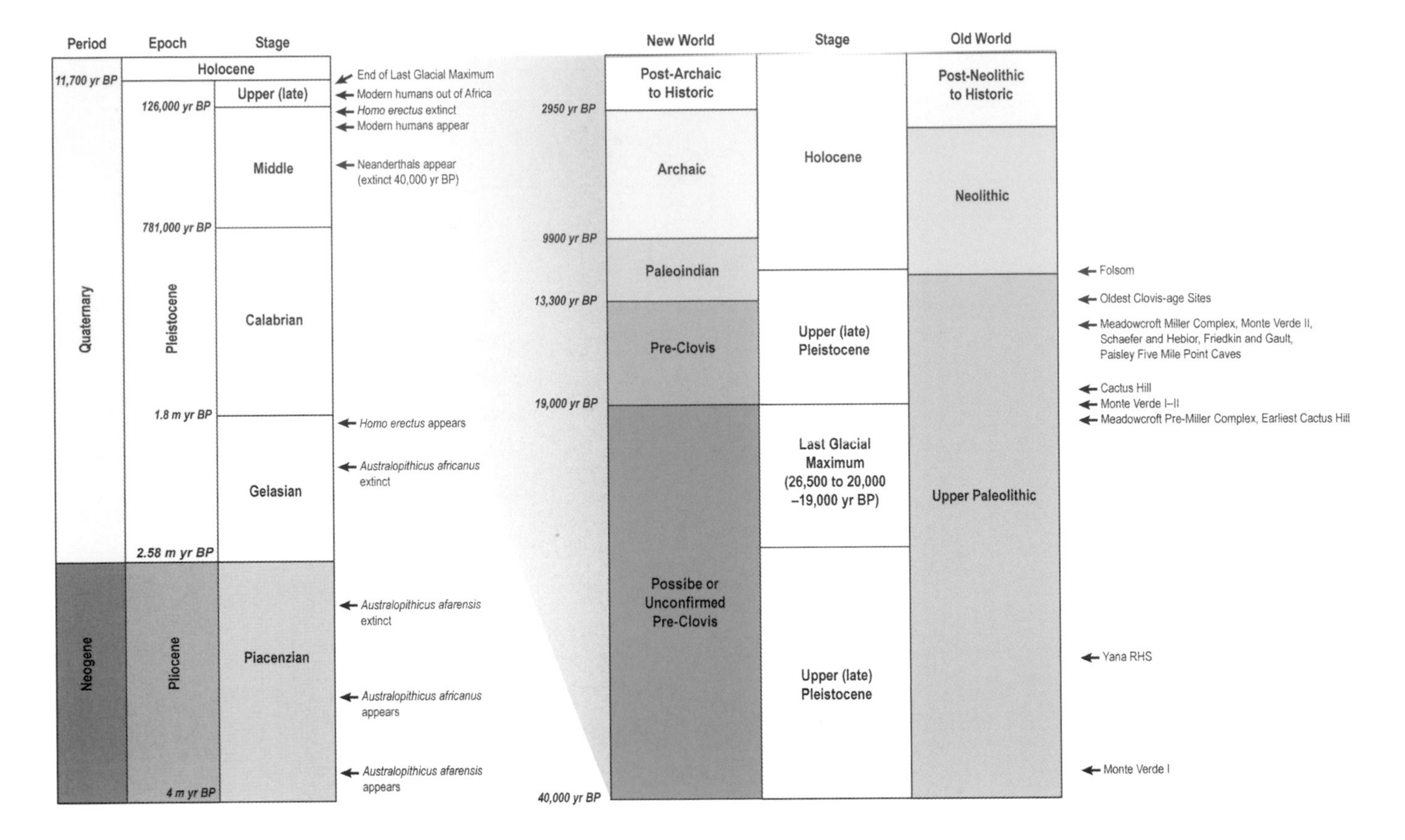

The International Commission on Stratigraphy has ascribed the geologic "history" of Earth to five time intervals or "eras" known as the Archean (~4,600–2,500 million yr BP), Proterozoic (2,500–540 million yr BP), Paleozoic (540–252 million yr BP), Mesozoic (252–66 million yr BP), and Cenozoic (66 million yr BP to present). All human evolution and history has occurred during the Pliocene (5.3–2.6 million yr BP), Pleistocene (2.6 million yr BP to 11,700 yr BP), and Holocene (11,700 yr BP to present) epochs, which are the most recent temporal subdivisions of the Cenozoic era. Our modern human species (*Homo sapiens*) is thought to have evolved in Africa sometime between 200,000 and 100,000 yr BP during the Cenozoic-era subdivision known as the Quaternary period, which subsumes the Pleistocene and Holocene epochs. Anatomically modern humans apparently began to migrate out of Africa to other parts of the world between 70,000 and 60,000 yr BP. As the earliest known archaeological site east of the Bering Strait in Siberia dates to about 33,000 yr BP, for the time being it appears safe to assume that the peopling of the New World could not possibly have occurred before then (despite the notable exception of the Monte Verde I site component). To provide a convenient context for readers to navigate the geologic, geologic, and cultural periods described in this book, the following chart is provided. The Old World temporal horizons are listed for comparison only, and are not to be considered parallel or consonant with those shown for the New World. Chart by David Pedler.

these materials have ranged about 1.8–3.2 million yr BP, though many others place the Diring-Yuriakh assemblage in the much more recent past.[25] While Mochanov has strenuously argued for an antiquity of about 1 million yr BP, others believe that the sediments at the site are no older than 270,000–260,000 yr BP.[26] Russian Academy of Sciences archaeologist Vladimir Pitulko and his colleagues maintain that even if the site's sediment dates are accurate, the artifacts probably post-date the end of the Last Glacial Maximum (20,000–19,000 yr BP) and are thus attributable to anatomically modern humans.[27] Whatever the true age of this or any other site alleged to be of great age in the Asian far north, most scholars agree that there is no incontrovertible evidence for a successful ancestral human colonization of this area before or during the Middle Pleistocene (780,000–126,000 yr BP).

The evidence for an even later human presence in the far north has, until recently, been equally controversial. Mochanov defined the Diuktai cultural complex on the basis of excavations at the northeastern Siberian site of Diuktai Cave as well as a series of open-air localities on the Middle Aldan River (a tributary of the Lena River)

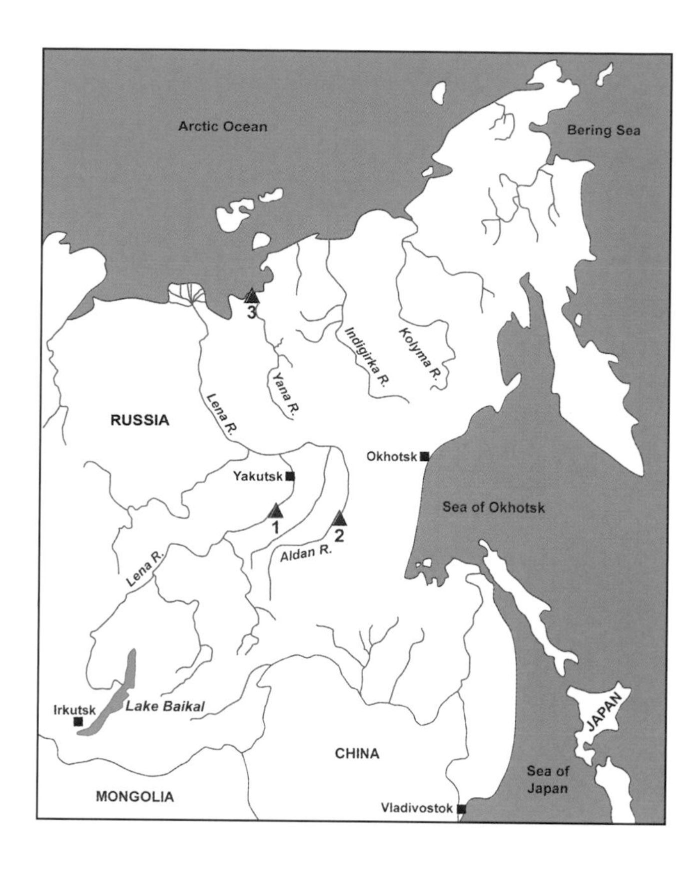

The eastern Siberian archaeological sites of Diring-Yuriakh **(1)**, Diuktai Cave **(2)**, and Yana RHS **(3)** are important for understanding the timing of human expansion into the far north, early adaptations to cold climates, and the peopling of the Americas. Map by David Pedler.

Typical unifacial choppers from Diring-Yuriakh. Courtesy of Michael Waters.

and nearby drainages in Yakutia.[28] On what have proven to be very questionable grounds, Mochanov has argued that all of the Diuktai complex sites he investigated are expressions of a unitary culture that began around 35,000 yr BP. These claims, like his earlier ones for a middle Pleistocene human presence in the far north, have been repeatedly challenged. The radiocarbon dating of the type locality, Diuktai Cave, indicates a probable age ranging between 17,360–16,620 yr BP and 15,960–15,320 yr BP. Additionally, virtually all of the putatively related open-air Diuktai localities are clearly disturbed, with indications of the transport, redeposition, and attendant mixing of artifacts and faunal remains. If Diuktai is a valid cultural construct, most contemporary verdicts on the age of this

Diuktai Cave bifaces. Courtesy of Yan Axel Gómez Coutouly and the Center for the Archaeology and Palaeoecology of Arctic People (Russian Academy of Sciences, Siberian Branch, Yakutsk, Russia).

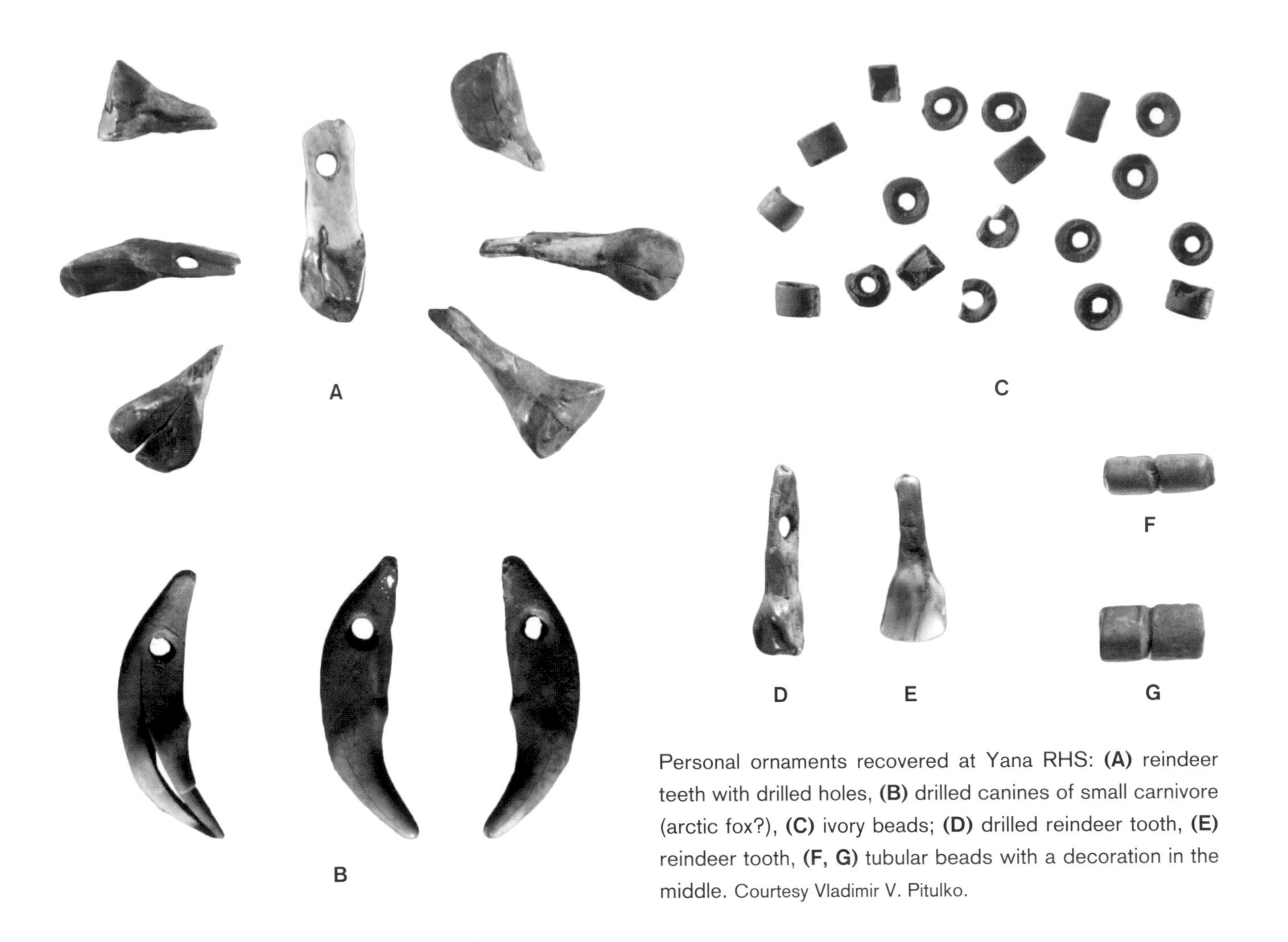

Personal ornaments recovered at Yana RHS: **(A)** reindeer teeth with drilled holes, **(B)** drilled canines of small carnivore (arctic fox?), **(C)** ivory beads; **(D)** drilled reindeer tooth, **(E)** reindeer tooth, **(F, G)** tubular beads with a decoration in the middle. Courtesy Vladimir V. Pitulko.

cultural complex concur with an assessment of an antiquity of no older than 18,000 yr BP.

The documented age for a human presence in the far north of northeastern Asia has been extended by more than ten millennia with the discovery and subsequent excavation of the Yana Rhino Horn Site (RHS) in Arctic Siberia by Vladimir Pitulko and his colleagues.[29] Located in the permafrost zone on the lower Yana River at 70°43' north latitude, Yana RHS is on the very western margins of Beringia. Excavations at this site have produced an extensive material culture inventory that includes lithic production debris and finished tools in the form of bifacial projectile points, a variety of unifacial forms, and so-called micro-tools apparently employed for working antler, tusk, and bone. The site also yielded worked bone, ivory, and antler in addition to a large suite of unmodified faunal remains. On the basis of multiple radiocarbon determinations, the primary visitation/utilization of the Yana RHS locality is dated as early as 33,130–31,720 yr BP, with episodic visits for some 3,000 years thereafter.

While the Yana RHS and the Diuktai sites contain lithic artifact assemblages with both bifacial and unifacial tools as well as blades, none of these items can be conclusively linked on technological terms to any early North American cultural manifestation. While isolated elements of the durable technology at both Yana RHS and Diuktai Cave are vaguely reminiscent of materials found in pre-Clovis and Clovis assemblages, in our opinion the connections are negligible and provide no solid artifact signature which can be back-tracked to help define a Native American Eden in northeastern Asia.

3. How Did They Get Here?

The Beringia Theory proposes that people migrated from Siberia to Alaska across a land bridge that spanned the current day Bering Strait. The first people to populate the Americas were believed to have migrated across the Bering Land Bridge while tracking large game animal herds. Photograph by Peter Bowers; courtesy of Mike Kunz.

Though superficially not quite as vexing as the "from where did they come?" question, the matter of "how did they get here?" has complications of its own. First, as we cannot presently specify a circumscribed area in eastern or northeastern Asia from which the first colonists emanated, we can only track their movements in a very general way. Second, tracking *any* population across the landscape using archaeological data alone is a process fraught with difficulty. Indeed, some of the great migrations of the more-recent past are essentially archaeologically invisible and would remain fundamentally unknown without written historical references.

Consider, for example, the westward expansion of the Hunnic Confederacy throughout Eastern Europe, the Caucasus, and Central Asia in the fourth and fifth centuries AD. With all of its myriad consequences, which were sufficiently profound to ensure that the name of their greatest leader Attila is still remembered 1,500 years later, the spread of the Huns occurred without an independently verifiable archaeological signature. While some graves of the period contain skulls with allegedly Hunnic cranial deformation,[30] metal work with putative Hunnic connections, and bone laths from the reinforcement of compound bows of alleged Hunnic manufacture, none of these items

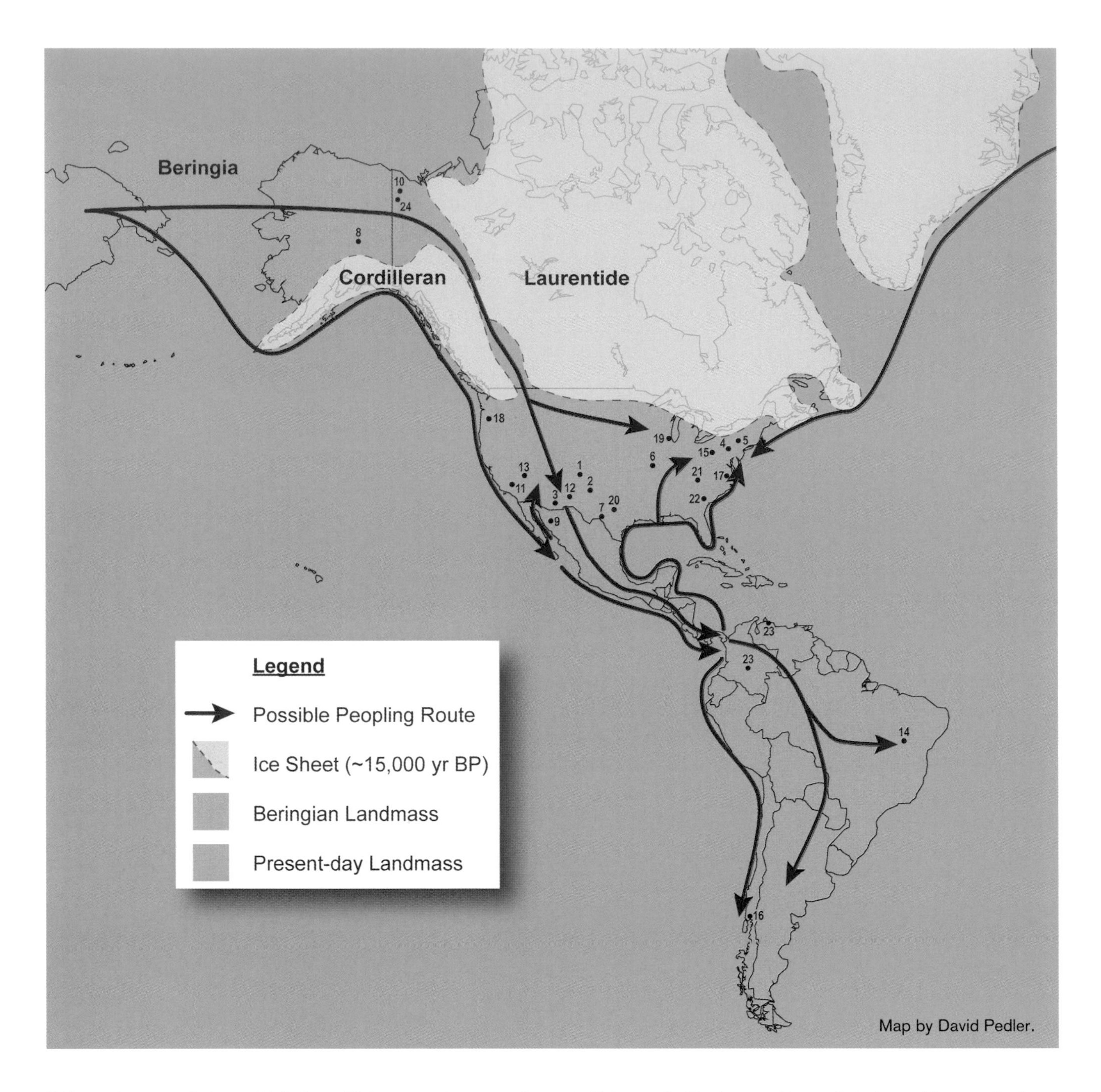

Schematic map of the multiple Late Pleistocene human migration routes (arrows) discussed in this book, showing the locations of sites (keyed using the chapter numbers in Part 2), ice sheets (as of about 15,000 yr BP), the configuration of Beringia as a dry land mass, and the present-day outline of the Americas. It is probable that various other routes and combinations of routes were taken, either contemporaneously or at broadly varying points in time. The ice-free corridor between the Laurentide and Cordilleran ice sheets was open and traversable by humans around 14,000–13,000 yr BP.

1. Folsom; **2.** Blackwater Draw; **3.** Lehner, Murrary Springs, and Naco; **4.** Shoop; **5.** Shawnee-Minisink; **6.** Kimmswick; **7.** Bonfire Shelter; **8.** Central Alaska; **9.** El Fin del Mundo; **10.** Old Crow; **11.** Calico; **12.** Pendejo Cave; **13.** Tule Springs; **14.** Pedra Furada; **15.** Meadowcroft Rockshelter; **16.** Monte Verde; **17.** Cactus Hill; **18.** Paisley Five Mile Point Caves; **19.** Hebior and Schaefer; **20.** Buttermilk Creek; **21.** Topper; **22.** Saltville; **23.** Taima-taima and Tibitó; **24.** Bluefish Caves.

are sufficiently sensitive as ethnic markers in the absence of the written historical record.

With these serious reservations aside, we believe the "how did they get here?" question is composed of two closely related elements: (1) via which route or routes did humans enter the New World and (2) what were the environmental circumstances and material/behavioral cultural mechanisms that helped and/or impeded their progress along the way? The immediate discussion addresses only the first element, while the second is considered later (see What Were They Doing?).

As is the case with the venerable, received-wisdom answer to the "when did they get here?" question, also to be addressed later, there has long been a single, time-hallowed answer to the "how did they get here?" query. Until recently, the exclusive avenue into the New World was considered to be the interior or "ice-free corridor" route, which could only be traversed on foot in the manner proposed by de Acosta over 400 years ago. In the most recent orthodox rendering of this scenario, a single colonizing group crossed the Bering-Chukchi shelf—popularly known as the Bering Land Bridge, but now referred to by many archaeologists as Beringia—during the Last Glacial Maximum (26,500–19,000 yr BP), when sea levels were dramatically lower. This reduction in sea level was on the order of 100–150 meters, exposing the huge Beringian landmass which at its widest point appears to have measured some 1,500 kilometers from north to south and essentially connected eastern Siberia and the Alaskan Peninsula. In contrast, the present-day Bering Strait separating the two continents consists of an 85 kilometer stretch of open water.

After their dry-shod crossing, this hardy group of migrants found their way into the ice-free interior of present-day Alaska. Apparently not being ones to linger, they promptly proceeded south through the McKenzie River Valley between the sinuous Cordilleran Ice sheet (which blanketed the mountainous backbone of western North America) and the far more extensive Laurentide ice sheet (which covered a vast section of Canada and the northeastern United States). There are several minor variations on this theme, the most commonly cited of which has human populations crossing Beringia, then proceeding south through the western interior plains of Canada, presumably during a warming phase when such movement would not have been impeded by glacial ice. The timing of this interglacial event would be crucial, however, and for obvious reasons it had to substantially pre-date all New World archaeological manifestations in order for this peopling scenario to be valid. The Clovis discoveries of the mid-twentieth century, and the subsequent discoveries of even earlier cultural entities throughout the length and breadth of the hemisphere, posed significant questions concerning the exclusivity of the ice-free corridor hypothesis.

In order to explain a very early human presence in archaeological localities situated over 5,000 kilometers to the south—and, in the case of Monte Verde in southern Chile, over 14,000 kilometers to the south—a decades-old coastal migration hypothesis has been resurrected by recent scholars. Initially advanced by Simon Fraser University archaeologist Knut Fladmark in 1979,[31] the notion that humans moved rapidly along the Siberian coasts, thence along the southern margins of Beringia and down the Pacific Coast to the Northwest United States and points south, has long been endorsed by archaeologists Alan Bryan,[32] Ruth Gruhn,[33] and James Dixon,[34] among others, and has been

The Pacific Coast Migration model proposes that people first entering the Western Hemisphere, beginning along the southern edge of the Beringia landmass, followed the Pacific coastline. Its most influential proponent, Kurt Fladmark, argued that the hypothesized narrow corridor between the Laurentide and Cordilleran ice sheets was likely to have been block or extremely difficult to travel through on foot. Paleoindian peoples traveling in boats along the shoreline and subsisting on marine resources would have had a much easier time reaching the unglaciated shores of Oregon and California and beyond. Photograph by Edward S. Curtis.

promoted most-recently by University of Oregon archaeologist Jon Erlandson[35] and his colleagues. This postulated movement is thought to have involved both limited pedestrian traverse and the extensive use of watercraft.

Several variations on the theme of coastal migration have also been proposed. Archaeologist David Anderson of the University of Tennessee, for example, has suggested that migration into interior North America first involved a southward coastal movement along the southern margins of Beringia to the tip of Baja California, followed by either a waterborne journey across a presumably narrow portion of the Gulf of California to mainland Mexico or a northward pedestrian movement up the east coast of Baja to the mouth of the Colorado River.[36] In either case, subsequent migration into different portions of the North American interior could have been facilitated. Alternatively, it has also been proposed that the coastal migration proceeded as far south as the Isthmus of Panama, which was then crossed at its narrowest point, followed by a northward movement along the Texas Gulf Coast and a bifurcation east to present-day Florida or northward through the Mississippi Valley. There is, of course, a distinct possibility that any combination of these routes could have been used simultaneously or sequentially by different pulses of colonists.

Part and parcel of various versions of these peopling scenarios, especially permutations of the interior entryway via the ice-free corridor hypothesis, is that the earliest populations journeyed on foot to the New World while following migrating herds of big game animals, especially mammoth. The implication is that humans entered the Western Hemisphere essentially by accident, not even realizing they were arriving in an unpopulated world. Moreover, the received wisdom of the greater part of twentieth century archaeological thought presumed that these first colonists were highly specialized hunters of big game who were utterly dependent on their highly focused lifeway. This adaptation is alleged to have been in place before the colonists left Siberia and employed throughout their traverse of Beringia, staying with them during and even after their dispersal into various parts of North and South America.

Naturally, the players selected by central casting for this ultra-predatory performance were prime-age males—clad in furs and muscularly wielding stone tipped spears—who successfully dispatched anything that moved across the picturesque landscape. This hackneyed focus on males and stone tools, as exciting as it may have been in that golden age of archaeological discovery, is the unfortunate legacy of over 175 years of paleoanthropological investigation in the Old and New Worlds. Conditioned largely by a temporally imposed tyranny of preservation which generally permits the recovery of *only* durable (notably stone) artifacts, and fostered by a one-time overwhelming preponderance of males in archaeological research institutions, vestiges of this antiquated conception of Paleolithic humans have persisted to this day.[37]

The underpinnings of this fanciful Late Pleistocene just-so story include, among other things: (1) an almost total ignorance of the role(s) of non-durable technology in the successful colonization of the New World, (2) an almost irrational resistance to appreciating the importance of watercraft in the peopling saga, and (3) a pervasive intellectual myopia concerning the critical roles played by people of all ages and both sexes in the peopling saga. As all of these issues more appropriately address the "what were they doing?" question than they do "how did they get here?" question, they will be addressed later.

Archaeological evidence confirming the specific routes that might actually have been followed by the New World's pioneer colonists is thin at best, as there are no signature artifacts yet known to be associated with their arrival, and none of the earliest widely accepted North American sites are located on the eastern fringe of Beringia (with the possible exception of the Bluefish Caves [see pages 306-313]). As presently known, the earliest well-dated sites and/or associated lithic industries in eastern Beringia or the western subarctic are no older than ca. 14,500 yr BP.[38] Moreover, none of these sites contain *any* diagnostic artifacts which can be conclusively linked to *any* northeastern Asian sites, broad resemblances notwithstanding.

Despite the discovery of submerged sites in coastal British Columbia[39] and the presence of Late Pleistocene people in California's northern Channel Islands,[40] the fact remains that none are as old as the oldest interior sites. Though the oldest evidence of humans in the far west may well be submerged on the inundated continental shelf, no known site conclusively demonstrates an early use of the coastal route. While we do not dispute the likelihood that maritime-adapted populations followed the so-called "kelp highway" into North and South America,[41] we reiterate that documentation of a very early use of this route is wanting. Similarly, no sites of deep antiquity have yet been associated with the postulated use of the Baja California or Panama-Gulf Coast route. While recent and, as yet, incompletely published research does suggest the potential for very old sites on the inundated continental shelf of the northeastern Gulf of Mexico,[42] none of these sites have been excavated to date.

In the most rudimentary terms, the only evidence that will ever conclusively demonstrate the use of any of the hypothetical avenues into the New World will be the indisputable archaeological presence of Native Americans along the way. While the linguistic and genetic data may lead one to confidently conclude that Native Americans ultimately came from somewhere in central or northeastern Asia, we still do not know precisely which route or routes were followed and whether any combination of them were used contemporaneously or sequentially.

Recently, two North American scholars have proposed yet another route for the early entrada into North America.[43] Dennis Stanford of the Smithsonian Institution and Bruce Bradley of the University of Exeter have resurrected Ronald Mason's 50-year-old hypothesis[44] that envisioned an ancient, maritime population pulse from, as some are wont to describe it, "Iberia not Siberia." Citing lithic materials from inundated and terrestrial sites on the coast and interior of eastern North America as proof of this seaborne entry, Stanford and Bradley claim that the technology and tools from several sites on Chesapeake Bay, Cactus Hill in Virginia (see pages 228-235), Meadowcroft Rockshelter in Pennsylvania (see pages 200-215), and several submerged locations off the coast of Virginia are congruent with Last Glacial Maximum-era Solutrean technology from Western Europe. They also consider the occurrence of a variant of Haplogroup X among eastern North American aboriginal populations as further evidence of a Solutrean connection, as this particular variant is known to have been represented in ancient Western Europe. Many scholars are critical of the supporting evidence for this hypothesis, however, and some have attacked it with a degree of vitriol reminiscent of Hrdlička in his Glacial Man-era prime. Despite the fact that some authorities find the evidence for a trans-Atlantic passage to be controversial, if not far-fetched, such a journey would at least have been technically feasible.

Of the questions we have considered so far, the issue of when Native Americans arrived in the New World has remained the most contentious. For a considerable duration of this lengthy debate, a migration earlier than de Acosta's estimate of 2500 yr BP was almost inconceivable for a variety of reasons, not the least of which was an ignorance of such basic facts as Earth's age and the antiquity of humankind. Since de Acosta's time and for centuries thereafter, the standard interpretations of Earth's age and its stages of development, notably the first appearance of humankind, were largely based on theological dogma rather than science. Indeed, establishing the antiquity of humans in the Old World would have to precede establishing the timing of their arrival in the New.

4. When Did They Get Here?

A turning point in our understanding of Earth's age and all living things would occur between 1830 and 1833. During these years, British geologist Charles Lyell published *Principles of Geology*, cited by many scholars as the most influential scientific book published in English in the first half of the nineteenth century. Building on the earlier work of James Hutton, William Smith, and others, in *Principles of Geology* Lyell formally articulated the concept of uniformitarianism. This elegantly simple notion holds that the forces which shaped the planet in the past were essentially those that still operate today, and as most of those forces worked slowly, Earth must be thousands, if not hundreds of thousands of years older than a literal reading of the Book of Genesis would have it. Lyell elucidated the idea of "deep time" within which scholars could more accurately consider some of the fundamental issues of life on Earth, and his work would prove to have a deep and abiding influence on Charles Darwin, Thomas Huxley, and many other scientists of his day.

The year 1859 proved to be even more pivotal in ways that would have far-reaching consequences. In that fateful year, not only did Darwin publish *On the Origin of Species*, with its profound implications for the antiquity of human life, but a more obscure yet equally portentous event also occurred in the Somme Valley of France. A group of respected British scientists visited the gravel beds that Jacques Boucher de Crèvecœur de Perthes, a French customs inspector and amateur archaeologist, had been excavating in the valley near Amiens since 1837. De Perthes' excavations recovered unmistakable, if somewhat crude stone tools and the remains of numerous Ice Age animals which he had long contended were contemporaneous with the artifacts. The British visitors confirmed the contemporaneity of the artifacts and the animal fossils, thereby simultaneously vindicating the maligned de Perthes and, more importantly, establishing that humans had lived in Europe for untold thousands of years.

One of the bifaces that Jacques Boucher de Crèvecœur de Perthes found in the gravels of Menchecourt-les-Abbeville, France, with remains of Ice Age elephants and rhinoceros. Photograph by Didier Descouens.

News of this sensational discovery and others that followed in France and England ultimately reached North America, setting off a flurry of research activity by amateurs and trained scientists who were soon scouring the landscape in search of "Glacial Man." In 1876, a New Jersey surgeon, Charles Conrad Abbott, published a widely read monograph which described what he claimed were Stone Age tools unearthed in gravel beds in Trenton. These discoveries and others by equally eager (if totally untrained) proponents of Glacial Man were actively promoted to an excited public and cited as incontrovertible proof that ancient humans had lived in the New World thousands of years before de Acosta's long-standing estimate.

While the idea of very early humans in the Americas was well received by the general public and even some scholars, it did not sit well with the archaeological establishment of the time, which included some formidable personalities such as John Wesley Powell, then the director of the Smithsonian Institution's Bureau of Ethnology and the U.S. Geologic Survey; William Henry Holmes, a geologist cum-archaeologist; and a bit later, one of the most daunting personalities to ever grace the American anthropological stage, Aleš Hrdlička, whom we have previously introduced.

Powell, and particularly Holmes, shredded Abbot's claims of Glacial Man by showing that the alleged Stone Age "tools" were not humanly made. Instead, they were the products of natural processes—"geofacts" rather than artifacts. By debunking the claims of Abbott and others of like mind, Holmes and company dealt Glacial Man a near-fatal blow. The *coup de grace*, however, was delivered by Hrdlička. A Czech-born scholar whose training at two medical schools made him thoroughly familiar with human anatomy, Hrdlička was also quite conversant in the paleoanthropology of early humans in Europe. Hired by Holmes in 1903 to head the newly created Division of Physical Anthropology at the Smithsonian Institution's National Museum, Hrdlička quickly became the ultimate arbiter of all ancient claims in the New World, or at least the northern third of it. From that perch he declared that none of the cases for Glacial Man were credible and, using criteria originally developed by his colleague Holmes, Hrdlička proceeded to demolish each new claim with cold scientific logic underscored with a condescending ferocity of withering dimensions.

While it may appear in hindsight that these quintessential "establishment" scholars were professional naysayers, there was nonetheless much substance to their skepticism. These Washington-based scientists formulated a set of simple yet undeniable probative criteria whereby the purported antiquity of a site had to be rigorously evaluated. Namely, one had to find: (1) artifacts of indisputable human manufacture or fossil bones that were unmistakably human within (2) a site whose geological layers could be unambiguously interpreted along with (3) appropriate chronological controls such as direct association with the remains of Ice Age plants and animals. None of the sites examined by Holmes and Hrdlička met these perfectly reasonable criteria, and with each passing year that was punctuated with more failed Ice Age candidate sites, establishment of an appreciation for the deep antiquity for humans in the New World seemed increasingly unlikely. A relatively recent arrival for the ancestors of Native Americans became dogma as if by default, and by the turn of the twentieth century it had become sacred writ that the first Native Americans arrived *after* the end of the last Ice Age.

A Folsom point. The Folsom complex dates to 12,750–11,700 yr BP and is thought by some researchers to have derived from the earlier Clovis complex. Folsom points have been found primarily in the Great Central Plains of North America. The point illustrated here was uncovered at the type site near Folsom, New Mexico, by a joint expedition of the American Museum of Natural History and the Denver Museum of Natural History during the 1927 and 1928 field seasons. Courtesy of the American Museum of Natural History.

Folsom

The man who would set into motion the events that would shatter the late arrival credo was not a professional scientist but, rather, an African American ranch foreman who was born into slavery. Drawn to the west after the Civil War, George McJunkin became quite proficient at his new occupation. One day in 1908 while checking fences and searching for stray animals in a dry wash near Folsom, New Mexico, McJunkin noticed bones eroding out of the wash's wall (see pages 36-45). He knew the bones were too large to be cattle and suspected that they even were too large for modern bison. After an incredibly circuitous path, the bones were eventually brought to the attention of Jesse D. Figgins, a paleontologist and the director of the Colorado Museum of Natural History, eighteen years after their discovery and four years after the death of McJunkin. The rest, as the cliché goes, is history...or perhaps more appropriately, prehistory. [45]

Figgins, a proponent of a late Ice Age migration, initiated excavations at Folsom and recovered not only additional large animal bones, which he recognized as extinct bison, but also directly associated delicately manufactured stone tools, the undeniable signature of Ice Age humans. With due fanfare, a delegation of prominent prehistorians including the leading field archaeologist of his day, Alfred Vincent Kidder, visited the site, examined the evidence, and pronounced it valid and acceptable. Curiously, Hrdlička never personally examined the site for himself, nor was he ever known to acknowledge acceptance of the Folsom discovery. Indeed, he went to his grave without recanting his recent arrival position.

The Folsom discovery and its acceptance marked a genuine watershed event in American archaeology. A Native American presence had been demonstrated before the end of the last Ice Age, and a new benchmark for a human presence in the New World was fixed at about 10,000 yr BP based on the geologically estimated extinction dates for *Bison antiquus*, the species identified at Folsom.

Simultaneously, and in a real sense equally momentously, the discoveries at Folsom provided the prologue to the "Paleoindians as lethal big game hunters" saga which has captivated generations of North American prehistorians.

Clovis

Shortly thereafter, another discovery in New Mexico, this time at a site called Blackwater Draw (see pages 46-57) near the town of Clovis, revealed evidence of what appeared to be an even earlier mammoth-hunting population that lived in the area before the Folsom bison hunters. Known and named for their trademark fluted lanceolate projectile points dubbed Clovis after the nearby community, additional evidence of these prehistoric North Americans began to be unearthed over much of unglaciated North America, both west and east of the Mississippi River.

Interestingly, though not yet recognized as such, Clovis points had been known to the scholarly community for more than 100 years *before* their excavation in New Mexico. In the fall of 1807, President Andrew Jackson dispatched William Clark, of Lewis and Clark fame, to the fossil quarry at Big Bone Lick in Kentucky to collect, if possible, a complete mastodon skeleton. In the course of excavating these fossils, five fluted points were recovered, three of which have been housed at the Cincinnati Museum of Natural History since 1817.

Because Clovis artifacts turned out to be so widely distributed, a picture gradually emerged of a population made up of small bands that spread with almost lightning speed throughout the hemisphere. Once radiocarbon dating was invented by Willard Libby (who won the Nobel Prize for this discovery) shortly after World War II, Clovis sites were dated to a very narrow range of 11,500–11,000 radiocarbon years BP. (When calibrated to calendar years like all of the radiocarbon ages mentioned in this book (see pages 318-327), this age for Clovis corresponds to about 13,300–12,800 yr BP.) This short time span correlated well with the notion of rapidly spreading colonists, and by the mid-1950s Clovis was viewed almost unanimously as the pioneer population in the New World. Moreover, because of the apparent congruity between the spread of Clovis fluted point makers and the extinction of some sixty-two species of North American mammals, it was presumed that the artificers of these points were directly associated with the disappearance of the Pleistocene bestiary.

Clovis point from Shawnee-Minisink, Pennsylvania. (see pages 78-89) Clovis points are bifacial (each face is flaked on both sides) and typically fluted on both faces. The origins and genesis of Clovis remain an open question. As a trail of progressively older Clovis technology does not appear along their presumed route of entry into unglaciated North America via the Canadian subarctic and northern Great Plains, it seems unlikely that the first Paleoindian pioneers brought Clovis with them. Many scholars nonetheless credit the relatively rapid and wide proliferation of Clovis to a single colonizing population, while others maintain that Clovis technology was instead communicated and shared among more-diverse, pre-existing peoples. Yet others have proposed that Clovis originated in the American Southeast, rather than at the classic locations in New Mexico and southeastern Arizona where Clovis was first identified. Courtesy Pete Bostrom, Lithic Casing Lab.

By the mid-1960s, the causal role of point-wielding Clovis hunters in the extinction of all sorts of Pleistocene fauna was formalized by University of Arizona geoscientist Paul Martin into the "overkill" or "blitzkrieg" hypothesis.[46]

According to a theory that to its acolytes would soon become as canonical as the New Testament gospels, the spread of Clovis point makers throughout North America occurred over the course of 500 years (in geologic terms, the blink of an eye) during which large Pleistocene creatures were slain with reckless abandon. The remarkable tenacity of this model, now subsumed under the broader rubric of Clovis First, is attributable to a number of interrelated factors. For one, Clovis First appeared to be supported by Greenberg's three-fold classification of native New World languages and Turner's tripartite dental chronology in a manner that equated the terms Clovis, Amerind, and Sinodont. Second, until relatively recently, there were no archaeological sites definitively older than Clovis. Third, and more subtly if not subconsciously, the overkill hypothesis appealed to male scholars in the same way that an earlier version of it had captivated the minds of nineteenth century European scholars.

The Specter of Pre-Clovis

Between the initial Clovis discoveries in the 1930s, through the subsequent refinement of the Clovis First model in the 1950s and early 1960s, to the model's apotheosis as an article of archaeological faith in the late 1960s and early 1970s, hundreds of sites in the New World were claimed to be older than the sacrosanct Clovis temporal benchmark. But for the most part, those sites enjoyed a brief, fleeting fame. Some of them are summarized in Part 2 of this volume, and they include Calico Mountain in Southern California (see pages 152-161), Old Crow in the Canadian Yukon (see pages 142-151), Valsequillo and Tlapacoya in Mexico, Sandia Cave in New Mexico, Tule Springs in Nevada (see pages 172-183), and, more recently, Pedra Furada in Brazil (see pages 184-197) and Pendejo Cave in New Mexico (see pages 162-171), to name but a celebrated few. These localities failed to convince most Clovis First advocates either because they believed the claimed artifacts were not true artifacts; the stratigraphy, context, and associations were ambiguous; and/or the chronological controls were lacking.

Despite decades of setbacks, a small core of staunch pre-Clovis proponents, notably including scholars Ruth Gruhn, Alan Bryan, Alex Krieger, and Richard Shutler, insisted that the Clovis First model was shot through with fatal flaws. First, no such rapid peopling event could be documented for *any* other part of the planet. Second, there were areas of North and South America where Clovis groups never went, or at least where their signature points were never found, and different cultural traditions clearly existed there. Third—and most damning—so much diversity existed in North and South America by the time of the Clovis horizon that Clovis could not possibly be the ancestral complex of these numerous, supposedly descendent cultures. So too did European scholars with near unanimity reject the Clovis First model as fundamentally flawed, as did their South American colleagues. The typical American response was "Those guys on the other side of the Atlantic don't know what they're talking about," or "Those guys in South America don't even publish in English!" But, squabbling aside, the question remained: where is the elusive proof?

Enter Meadowcroft Rockshelter

In the spring of 1972, another chain of events about as unlikely as the Folsom saga was kicked off when co-author Adovasio was approached by the University of Pittsburgh about a position in its anthropology department. One of the responsibilities of the new position was the resurrection of that institution's then-moribund North American archaeology program. Adovasio couldn't join the faculty until the fall of 1973 due to research commitments on Cyprus, but took the appointment anyway.

Before leaving for the Mediterranean, Adovasio circulated word that he would like to find a place suitable for a student field school that met various criteria: it should be within about 40 kilometers of Pittsburgh; it should have received little or no previous archaeological attention so that the students would be uncovering new information about the prehistory, geology, and environment of the area as they learned to excavate; and it should include a cave or rockshelter, as these were the kinds of sites with which Adovasio was most familiar.

Somehow, in the spring of 1973, the late Phil Jack, a historian at what was then California State College (now California State University) in California, Pennsylvania, heard about Adovasio's quest and contacted him about a site on the property of two local landlords, one an enthusiastic student of history, Albert Miller, and his brother Delvin, an illustrious harness racer. The property was called Meadowcroft. As soon as Adovasio saw the site later that spring, he knew it was ideal for his purposes. Like Albert—who suspected its potential as a child and who deserves credit for discovering the site and ensuring that it was never despoiled—Adovasio knew that generations of people throughout history would have seen it, as both he and Albert did, a perfect place to camp for a night or two or even a couple of weeks.

With the first publication of radiocarbon ages from Meadowcroft in 1975, the site became unarguably the most controversial (if not notorious) locality ever advanced for a pre-Clovis human occupation in the New World. As eminent scholar of all things Clovis and pre-Clovis David Meltzer has observed, Meadowcroft Rockshelter has remained in the vortex of controversy to the present day. This vortex was set into motion by and continues to swirl around the radiocarbon ages from the site. Taking into account the radiocarbon age of the site's youngest or upper strata, its earliest age is on the order of 13,800–11,500 yr BP, which is well within or just slightly older than Clovis. But the eye of the vortex has been the site's Miller complex artifacts, recovered from living surfaces in the site's lowest culture-bearing stratum that are bracketed above and below by radiocarbon ages of 15,160–11,250 yr BP and 17,580–13,060 yr BP, respectively. The mean of these ages alone pre-dates Clovis by almost 1,800 years, and the mean ages of hearths and fire features underlying the later Miller levels extend this age range by at least an additional 4,700 years.

As is usually the case when an entrenched theoretical perspective is challenged in any field of enquiry, the counter-attack was swift. Because even the critics acknowledged that the fieldwork at Meadowcroft and subsequent analyses were done to the most meticulous archaeological standards, these attacks boiled down to the peculiar claim that any date older than that of the semi-mystical Clovis benchmark must be wrong. Specifically, it was suggested that despite a stratigraphically coherent series of radiocarbon dates from the site's Holocene deposits—which actually correspond to those obtained from contemporary components elsewhere in the upper Ohio Valley—all of the radiocarbon samples from Meadowcroft's older deposits were somehow contaminated and thus rendered artificially older than their actual ages. It was also suggested that the associated animal and plant remains were inconsistent with a Late Pleistocene occupation. (No contamination by at least five external analyses has ever been detected, and data from other contemporary localities have demonstrated that the site's animal and plant remains are not anomalous.)

A Miller point from the Meadowcroft Rockshelter. Dated to 17,160–15,500, it is at least 2,000 years older than and quite dissimilar to Clovis. Miller populations were apparently broad spectrum forgers rather than hunters focused on the predation of megafauna. Courtesy of J.M. Adovasio.

As the vortex of controversy continued to swirl on the archaeological horizon, its radius of gyration came to include the question: if Meadowcroft Rockshelter is really that old, where are the other Meadowcrofts? Apparently having grown weary of focusing on the issue of age, the Clovis-First skeptics turned their attention to the fresh concepts of high visibility and replicability. In other words, overwhelming and widespread evidence was insisted upon with no room for exceptional cases—such as de Perthes' excavations in the Somme Valley or the discoveries at Folsom, for example—and their demands were clear: give us more Meadowcrofts.

End Game for Clovis First

Perhaps apropos of our personal take on the old saw, be careful what you wish for and especially careful of what you ask for, other "Meadowcrofts" have indeed appeared, most spectacularly in the form of Monte Verde in southern Chile (see pages 216-227). Since 1976 this site has been intensively investigated and authoritatively reported on by a

A projectile point recovered from a Late Pleistocene level at Monte Verde, Chile. New evidence of stone artifacts, faunal remains, and burned areas suggests human activity at this site 15,000 kilometers south of Beringia by at least 14,500 yr BP. Courtesy of Tom Dillehay.

multi-disciplinary, multi-national team directed by archaeologist Tom Dillehay of Vanderbilt University. Multiple settlements have been identified at Monte Verde, the most extensive of which (Monte Verde II) consists of two clusters of stream-side architectural structures. Scattered in and among these structures were a few stone tools very different from those found at Clovis or Meadowcroft and a host of wood and fiber artifacts.

Most telling is the fact that the Monte Verde II component has been securely dated to between 15,120–14,030 yr BP and 14,590–13,400 yr BP, which makes it as many as 2,000 years older than Clovis while being almost 9,000 kilometers further south of Beringia than Blackwater Draw, as an exceptionally sturdy crow might fly. Despite even more relentless—and frankly, occasionally quite breathtakingly ludicrous[47]—attacks than those directed at Meadowcroft, Monte Verde is now almost universally viewed as the site that shattered the Clovis-First model. Very recently, Dillehay and his colleagues have found convincing evidence of even older, episodic, and ephemeral human presence at the site that may date to as early as 22,860 yr BP.

The pre-Clovis sites have continued to pile on: ongoing investigations at Cactus Hill in Virginia (see pages 228-235), the Paisley Five Mile Point Caves in Oregon (see pages 236-245), Gault and Debra L. Friedkin in Texas (see pages 258-271), as well as several other sites in North and South America have definitively demonstrated that the Clovis-First model for peopling of the New World is finished. Humans were clearly in the New World before Clovis, and the mounting evidence shows an emerging picture of lifeways that are also quite different from Clovis.

The question of precisely when humans arrived in the New World, on the other hand, like most of the other questions we have posed so far, is presently unknown. Assuming a rather more leisurely movement through the Americas than initially imagined by the "blitzkrieg" proponents, the first population pulses *could* have occurred before the end of the Last Glacial Maximum, now placed at about 20,000–19,000 yr BP. Conversely, if the dispersal schedule is speeded up, the initial colonization movements *may* post-date the Last Glacial Maximum. As somewhat of an outsider to this controversy, the linguist Johanna Nichols once suggested that her reading of the linguistic diversity data supported an initial human spread into the hemisphere at least some 30,000 years ago. An arrogant American archaeologist's response was reportedly "Where are the 30,000 year old sites?" Nichols is alleged to have responded "That's not my problem."[48] She was, and of course still is, correct.

The "when did they arrive?" question is indeed our problem as archaeologists, and it is one that in all probability will never be perfectly resolved. We may very well find 20,000 or 30,000 year old sites, but the likelihood of finding the first campsite or landfall in the Western Hemisphere would be akin to finding a winning mega-trillions lottery ticket in a haystack during a tornado. What we will find, however, are a few of the scattered and ephemeral sites of slightly later movements whose age and geographical locations will enable us to provide a hopefully better—if not final—answer to the "when did they arrive?" conundrum.

5. What Were They Doing?

However they got here, via whatever routes they traveled, and whenever they arrived, the First Americans must have soon realized that they were indeed strangers in a new land. The smoke from campfires was literally behind them, and there was no sign of it on the forward horizon. The manner in which they populated this new land and successfully spread through two environmentally and ecologically diverse continents—essentially the answer to the "what were they doing?" question—is one that Columbus and his entourage apparently never thought to ask. But before discussing what they were doing, we should address what they *were not* doing. Certainly, they were not just walking or running through the hemisphere, nor were they simply and exclusively making stone tools. And most assuredly, they were not just killing and eating large animals while driving some of them to extinction.

Determining the rate of human dispersal through the New World is not presently possible because we still do not know exactly when the first migrants arrived, but it seems clear that their spread occurred much more slowly than the proponents of the overkill hypothesis ever imagined. New lands with new environments and new microclimates, new plants, and new animals obviously necessitate a period of adjustment on the part of any pioneer population. In this regard, it might be instructive to consider the research of Quaternary scientist and archaeologist Rupert Housley and his colleagues regarding the initial dispersal of humans into the northern reaches of deglaciated Europe.[49] Housley's research indicates that it took some 4,000 years for humans to travel only 1,300 kilometers northward, despite the fact that they were entering a landscape that was already known to exist. As also appears to have been the case for the New World, the earliest sites of Housley's Pioneer phase of this movement are few in number and far between, diminutive, and essentially invisible in archaeological terms. Crucially, though this northward dispersal was somewhat imperfectly marked by a trail of stone tools, lithic artifacts are not thought to have been the most important component of the migrants' tool kits.

Paleoindian hunters versus a mammoth in the swamps of what is today Georgia. Several scenarios have been advanced to account for extinction of many species of North American megafauna at the end of the Pleistocene: (1) overkill by human hunters, (2) climate change, (3) disease, and (4) extraterrestrial impact event resulting from comets. From Richard L. Thornton Ancient Roots IV: The Architectural Heritage of the Muskogean Peoples.

For some incomprehensible reason, many New World archaeologists have maintained longstanding biases against the potential role of watercraft in the initial dispersal of the First Americans and the crucial role of non-durable technologies—especially plant-fiber-based industries—in the successful colonization of new lands. This is particularly perplexing given the ample ethnographic evidence that perishable technologies form the bulk of hunter-gatherer material culture, even in Arctic and subarctic environments.[50] Moreover, archaeologists working with materials recovered from environmental contexts with ideal preservation confirm that the ubiquity of non-durable artifacts observed in the more-recent ethnographic record also holds for the past.[51]

It is now abundantly clear that contrary to the received wisdom of mid-twentieth century archaeological thought, the production of plant-fiber artifacts extends well back into the Pleistocene. Once defined as the hallmark of the Mesolithic and Neolithic periods of the Old World or the Archaic and Formative periods of the post-10,000 yr BP New World, the manufacture of cordage (*i.e.*, string or rope), cordage byproducts (like nets and snares), baskets, textiles, sandals, and their kin is now widely recognized as a Paleolithic innovation.

Research on fired and unfired clay impressions from the Upper Paleolithic sites of Dolni Vestonice I and II and Pavlov I in Moravia, for example, has conclusively documented the manufacture of a wide range of plant fiber derived artifacts in the Late Pleistocene of Europe. The plant-fiber artifact inventory from these Moravian sites, which date to 29,000–24,000 yr BP, includes numerous varieties of cordage, knotted netting, plaited wicker-style basketry containers, and a surprising range of twined and plain-woven textiles. The textile suite represents a diversity of forms including mats, wall hangings, blankets, flexible bags, and an array of apparel forms such as shirts, shawls, skirts, and sashes.

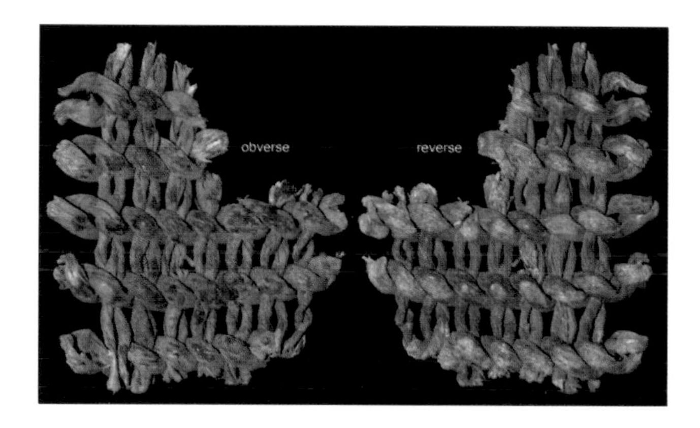

The obverse and reverse sides of a fragment of a twined mat or basket container dated to about 11,200 yr BP from the Gultarrero Cave, Callejón de Huaylas, Yungay Province, Peru. Black grimy residue (left) and wear from use (right) are visible. Photograph by Edward Jolie.

Ample additional evidence in the form of even more impressions, actual specimens, iconographic representations, and tools used in the production of fiber artifacts have conclusively confirmed both a widespread distribution of and very deep antiquity for this technology.[52] By no later than about 15,500 yr BP, such technologies are evident as impressions on the world's oldest ceramic artifacts and clearly have reached the very threshold of Beringia in the Russian Far East.[53] As we have emphasized many times in the past, perishable technologies *must* have accompanied the first travelers to the New World.

The ~30,000 years previous to Columbus' landfall at Guanahani represents 99 percent of human history in the Americas.

The elaboration of non-durable technologies in the material cultural repertoire of the First Americans has numerous and profound implications for our understanding of their behaviors. The early development of cordage and netting, in particular, illuminate aspects of social organization which have been virtually ignored in discussions of the first arrivals in the New World. In fact, the production and use of both knotted and knotless nets for hunting, storage, and transport may well have been a more critical element in the successful mastery of diverse New World environments than any kind of stone projectile point or lithic tool.

Hunting with nets is a transformational subsistence technique on a number of levels. At one level, hunting or fishing with nets is entirely different from the taking of individual animals or even the systematic predation of large aggregates of animals via lances, spears, leisters, spear-throwers, and darts. At another level, cross-cultural research indicates that net-hunting is often a communal effort which, because of the relative lack of expertise necessary for success as well as the minimal danger and low physical demands involved, often employs the entire co-residential social unit—that is, the young and old of both genders. At yet another level, net-hunting and net-fishing are also associated with mass harvests in short periods of time and, thus, with the production of a surplus. Significantly, the recovery of very early nets from inland and coastal contexts indicates that net-hunting and net-fishing were practiced by the earliest inhabitants of the New World.[54]

The implications of the production and use of watercraft, and therefore the movement of populations by boat, are at least as far-reaching as the adaptive consequences of non-durable technology. It is now clear that open-water navigation is far more ancient than had previously been imagined, and it was not necessarily the sole province of anatomically modern humans. Current data suggest that *Homo erectus* was utilizing boats to populate islands off southeastern reaches of Asia's continental shelf by 880,000–780,000 yr BP,[55] and there is gathering evidence that some Mediterranean islands such as Cyprus and Crete were colonized via boats much earlier than previously believed.[56] Given the probable existence of vessels that were capable of plying open water by the early reaches of the Middle Pleistocene, it is certainly not a surprise that Australia was peopled by anatomically modern humans using boats perhaps as early as 60,000–50,000 yr BP and certainly by 40,000 yr BP. Indeed, there is no other way to cross the open water between Asia's continental shelf and Australia, except by boat.

If watercraft were indeed operational in the open waters of Southeast Asia by at least 40,000 yr BP, it is virtually certain that ancient coastal populations further to the north also had access to boats.[57] Obviously, if boats of any kind (but most probably umiak-style, skin-covered vessels) were accessible to the first migrants to the New World, any discussions of the timing of the exposure or inundation of Beringia or the opening and closing of the ice-free corridor are more or less moot. The use of boats also would have permitted a coastal entry anywhere south of Beringia and the Cordilleran glacier, transits across the Gulf of California and up the Colorado River, or circumnavigation of the Gulf of Mexico and the potential to travel up the Mississippi River. Needless to say, without watercraft the trans-Atlantic peopling scenario posited by Stanford and Bradley would not have been possible.

By emphasizing the multifaceted roles of plant-fiber derived artifacts and the possible use of watercraft in the peopling of the New World, we certainly do not intend to de-emphasize other innovations that would have facilitated the successful colonization of the hemisphere. Items and behaviors such as tailored clothing, the durable construction of weatherproof housing, the development of complex food preservation and storage capabilities, and the invention of atlatls and projectile weaponry, among myriad others, would have been key elements for existing, spreading, and flourishing in a new land and all waypoints encountered on the journey there.

Our intent, rather, is to counter the modern temptation to flatten the picture of life in the waning years of the Pleistocene by attempting to tease out the rich, everyday details that lie behind the tired tableau of fur-wrapped, spear-wielding males as players in a B-movie colonizing adventure. This is not to say, of course, that the contributions of men, stone tools, and the hunting of game both big and small were not crucial aspects of the First Americans' lifeway. Our point is that they were not the *only* crucial aspects, and that any culture's survival in any circumstance—much less a hazardous trek into parts

The First Crossing by Greg Harlin, Wood Ronsaville Harlin, Inc.

unknown—depends on the direct involvement of all of its constituents, man or woman, young or old.

In the end, perhaps the fullest, most comprehensive answer to the "what were they doing" question lies in Part 2 of this book, which catalogues the archaeological history of the First Americans from the perspective of key, well-published sites they left behind, placing them in their natural and cultural landscapes. Starting with the Folsom and Clovis discoveries of the early twentieth century, we mark the definitive unearthing of the New World's human antiquity and the gradual realization of an apparently uniform, continent-wide cultural complex that was believed to have evolved essentially in place. We then move on to describe the subsequent mid-century explorations of putatively even more-ancient sites throughout the length and breadth of the Western Hemisphere, from the Canadian Yukon to eastern Brazil, and their ultimate debunking by an increasingly wary community of archaeological scholars. Next, we arrive at the archaeological discoveries of the later twentieth century, which advanced sites demonstrably older than, and distinctly different from, those of the Clovis horizon. Part 2 concludes with a tour of several localities for which scholarly consensus has not yet been (or may never be) reached, but which nonetheless hold the promise to further expand or refine our understanding of what is probably humankind's last great terrestrial migration.

Folsom point found in association with extinct bison at Folsom, Arizona in 1927.

CLOVIS AND FOLSOM AGE SITES

The discovery and acceptance of the Folsom site in 1927 was a pivotal event in American archaeology that opened the way for all of the ultimately more ancient archaeological discoveries that followed. Appearing at a time when the American archaeological community had essentially concluded that Native Americans could not have been present during the last Ice Age, this exception to the rule was eventually discerned to be part of a cultural horizon that spanned a vast portion of the Great Plains, and soon discovered to have been preceded by an even older Clovis horizon whose reach covered virtually all of unglaciated North America. This era of American archaeology saw the development of key analytical and site evaluation criteria that are still in place today (see Disputed Pre-Clovis Age Sites, pages 140-197), while also setting the stage for refinements in field and laboratory methods that have defined modern archaeological practice. The sites described below include two well-known Folsom localities (Folsom and Bonfire Shelter), a variety of Clovis sites that span thousands of miles and diverse environments, and localities from a remote focal point of Clovis-era human activity in central Alaska whose sites are decidedly dissimilar from their southerly contemporaries.

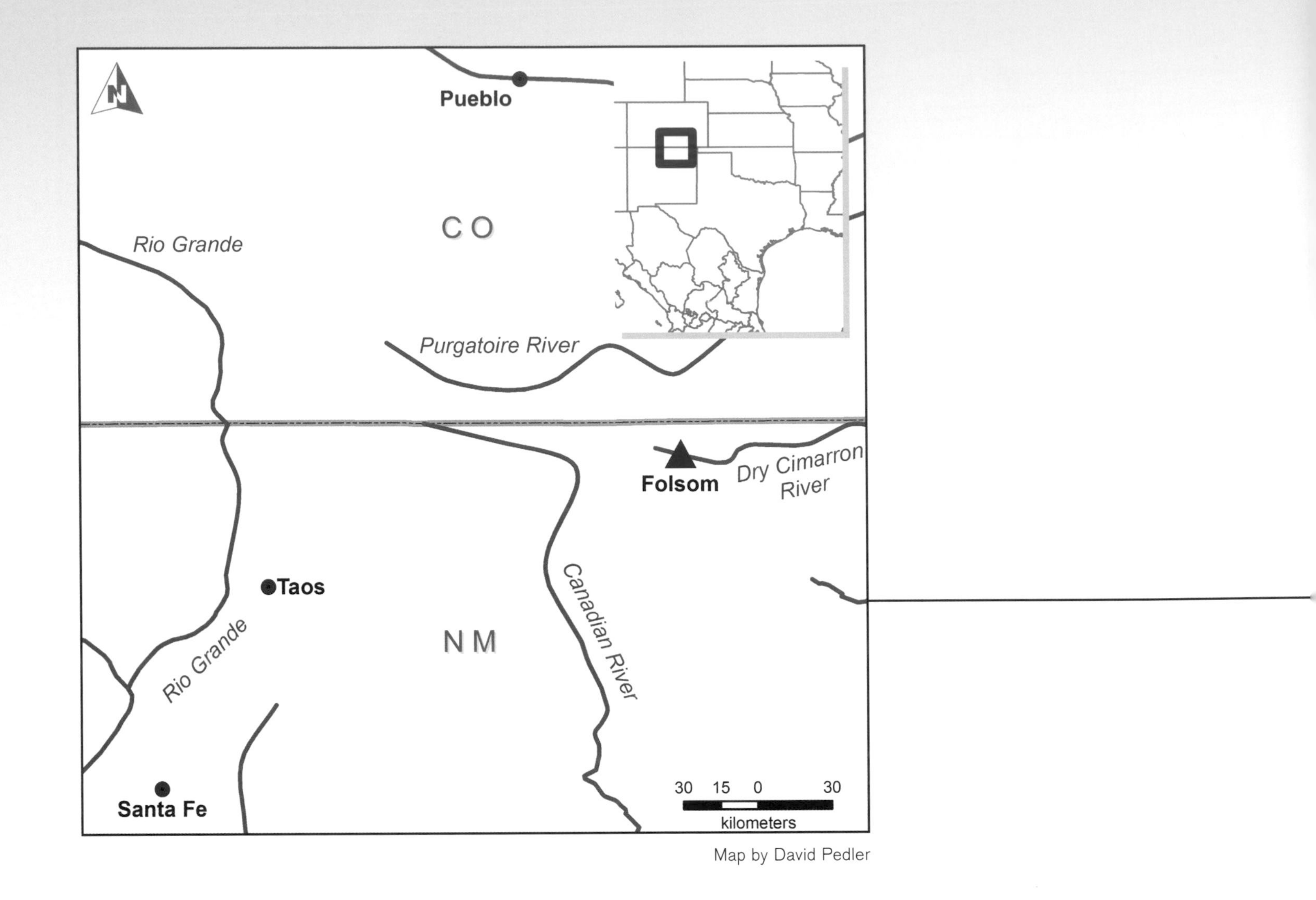

Map by David Pedler

While Folsom is one of the best-known Paleoindian sites in American archaeology, after over eighty years of archaeological investigation much remains unknown about the Paleoindians who hunted now extinct bison there.

1 FOLSOM

LOCATION UNION COUNTY, NEW MEXICO, UNITED STATES

Coordinates 36°52'54.05"N, 104° 4'16.13"W.

Elevation 1,948 meters above mean sea level.

Discovery George McJunkin in 1908.

The Folsom site provided the first definitive evidence that humans shared the New World landscape with and hunted extinct Late Pleistocene animals, in this case *Bison antiquus*, and has been called "the place that forever changed American archaeology" by noted authority on the site David Meltzer. Folsom also serves as the complex's type site, defined by the *Oxford English Dictionary* as "the site where objects or materials regarded as defining the characteristics of a particular period were found." The Folsom complex was a cultural manifestation that endured for about 700 years and whose range extended from the Rocky Mountains to the central Great Plains, from central Texas to just south of the Canadian border.

The Folsom site is located in the northeastern corner of New Mexico near the Colorado border, about 15 kilometers west of the village of Folsom and about 200 kilometers northeast of Santa Fe, on the southwestern reaches of the Great Plains about 100 kilometers east of the Rocky Mountains. The site lies principally on the southern bank of Wild Horse Arroyo (also known as Dead Horse Arroyo), a narrow, ephemeral tributary of the Dry Cimarron River which in turn is a tributary of the Arkansas River and, ultimately, the Mississippi. The regional physiography is dominated by volcanic landforms and large flat-topped mesas that are divided by several major river systems and interspersed with grasslands and open woodlands. [Figure 1.1] The river valleys generally trend from west to east and, along with several mountain passes currently used by modern highways, would have provided ready human and animal access to the region from the High Plains to the east.

During Folsom times, the climate appears to have been drier and cooler than today with fairly snowy winters. Although tree species were probably similar to

FIGURE 1.1 High plains grassland-desert scrub in the Folsom area. Johnson Mesa is to the left, the Folsom site is in the right foreground. At the time bison were hunted by Paleoindians, it appears that the climate at Folsom was drier and cooler. Photograph by David Meltzer.

those present today, it appears that vegetation was more open with abundant grass- and shrub-covered parkland, scattered wetlands, ponds, and lakes. These conditions would have made the region attractive to bison, but apart from the prospects for hunting, not altogether attractive to humans. There were no sources for nearby toolstone, few springs, and few edible plants to distinguish the site's immediate area from adjacent, more ecologically diverse places. The site's attraction, in other words, was primarily its position in a landscape dominated by a topography that helped to channel people and animals around impassable, steeper areas such as mesas and volcanoes.

The Folsom site is widely believed to have been discovered by a cowboy named George McJunkin sometime after a destructive and deadly 1908 flash flood of Wild Horse Arroyo exposed bison bones on the eroded arroyo floor. [Figure 1.2] The site had remained known only to avocational naturalists until early in 1926, when Jesse Dade Figgins and Harold Cook of the Colorado Museum of Natural History in Denver visited the site in the interest of recovering a bison

FIGURE 1.2 The site at the head of Wild Horse Arroyo contains no special resources that would distinguish it from its surroundings. The chain of events that led to its discovery as a major Paleoindian archaeological site began with a late summer 1908 flash flood through the arroyo that exposed fossil bones. George McJunkin, foreman at a nearby ranch, spotted the large bones while looking for stray livestock and checking fences. He reported his find to Carl Schwachheim and Fred Howarth, amateur naturalists in the area, who in turn contacted Jesse Dade Figgins and Harold Cook at the Colorado Museum of Natural History in Denver. Excavation at Wild Horse Arroyo began in February 1926. Courtesy of the Denver Museum of Nature and Science.

skeleton for display. After about a month's excavation, this presumed exclusively paleontological site became an archaeological locality with the discovery of a Folsom point, followed by a second point a month after that. Neither of these artifacts was found actually embedded in bison remains. Upon presenting Smithsonian curator Aleš Hrdlička with his finds in early 1927, Figgins was advised that he should halt excavation and invite the inspection of "outside scientists" to confirm the recovery of any embedded (and therefore indisputably associated) artifacts that might be encountered in the future.

Several Folsom points embedded between the ribs of extinct bison were subsequently recovered from the site during the 1927 and 1928 field seasons (the latter conducted in collaboration between the Colorado Museum of Natural History and the American Museum of Natural History). Visits by elite scholars ultimately confirmed the site as, minimally, a very Late Pleistocene bison kill. [Figures 1.3 and 1.4] But some significant questions remained. In the absence of radiocarbon dating, which would not be available for another twenty years, the site's age remained unknown and only broad estimates (ranging from "thousands of years" old to 20,000–15,000 yr BP) could made by the excavators. Additionally, because the excavation methods of the time employed only very crude measurement techniques and the project was primarily focused on removal of the bones rather than precisely documenting the site, the lay of the land at the time of the site's formation also remained unknown. [Figure 1.5] Had the site been a streambed, marsh, pond, lake during the Late Pleistocene? No one could say.

Despite its international acclaim and pivotal role in a watershed moment of American archaeology, the Folsom site remained curiously absent from scholarly publications following the brief flourish of mostly superficial treatments that appeared immediately after its discovery. The reporting of subsequently discovered Folsom sites—there are now at least forty-five localities recorded on the Great Plains and in the Rocky Mountains—had essentially eclipsed the type site until 1997, when archaeologists at Southern Methodist University in Dallas returned to the site with a major multidisciplinary research project.

The Southern Methodist University work was conducted over the course of three field seasons (1997–1999), periodic brief site visits thereafter (2000–2004), and several years of laboratory investigations, culminating in the publication of an outstanding monograph in 2006, *Folsom: New Archaeological Investigations of*

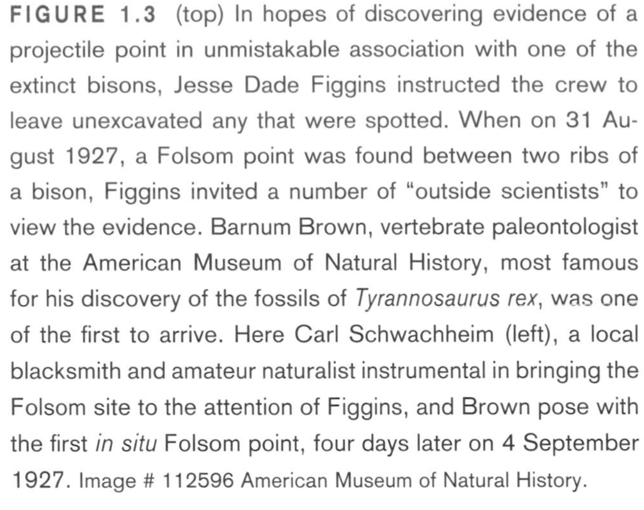

FIGURE 1.3 (top) In hopes of discovering evidence of a projectile point in unmistakable association with one of the extinct bisons, Jesse Dade Figgins instructed the crew to leave unexcavated any that were spotted. When on 31 August 1927, a Folsom point was found between two ribs of a bison, Figgins invited a number of "outside scientists" to view the evidence. Barnum Brown, vertebrate paleontologist at the American Museum of Natural History, most famous for his discovery of the fossils of *Tyrannosaurus rex*, was one of the first to arrive. Here Carl Schwachheim (left), a local blacksmith and amateur naturalist instrumental in bringing the Folsom site to the attention of Figgins, and Brown pose with the first *in situ* Folsom point, four days later on 4 September 1927. Image # 112596 American Museum of Natural History.

FIGURE 1.4 (bottom) The first Folsom point recovered *in situ* at Folsom on 31 August 1927. An intact block of sediment containing the point and ribs was carefully removed from the south bank where it was found and is now on display at the Denver Museum of Nature and Science. Courtesy of the Denver Museum of Nature and Science.

FIGURE 1.5 A Colorado Museum (now the Denver Museum of Nature and Science) truck is used to haul a plaster-jacketed bison skull up the north flank wall at Folsom, 1927. Courtesy of the Denver Museum of Nature and Science.

Bellisle Lake
JOHNSON MESA
Dry Cimarron River
Wild Horse Arroyo

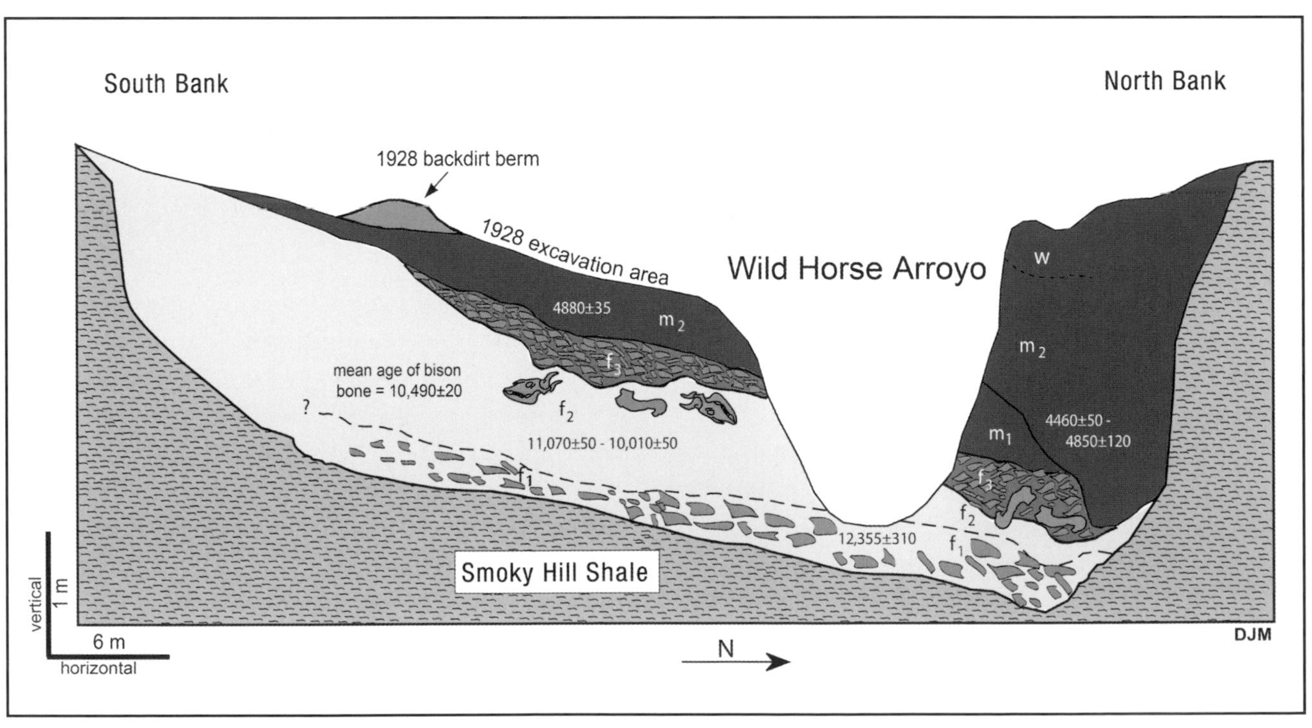
South Bank
North Bank
1928 backdirt berm
1928 excavation area
Wild Horse Arroyo
w
4880±35
m2
f3
mean age of bison bone = 10,490±20
f2
11,070±50 - 10,010±50
m1
4460±50 - 4850±120
12,355±310
f1
Smoky Hill Shale
vertical
1 m
6 m
horizontal
N
DJM

FIGURE 1.6 (left, top) Aerial photograph of the Folsom site (in the white circle) looking west-northwest, taken in 1934 as part of Barnum Brown's American Museum of Natural History Dinosaur Expedition sponsored by the Sinclair Oil Company, which had chosen a brontosaurus as its company logo. Image # 126768 American Museum of Natural History.

FIGURE 1.7 (left, bottom) Stratigraphic profile of the south and north banks of Wild Horse Arroyo. The ages shown in the profile are reported in radiocarbon years before present. Selected calibrated ages from Folsom and the other sites feature in this book are reported in the text and Radiocarbon Dating pages 318-327.

a Classic Paleoindian Bison Kill. The project team, led by David Meltzer, sought to (1) determine whether intact bone deposits remained, (2) determine whether a Folsom-age camp or habitation site was present, and (3) elucidate the site's stratigraphy, chronology, and paleoenvironment. The project also sought to integrate its approach, methods, and findings with those carried out at the site some seventy years earlier and by Vance Haynes and his colleagues in the 1970s under the auspices of the Folsom Ecology Project. Newly excavated portions of the site covered about 375 square meters over the course of the Southern Methodist University project, and questions concerning the site's stratigraphy and formation were addressed via extensive sedimentological coring, augering, and geophysical studies.

The new work at the site demonstrated that the Folsom site landscape had undergone considerable geological alteration since late glacial times, and that a thick layer of more-recent sediments overlays and masks the former Late Pleistocene landscape. [Figure 1.6] The site's stratigraphy is composed of three discrete formations which include, from top to bottom, the Wildhorse (a relatively thin, recent deposit that is no older than 700 years), the McJunkin (a 2-meter thick Holocene deposit that dates from about 7950–7580 yr BP to 5320–4860 yr BP), and the Folsom (a 2-meter thick Holocene through Late Pleistocene deposit that dates from about 15,150–13,790 yr BP to 10,510–10,250 yr BP). The bison bones and Folsom artifacts recovered from the site occurred within the middle of three Folsom formation subdivisions, named f2, which was radiocarbon dated to an unhelpfully broad range of about 13,440–13,270 yr BP to 11,720–11,280 yr BP, a span which is about three times the presently known duration of the Folsom complex.

Obtaining a more precise date for of the Folsom-age bone bed within the f2 deposit was partly achieved by modeling the site's topography as it appeared in the Late Pleistocene. [Figure 1.7] Sediment core data collected by the Southern Methodist University researchers indicated that a high bedrock wall lines the southern margin of Wild Horse Arroyo on both sides of the site, but not at the site proper, where there exists a gap. This gap was apparently created by an ancient north-flowing, two-pronged tributary (termed the "paleotributary" by Meltzer) whose channel contained a large portion of the bison bone bed and was surrounded by higher ground on both its west and east sides. This site configuration would have been well-suited for hunting, as bison attempting to avoid hunters by fleeing into the lower ground of the present-day arroyo (a former, ancestral valley termed the "paleovalley"), and possibly even being trapped there, would have presented themselves as easier prey for hunters positioned on the surrounding higher

FIGURE 1.8 (left) The first projectile point recovered from the Folsom site. Other than lithic artifacts and cut marks on bone, no additional evidence of human activity at Folsom has been discovered. Courtesy of the Denver Museum of Nature and Science.

FIGURE 1.9 (below) The original mounted bison (*Bison antiquus*) from Folsom at the Denver Museum of Nature and Science. Evidence indicates that at least thirty-two bison were killed at Folsom. Their bones have been dated to mean age of about 12,450–12,400 yr BP. Courtesy of the Denver Museum of Nature and Science.

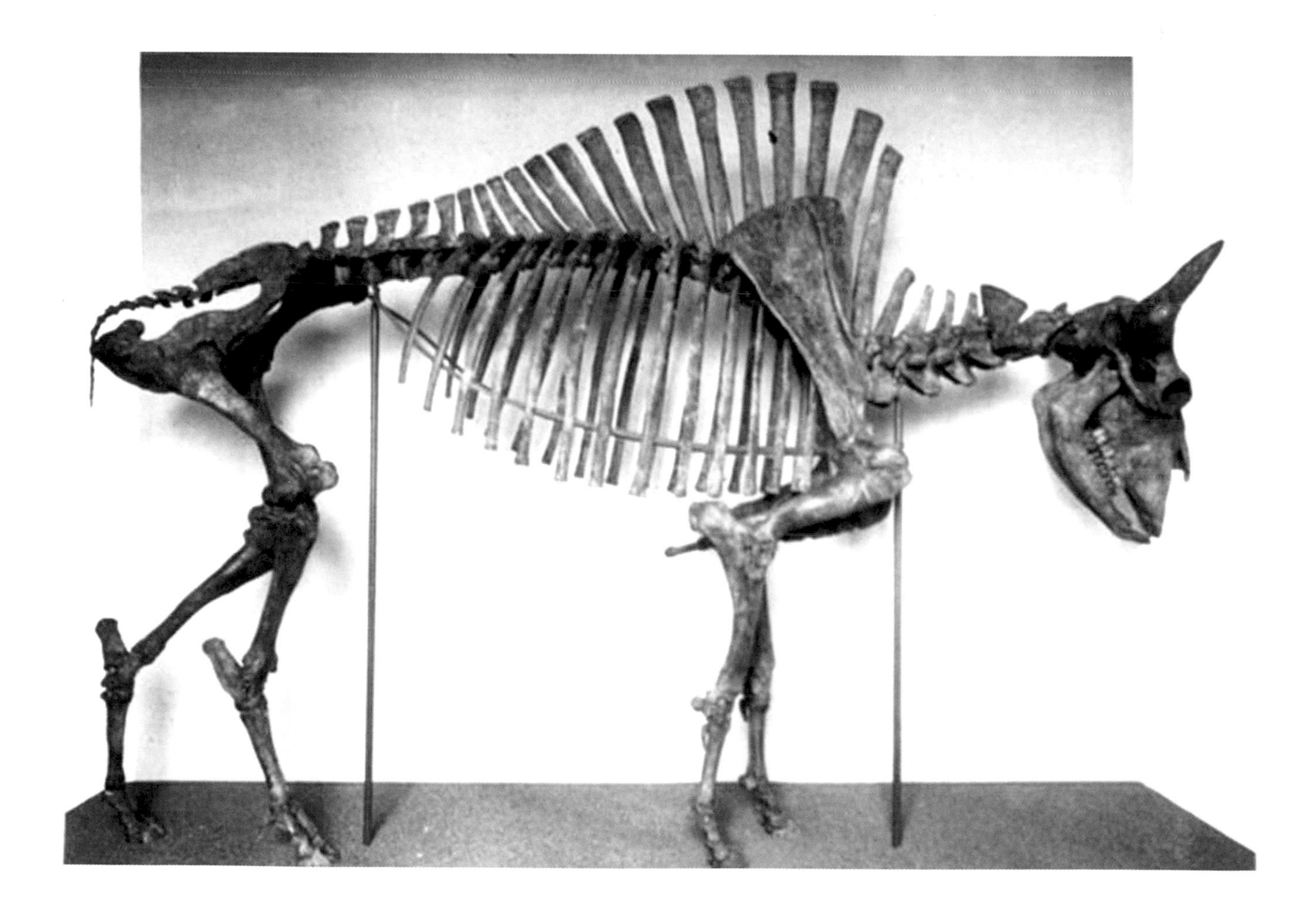

ground. [Figure 1.8] The radiocarbon dates from the two site areas, however, were problematically discrepant.

Twenty-four radiocarbon samples, six of which were run directly on bison bones, were employed to refine the chronology of the site's Folsom material. The eighteen dates obtained from charcoal within the f2 sediment package appeared to indicate that deposition began and ended earlier in the paleovalley, which was determined to be significantly older than the paleotributary, and that the two portions of the site shared only a 500-year overlapping interval. As no excavators have identified cultural features (such as fire pits) apart from the bone deposit and the dated charcoal was from naturally occurring fires over a very broad time range, the Southern Methodist University research team employed advanced radiocarbon dating techniques to process the organic fractions of six bison bone samples, three from each site area. The results from across the site proved to be statistically undistinguishable and produced a mean age of about 12,540–12,400 yr BP for the Folsom bison kill, which places it in the virtual midpoint of the age range obtained for the entire extent of the Folsom formation at the site.

The Southern Methodist University research concluded that the bison kill was a single autumn event that took at least thirty-two individual animals in both the paleotributary and paleovalley portions of the site, distributing several thousand bones across a well-defined archaeological surface. The animals clearly did not die a natural death, as the herd appears to have been relatively healthy and free from animal predation, but unequivocally showed signs of being hunted (the presence of a large number of projectile points) and butchered (dismemberment and the presence of occasional cutmarks). [Figure 1.9] Moreover, the specific bison bones recovered (and hence, discarded on site) consisted mainly of so-called low-utility skeletal elements such as lower limbs, mandibles, and crania. High-utility elements such as ribs, vertebrae, and upper limbs were for the most part removed from the site and transported to an unknown location.

The small assemblage from the site is composed entirely of flaked stone artifacts and dominated by the eponymous Folsom projectile point, the vast majority of which appear to have been broken as a consequence of the hunt. [Figure 1.10] The entire artifact count from all excavation seasons at the site stands at twenty-eight projectile points, though obviously more may have been recovered over the decades in which the site remained open, along with several flake tools and a quartzite skinning knife. While the raw material source for the quartzite knife is unknown, the stone employed in the manufacture of the site's Folsom points—mostly Alibates agatized dolomite and Tecovas jasper from the High Plains of the Texas Panhandle—appears to have been obtained from at least five discrete sources that are encompassed by an area of about 90,000 square kilometers. This mix of lithic raw material is not unusual for a Folsom complex assemblage, and the narrow array of artifact forms is typical of kill localities. The lack of fire

FIGURE 1.10 In the 1950s Homer Farr, longtime Superintendent of the nearby Capulin Volcano National Monument, collected three projectile points at Folsom, two of which are seen here. Folsom points are characteristically bifacial, fluted, and have a symmetrical, leaf-life shape and concave base. Photograph by David Meltzer.

FIGURE 1.11 Folsom projectile *in situ* with a bison (*Bison antiquus*) rib and vertebra. These extinct Late Pleistocene bison were significantly larger than living bison (*Bison bison*), and could reach a shoulder height of about 2 meters, a body length of almost 5 meters, and an estimated body weight of over 1,000 kilograms. They were formidable animals. Courtesy of the Denver Museum of Nature and Science.

pits or other cultural features clearly suggests that prolonged residential or domestic activity did not occur at the site. There probably was a nearby camp, as it would have taken the hunting party several days to process the remains of over thirty bison. To date, however, such a locality has not been found despite considerable efforts to locate one. [Figure 1.11]

Prior to the Folsom discovery, an indomitable consensus had held that human occupation of the Americas was quite recent, certainly occurring during Holocene times, and all claims for Pleistocene antiquity were met with withering criticism and denunciations from an uncompromising scholarly elite. In the absence of rigorous field methods, modern dating techniques, and later twentieth century refinements in geologic science, the fortuitous discovery of a Folsom point embedded in the ribs of an extinct Pleistocene species proved that antiquity beyond all doubt.

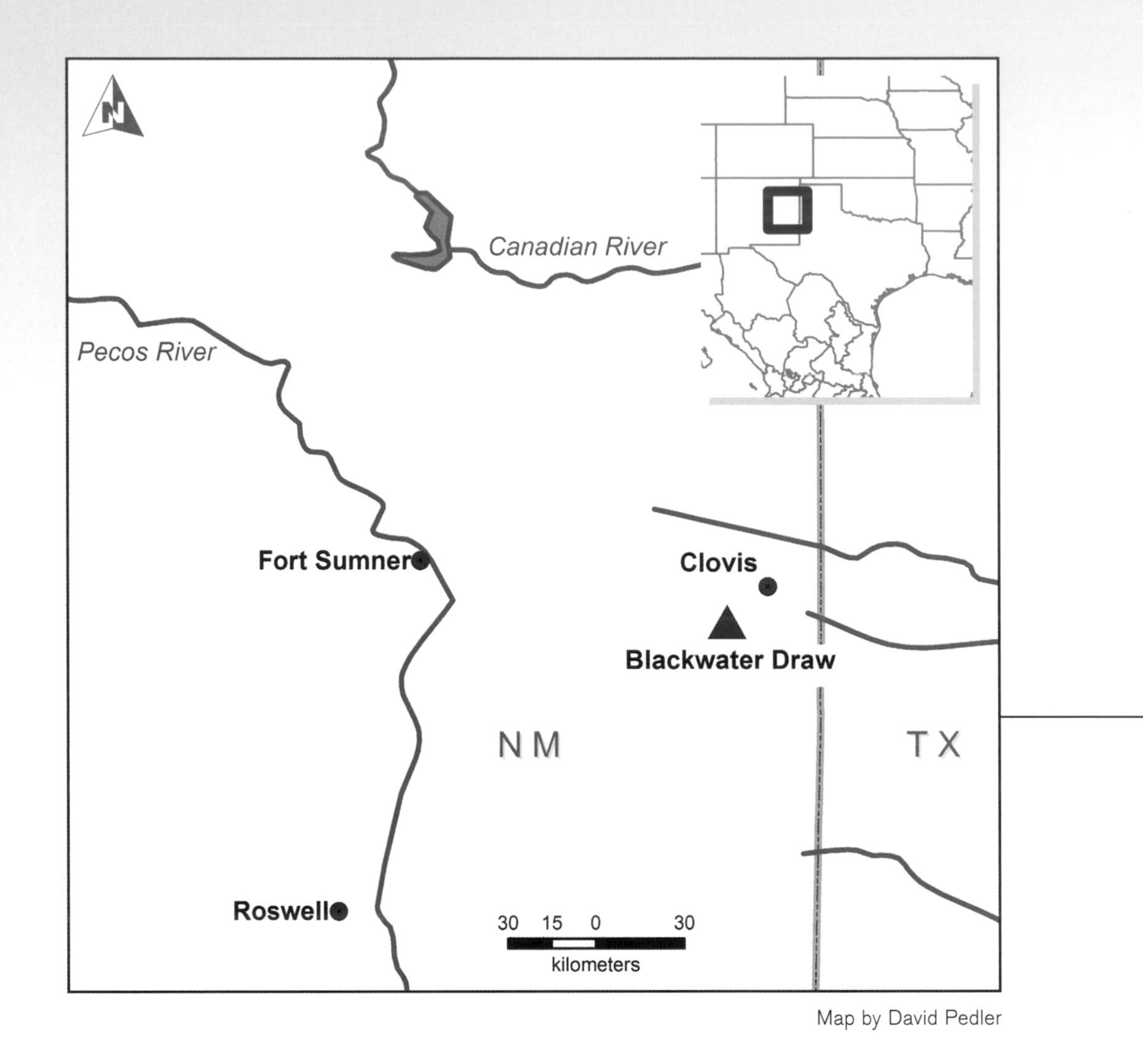

Map by David Pedler

Geologically, the landform known as Blackwater Draw is a wide, shallow valley that extends nearly 200 kilometers southeastward from Fort Sumner, New Mexico, to Lubbock, Texas, where it joins with Yellowhouse Draw to become the Brazos River. Paleoindian sites are common along the entire length of the draw.

2 BLACKWATER DRAW

LOCATION ROOSEVELT COUNTY, NEW MEXICO, UNITED STATES

Coordinates 34°15'0.38"N, 103°14'58.40"W.

Elevation 1,235 meters above mean sea level.

Discovery James Ridgely Whiteman in 1929.

Blackwater Draw is renowned as the Clovis horizon's type site and thus is considered the model for Clovis culture. The initial 1932 finds and subsequent work at the site have been instrumental in refining our understanding of Late Pleistocene environments and human ecology, and were crucial for the interpretation of findings at the Folsom site, discovered about 300 kilometers to the north and formally excavated just a few years earlier. The investigations at Blackwater Draw soon established Clovis rather than Folsom as the Western Hemisphere's pioneer human population, an interpretation that would persist and guide the discipline for decades until the gradual unraveling of the Clovis-First model at the turn of the twenty-first century.

The Blackwater Draw site is located in the northwestern reaches of the Llano Estacado in east-central New Mexico, about 10 kilometers north of Portales and 25 kilometers west of the Texas border. The site lies in its namesake geographic feature, Blackwater Draw, a broad and shallow, dry river valley that cuts through the vast Llano Estacado mesa and into the High Plains of Texas, ultimately connecting with the Brazos River basin and the Gulf of Mexico. [Figure 2.1] The Late Pleistocene climate and physical environment of this portion of the southern High Plains were quite different from today. The climate was cooler and wetter in Clovis times, and the landscape appears to have been dominated by savannah with patches of gallery forest and riparian (river bank) pine woodlands lining perennial streams in some of the region's draws. Wetter climatic intervals permitted the development of a patchwork of shallow, upland lakes and ponds and perhaps even short-lived spruce-parkland occurring between the draws. This mosaic environment and available water year-round

FIGURE 2.1 View looking southwest across the stretch of Blackwater Draw that has undergone the most sustained and intensive archaeological investigation. The Interpretative Center building on the right houses an ongoing excavation open to the public throughout the summer. Photograph by George Crawford, Director Blackwater Draw National Historic Landmark.

would have provided attractive habitats for now-extinct herbivores, other smaller mammals, invertebrates such as mollusks, and a wide variety of plant resources.

The site is composed of several discrete localities dotting a 15–20 kilometer stretch of Blackwater Draw that have undergone varying degrees of archaeological investigation. In the vicinity of the site, the draw is about 2 kilometers wide and 8 meters deep. The more prominent localities include Anderson Basins No. 1 and No. 2 on the east, the McCullom Ranch site on the west, and the more or less centrally located Blackwater Draw Locality No. 1, which lies within a natural depression that is interpreted to have been a spring-fed pond during Late Pleistocene times. Blackwater Draw Locality No. 1, also known as the "Gravel Pit," was the first locality to be discovered and shall serve as the focus here. [Figures 2.2 and 2.3]

The Blackwater Draw archaeological site was discovered in 1929 by a teenaged Eagle Scout named James Ridgely Whiteman of nearby Portales, who on one of his frequent hikes in the sand hills south of Clovis encountered large bones in association with stone artifacts in a wind-eroded hollow. Whiteman spoke of his findings to a local journalist, printer, and artifact collector named A. W. Anderson, who in turn notified Edgar B. Howard of the University Museum and the Academy of Natural Sciences in Philadelphia. Howard had recently concluded his work at Burnet Cave 250 kilometers to the south, where he had been excavating Basket Maker (3450–2450 yr BP) archaeological materials and inadvertently discovered a Folsom point in association with the remains of extinct animals. Howard first visited Blackwater Draw in August 1932, and prior to his return to Philadelphia obtained from Anderson and other collectors what he described as "some of the best artifacts" from the site for later examination.

Howard hastily returned to Blackwater Draw in November of that same year after a road construction crew quarrying for gravel had uncovered large quantities of bison and mammoth remains in a large excavation pit at the site, whereupon it was decided to commence formal field investigations in the coming year. The 1933 investigations concentrated on the gravel pit and a series of adjacent surface features that were locally known as "blow outs" and described by Howard as basin-like depressions resembling old lake beds. The gravel pit excavation uncovered an extensive and dense deposit of bison remains along with the remains of at least three mammoths, and the blow outs yielded mammoth and bison bones, some of which had remained articulated (*i.e.*, in anatomical position) for millennia. Howard returned to the site in 1934, and while these first two years of research successfully documented the contemporaneous presence of humans and extinct megafauna and accurately described the geological processes at work there, they did not succeed in determining the timing of that event or in demonstrating a firm association between the site's artifacts and the ancient animal remains.

Howard and his field crew returned to Blackwater Draw in July 1936, inspecting the margins gravel pit for eroding bone and cultural material while also commencing the excavation of a large 3,500 square meter rectangular excavation trench. The excavators soon shifted their focus to undisturbed deposits identified on the gravel pit walls in order to better define their context and further examine the associations between the artifacts and extinct faunal remains. This phase of the excavation yielded what arguably are the most spectacular and important results of the entire project—conclusive proof of the coexistence of humans and extinct animals—which overshadowed those from the following and final 1937 season conducted at the site by the University Museum and the Academy of Natural Sciences of Philadelphia. The site has been revisited several times since this original work, notably by Elias Sellards and Glen Evans in 1949–1950 (the first time radiocarbon dating was employed to date the site deposits), Fred Wendorf and Alfred Dittert in the mid-1950s, and numerous Clovis scholars such as Arthur Jelinek, C. Vance Haynes, and Dennis Stanford, among many others, who have made important contributions to the massive body of literature concerning the site and its implications for our understanding of the Clovis horizon. [Figures 2.4, 2.5, and 2.6]

A key development in the site's interpretation was gaining an understanding of its stratigraphy and how the strata were deposited. Howard's original research team and subsequent scholarly visitors to the site have generally noted the presence of several discrete strata. The top two or three strata appear to represent a Holocene layer of sands underlain by silt that contains Archaic and Late Paleoindian cultural materials along with biological evidence of a drier, hotter climate. This zone is underlain by a layer containing Folsom artifacts, the remains of bison and horse, and invertebrate remains that suggest the presence of dry to damp woods and grassland habitats. This layer is in turn underlain by what Howard and his colleagues called the "brown sand wedge," which dates to about 13,300 yr BP and contains Clovis artifacts in association with the remains of extinct faunal species (bison, pronghorn antelope, and horse) and smaller animals (opossum, squirrel, shrew, and vole) suggesting the presence of forests, a moist climate, and stream environments. Positioned between this "brown sand wedge" and the site's basal gravel layer is the so-called "gray sand," which contains Clovis artifacts in association with the remains of extinct large vertebrates (mammoth, camel, bison, and horse), small carnivores (saber-toothed cat, fox, and wolf), and other

FIGURE 2.2 (above) Aerial view to the northeast of the Blackwater Draw National Landmark where archaeological work continues. The overall site is managed, owned, and administrated by Eastern New Mexico University in cooperation with the State of New Mexico. This main portion of the site measures about 800 meters on each side. Courtesy Blackwater Draw Archives, Eastern New Mexico University.

FIGURE 2.3 (right) This map shows the locations of several major cultural areas within the main portion of Blackwater Draw National Landmark, including Isequilla's Pit, the South Bank, blade caches, and the Mitchell Locality. Courtesy George Crawford, Director Blackwater Draw National Historic Landmark.

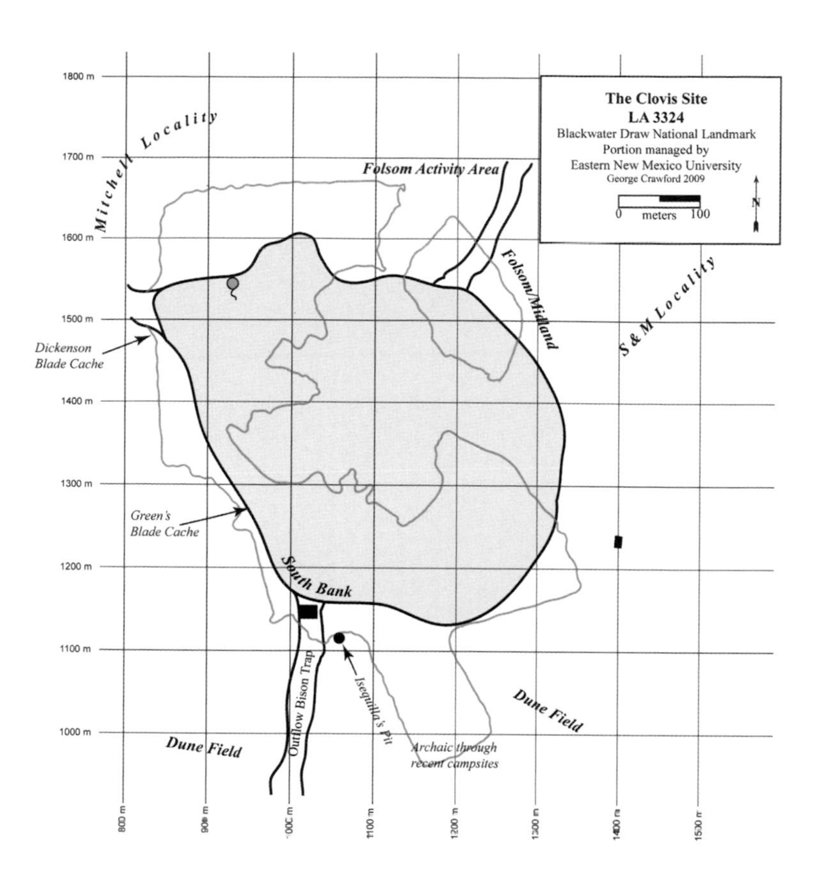

smaller faunal species (muskrat and turtle) suggesting the presence of permanent water, evenly distributed annual rainfall, wooded areas, and a warmer climate. This lower cultural layer hosted the site's earliest Clovis occupants. [Figures 2.7 and 2.8]

Perhaps the most important find of any visit to Blackwater Draw Locality No. 1 was uncovered by the Mammoth Pit excavation undertaken during Howard's 1936 field season, when Howard located a mammoth cervical vertebra that had been exposed in the south-western corner of the large excavation pit. Further efforts to expose this element revealed additional bones and a large lanceolate (*i.e.*, shaped like a lance head) point that did not quite correspond to the Folsom type and, crucially, occurred much deeper in the deposits than other identified Folsom artifacts and bison remains. The uncovering of yet more mammoth bone led the team to systematically excavate a 450 square meter trench that ultimately revealed the remains of what appeared to be a single mammoth, followed by those of a second in direct association with artifacts of indisputable human manufacture. It remains unknown whether the dozen or so mammoths ultimately identified at the site were killed by humans or simply died there. [Figure 2.9]

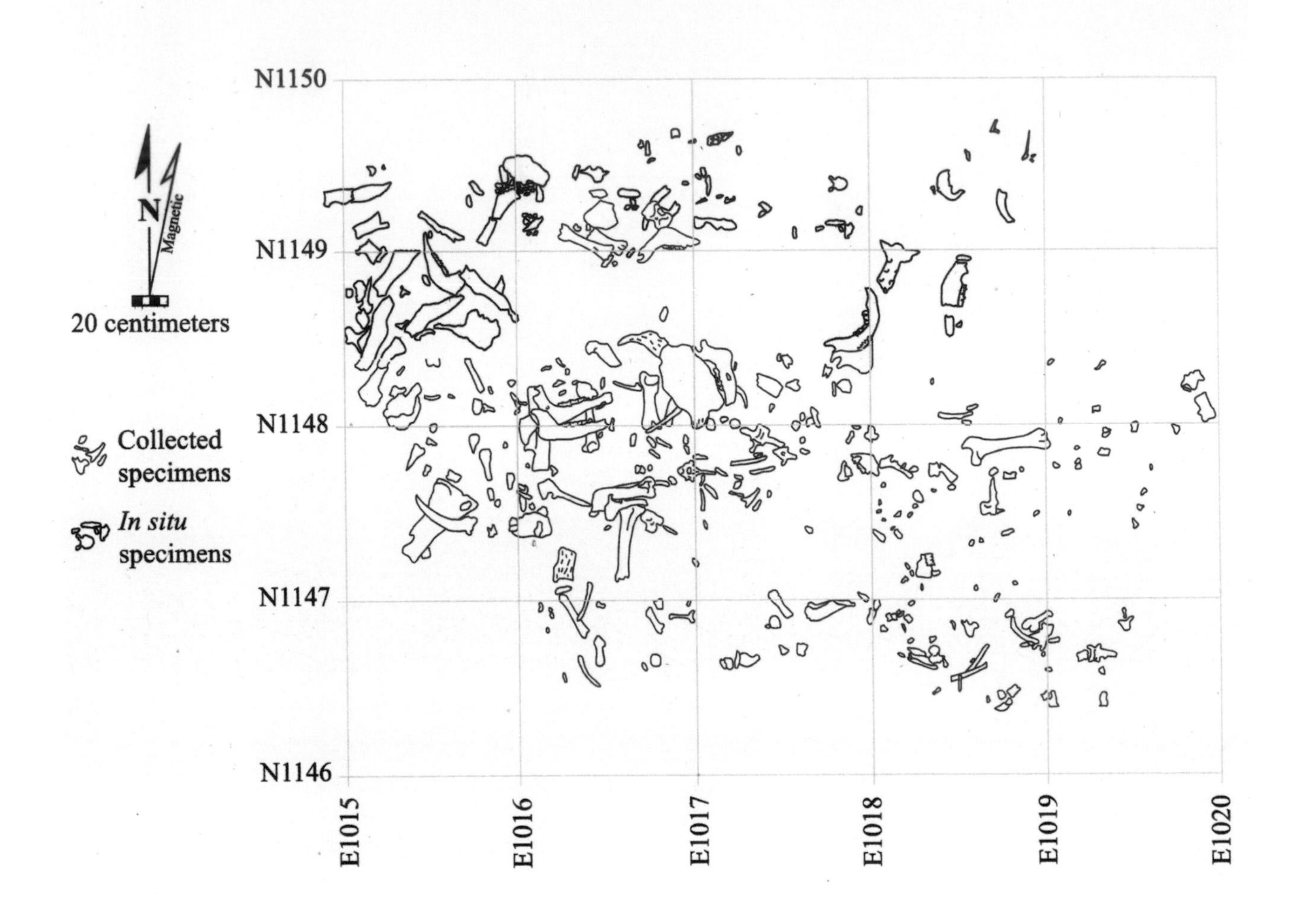

FIGURE 2.4 A detailed 1984 site map of the placement of mainly bison bones in what is thought to be a late Paleoindian butchering area on the south bank of Blackwater Draw. The remains of at least six bison, two turtle carapaces, and the fractured thighbone of a rabbit-sized animal were recovered. Courtesy Stacey D. Bennett.

FIGURE 2.5 (top) Part of the Clovis age mammoth kill site on the north bank of Blackwater Draw excavated in 1964 under the supervision of the late George Agogino, long-time Chair of the Department of Anthropology at Eastern New Mexico University and founding director of the Blackwater Draw Museum. Courtesy Blackwater Draw Archives, Eastern New Mexico University.

FIGURE 2.6 (bottom) Two of the many Clovis points recovered in 1964 from the north bank mammoth kill site, illustrating their broad range in size. Photograph by George Agogino; courtesy Blackwater Draw Archives, Eastern New Mexico University.

FIGURE 2.7 (above) "Isequilla's Pit," an area of Blackwater Draw Locality #1, during its 1967–1969 partial excavation under the direction of Alberto Isequilla. He abruptly abandoned this fieldwork, leaving a 10 meter long profile that documents environmental change from about 13,500 yr BP. Courtesy Blackwater Draw Archives, Eastern New Mexico University.

FIGURE 2.8 (right, top) In 2009 Isequilla's Pit was reopened for excavation by the Eastern New Mexico University, during which two overlaying bison bone beds were uncovered dating to the late Pleistocene. Photograph by George Crawford, Director Blackwater Draw National Historic Landmark.

FIGURE 2.9 (right, bottom) *Bison antiquus* bones uncovered in late Paleoindian strata at Isequilla's Pit in 2010 by the Eastern New Mexico University under the direction of J. David Kilby, Assistant Professor of Anthropology and Applied Archaeology, Eastern New Mexico University, and George Crawford, Director of the Blackwater Draw National Landmark. Even though these bones are extremely old, the arid environment of the area preserves them in a relatively pristine state (*i.e.*, not fossilized), making them extremely fragile. Courtesy Blackwater Draw Archives, Eastern New Mexico University.

According to the original report on the 1936 excavation by Howard's assistant, John Cotter, this trench yielded at least four Clovis points (three whole and one fragmentary), two uni-beveled bone artifacts (representing either toggles or foreshafts for propelling or mounting the points on spears), one scraper, three flake tools, and three modified flakes in exceptionally tight association with the mammoth remains. In the subsequent 1937 field season, the excavation in this portion of the site was expanded considerably, resulting in the recovery of an additional two Clovis points and additional mammoth bones. The lithic raw materials identified in the Clovis point assemblage include Edwards chert from the Edwards Plateau of central Texas, about 450 kilometers southwest of the site, and Alibates agatized dolomite from the High Plains of the southern Texas Panhandle, over 200 kilometers to the east. [Figure 2.10]

ENMU
LA 3324
BWD
SITE
N 1131 E1045
FS 636
THRU
645
LEVEL 3 STRATUM E
06 21 2010

FIGURE 2.10 A Clovis-age biface preform recovered in the northwest portion of the Blackwater Draw National Landmark in what was probably a storage cache. The flaking on this artifact is classic soft-hammer Clovis technique. Photograph by George Crawford, Director Blackwater Draw National Historic Landmark.

The report on the 1936 excavation at the site concludes with the observation that although "...there seems to be little doubt that man lived in America contemporaneously with certain types of animals that later became extinct... it seems to be increasingly clear that the time of this extinction is hard to fix, and that other methods will have to be used to date such finds." Those other methods would not arrive until several years after Howard's death, when physical chemist Willard Libby developed the radiocarbon dating technique in the late 1940s. Thanks to this new dating tool and the efforts of Elias Sellards and Glen Evans, and particularly the definitive work of geoarchaeologist C. Vance Haynes over the course of decades, the site's Clovis component has been dated to about 13,300–13,000 yr BP.

Just as it is astounding that some of the ephemeral pre-Clovis sites discussed later in this book were discovered, so too is it astounding that any of the data from Blackwater Draw was ever recovered in the first place. The archaeological project was spurred on by and conducted in the face of an extensive, decades-long gravel quarrying operation that destroyed a considerable portion of its deposits and has led to some confusion over interpretations of its earliest and seminal occupation. [Figure 2.11] Fortunately, quarrying ceased when the site was purchased by Eastern New Mexico University in 1978 in the interests of security, access, interpretation, and conservation.

FIGURE 2.11 Excavations 1962–1964 pitted archaeologists against a mine operator to save what they could of the north bank of the Blackwater Draw from destruction and looting. Miners used the bulldozer seen here to dig out enormous quantities of artifacts for private collections and possible commercial sale. The white oblong object to the right of the bulldozer is a plaster jacketed remnant of a Folsom bison kill wrapped by the archaeologists for preservation. Had this area been excavated with proper methods, this portion of the site alone would stand as one of the most significant Clovis mammoth kills in the archaeological record. Courtesy Blackwater Draw Archives, Eastern New Mexico University.

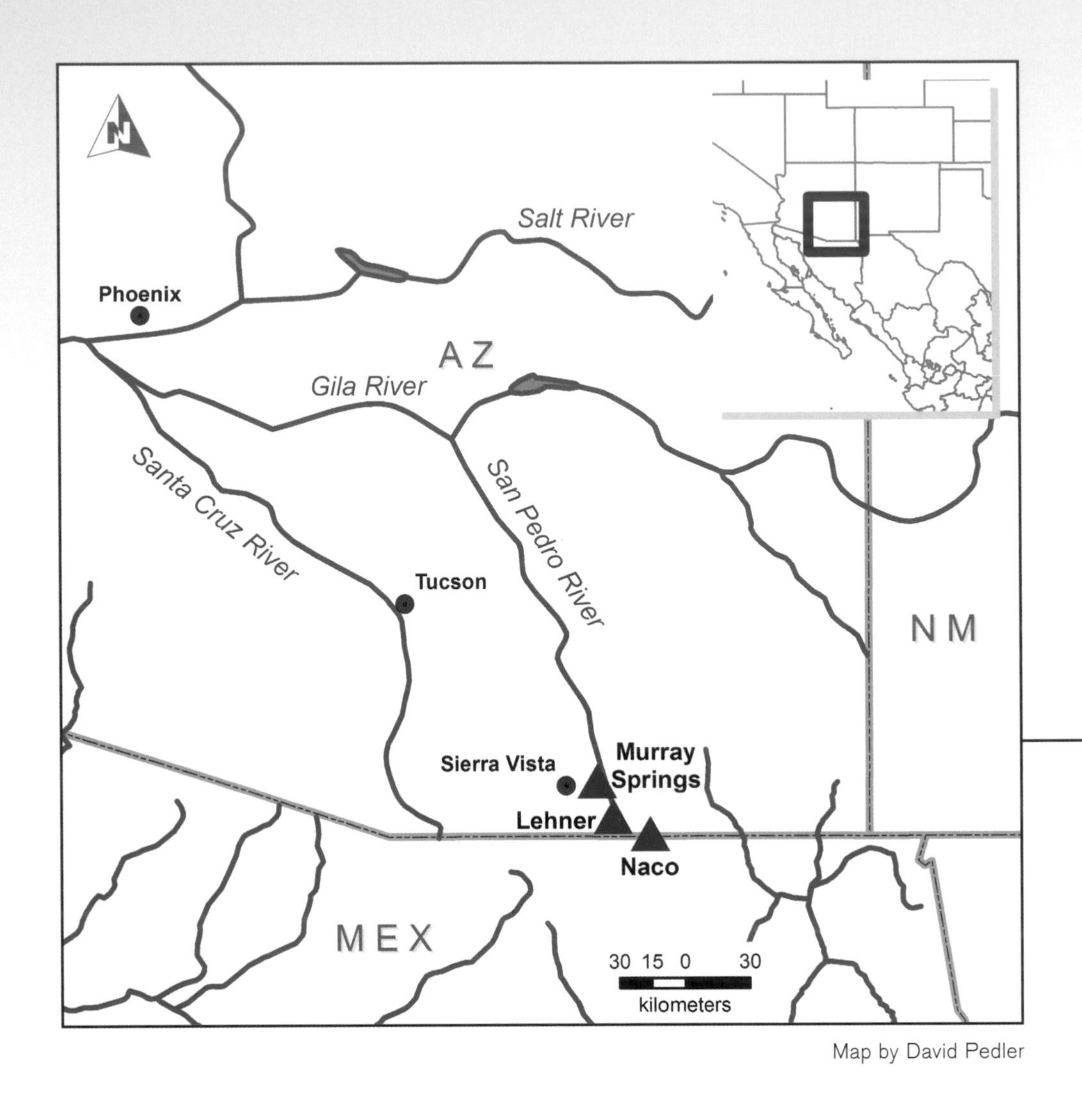

Map by David Pedler

The Lehner, Murray Springs, and Naco sites were the first archaeological localities to produce Clovis points in direction association with mammoth kills and the first to contain demonstrable Clovis hearths. They were also among the first archaeological localities to have been radiocarbon dated.

3 LEHNER, MURRAY SPRINGS, AND NACO, ARIZONA

LOCATION COCHISE COUNTY, ARIZONA, UNITED STATES

Coordinates Naco, 31°21'0.21"N, 109°57'36.42"W; Lehner, 31°25'22.23"N, 110°6'44.52"W; Murray Springs, 31°34'14.26"N, 110°10'43.91"W.

Elevation Lehner and Murray Springs, 1295 meters above mean sea level; Naco, 1,390 meters above mean sea level.

Discovery Lehner, Edward Lehner in 1952; Murray Springs, C. Vance Haynes, Jr., in 1966; Naco, Fred and Marc Navarette in 1951.

The Naco Mammoth Kill Site has the distinction of being the first buried Clovis site identified west of North America's Continental Divide at a time when only three other buried Clovis sites had been known to exist. The nearby Lehner Mammoth Kill-Site was investigated by the Naco site researchers three years after the Naco investigation, and their work provided some of the first radiocarbon ages ever reported for the Clovis complex. The only slightly more distant Murray Springs site was investigated ten years after the excavations at Lehner. Together, these megafauna hunting and processing sites have served as the baseline for the study of a region which by any measure contains a remarkable concentration of Clovis archaeological sites.

The three sites are located in southwestern Cochise County, southeastern Arizona, just a few kilometers north of the Mexican border, about 285 kilometers southeast of Phoenix, 115 kilometers southeast of Tucson, and 10–35 kilometers east of Sierra Vista. They lie within the southeastern Arizona's and northwestern Mexico's upper San Pedro Valley watershed, which covers an area of about 6,400 square kilometers. The entire San Pedro River watershed from its headwaters in Mexico to its confluence with the Gila River about 200 kilometers northwest of the three sites covers an area of about 12,200 square kilometers. The greater watershed of the Gila River, which in turn flows into the Colorado River basin at Yuma, Arizona, and thence into the Gulf

FIGURE 3.1 An arroyo, or draw, at Naco. Mammoth and bison bones that were washed out of arroyos by seasonal rains at these sites caught the attention of archaeologists. Courtesy Arizona State Museum.

of California, drains an area totaling more than half of Arizona's landmass.

The upper San Pedro Valley landscape is dominated by steep, linear mountain ranges interspersed with flat, semi-arid valleys containing highly variable ecological zones. Essentially defining a transitional zone between the Sonoran Desert on the west and the Chihuahuan Desert on the east, this biodiverse region supports Earth's second highest land-mammal diversity and almost 400 bird species. The region's vegetation is predominantly desertscrub, grassland, and mesquite woodland, with less commonly occurring patches of oak woodland-savannah, riparian (river bank) forest, and coniferous forest at higher elevations. At the time of the region's Clovis occupation, as appears to have been the case from the Late Pleistocene through much of the Holocene, the valley was predominantly covered in grassland. Beginning sometime before European contact—but intensified by human settlement, cattle-grazing, wood-clearing, and drought throughout the eighteenth century—this rich grassland has been impinged upon by desert scrub more characteristic of the Chihuahuan Desert.

The three sites are relatively closely spaced, with only about 17 kilometers separating easternmost Naco from centrally located Lehner, and roughly the same distance separating Lehner from westernmost Murray Springs. The latter two sites are close (800 meters and 3.6 kilometers, respectively) to the San Diego River's present-day west bank at roughly the same elevation (around 1,275 meters above sea level), while Naco lies about 15 kilometers east of the river at an elevation of 1,390 meters above sea level. The sites' circumstances of discovery were broadly similar, in that all three were first identified in the eroding walls of arroyos (*i.e.*, predominantly dry stream beds that flow only temporarily or seasonally), also known as "draws," thanks to erosional processes that are thought to have begun in the 1920s in the cases of Naco and Lehner, and as late as the 1950s or early 1960s at Murray Springs.

The Naco site was discovered in September 1951 by father and son avocational paleontologists Fred and Marc Navarette, who recognized two large projectile points and associated mammoth bones in an arroyo wall on the south bank of Greenbush Creek, a major tributary of the San Pedro River. [Figure 3.1] The Navarettes approached staff of the Arizona State Museum with their find, and aroused sufficient interest that the site was excavated in the following spring by an interdisciplinary team of archaeologists, geologists, and paleontologists under the direction of University of Arizona archaeologist Emil Haury. [Figures 3.2 and 3.3] While Haury and his colleagues were investigating Naco over 14–18 April 1951, local resident Edward Lehner conveyed his own discovery of mammoth bones in an arroyo wall near the west bank of the San Pedro River to personnel of the Arizona State Museum. This second discovery led to a visit to the Lehner site by Haury, who was struck by the similar deposit that he had just observed at Naco. After the especially heavy summer rains of 1955 further exposed the Lehner mammoth deposit, Haury and his former Naco research team moved on to Lehner, where they conducted excavations in December 1955 and February 1956. [Figure 3.4]

FIGURE 3.2 A Clovis point *in situ* with bones of a Columbian Mammoth (*Mammuthus columbi*) at the Naco Mammoth Kill site, 1952. The Columbian mammoth was one of the largest of the mammoth species and also one of the largest probiscideans to have ever lived, measuring up to 4 meters tall and weighing up to 10 tonnes. It was one of the last American megafaunal species to become extinct. Courtesy Arizona State Museum.

FIGURE 3.3 The eight Clovis points Emile Haury recovered in association with the remains of a Columbian Mammoth (*Mammuthus columbi*) at Naco. Clovis points have been found at over 1,500 sites across North America. Courtesy Arizona State Museum.

FIGURE 3.4 A Clovis point *in situ* near a bison mandible and mammoth bone at the Lehner site, 1955. Thirteen fluted Clovis fluted projectile points, butchering tools, chipped stone debris, and hearths were recovered at the site. Courtesy Arizona State Museum

The Murray Springs site was discovered a decade later in the spring of 1966 by C. Vance Haynes, Jr., a newly minted University of Arizona PhD and one of Haury's former graduate students, while he was mapping late Quaternary geological deposits in the San Pedro Valley. While examining geologic profiles in Curry Draw, Haynes attention was drawn to the presence of a distinct layer of black, organic-rich clay that was identical to a layer he had observed at Lehner some years before. Dubbed the "black mat" by Haynes and "black swamp soil" by the Lehner research team, this stratum was observed immediately above the megafaunal remains and artifacts recovered from Lehner. Almost immediately after relating that observation to his colleague, Haynes spotted mammoth bones in the wall of the draw, directly beneath the black mat layer. Haynes hastily proposed an excavation project to the National Geographic Society's Committee for Research and Exploration, and work commenced at Murray Springs shortly thereafter in June 1966. Work was carried out at Murray Springs by Haynes and his colleagues, in collaboration with Emil Haury and others, in summer field seasons from 1966 to the project's conclusion in 1971. The Murray Springs research team returned to the Lehner site in 1974 and 1975 in an attempt to locate an associated Clovis hunting camp, but was unsuccessful.

The largest and most extensively excavated of the three sites, to an exponential degree, is the Murray Springs site. Covering an area of over 4 hectares, Murray Springs is composed of three primary archaeological concentrations: a megafaunal kill and processing zone (Areas 1–5) located in the site's northeastern portion and bisected by the south branch of Curry Draw, a briefly occupied hunting camp (Areas 6 and 7) directly to the south, and a smaller area that apparently represents an ancient springhead (Area 8) on the site's western margin. [Figure 3.5] The Lehner site, which is the smallest of the three localities, is composed of two zones along a northwest–southeast oriented, 35 meter stretch of Curry Draw: a bone bed on the southeast, and a hearth area on the site's northwestern margin. The somewhat larger Naco site is composed of a single zone

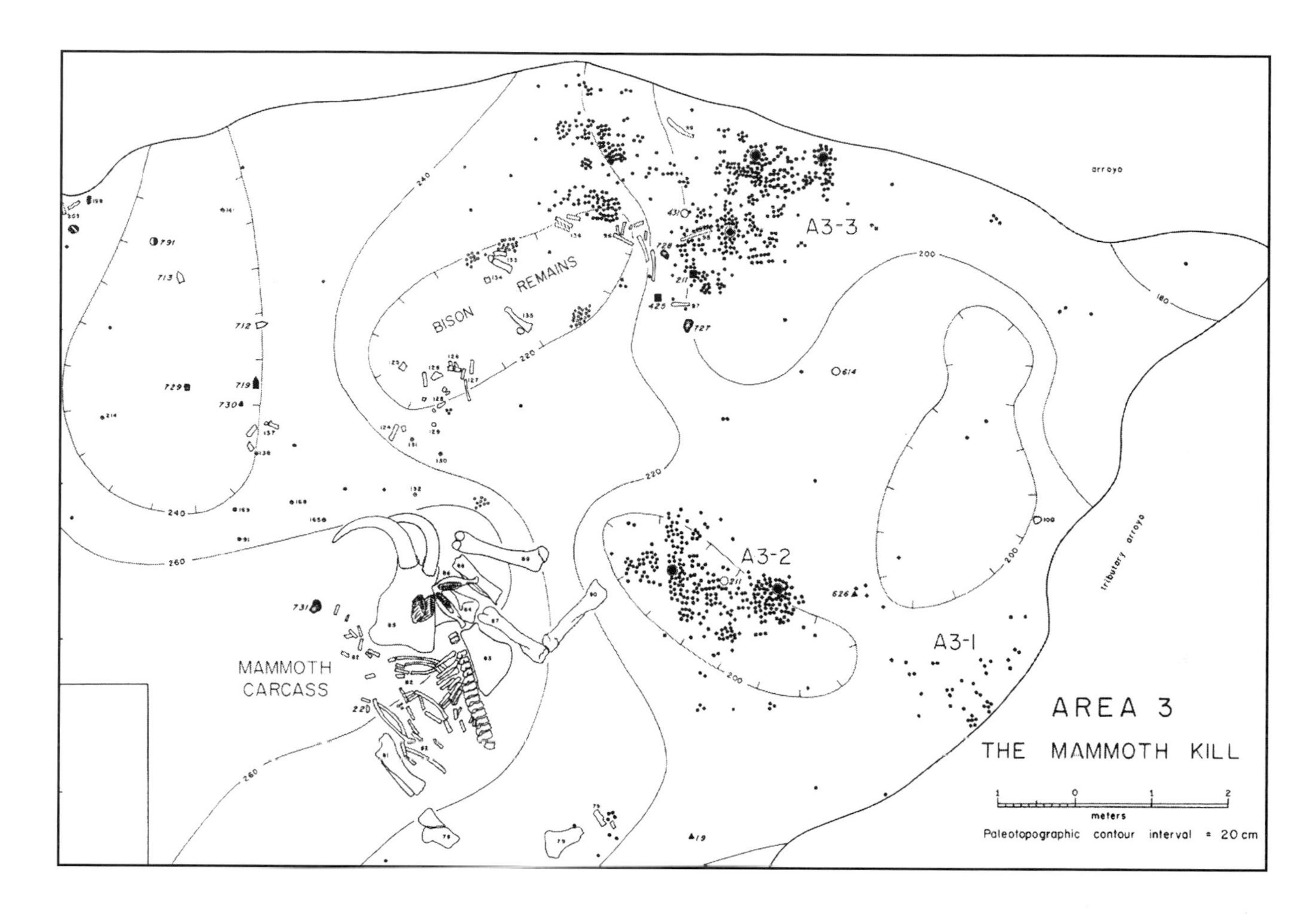

FIGURE 3.5 Map of Murray Springs Area 3 mammoth kill site. Area 3 comprises 350 square meters of buried occupation surface excavated primarily in 1967 and extended to the west and south in 1968 and 1969. This area was the central location for kill and processing activities based on the density and variety of cultural debris. Courtesy of the University of Arizona Press.

of mammoth bones and artifacts in a 40 meter by 30 meter area on Greenbush Creek.

The silt, sand, and gravel stratigraphy at Naco is not readily comparable to the broadly similar Lehner and Murray Springs sites, which is understandable given Naco's position on a tributary channel some 15 kilometers east of the San Pedro River and 1,000 meters above the valley floor. The complex Murray Springs stratigraphic sequence masterfully described by Haynes, a subject which after all is what first drew him to Curry Draw, stretches back to earlier than 30,000 yr BP. But what is of particular interest here is the Clovis-age boundary that separates the Pleistocene and Holocene epochs: the lower Murray Springs Formation and the overlying Lehner Ranch Formation, which contains the black mat at its base. Through time, the black mat was covered by additional sedimentation measuring about 1–2 meters thick at Lehner and about 2–4 meters thick at Murray Springs. [Figure 3.6]

Haynes postulates that prior to the arrival of Clovis people and long before the deposition of the black mat, the Murray Springs area was a wet spring field covered in grasses and sedges. During the site's Clovis occupation and the associated killing and butchering

FIGURE 3.6 Precisely how the "black mat" layer was deposited at the Murray Springs (seen here) and Lehner sites has never been fully explained. But it is clear that soon after their Clovis occupations, the sites' low areas became submerged. Pond and marsh conditions prevailed for the next millennium, suggesting an algal origin for the black mat deposit. Courtesy of C. Vance Haynes.

of megafauna, the region appears to have experienced episodes of drought that were soon followed by the Clovis abandonment of the site and the megafaunal extinctions emblematic of the terminal Pleistocene. The water table then appears to have risen again gradually, inundating the site's lower elevations to form ponds and marshes that existed for about 1,000 years after the Clovis occupation. The most significant net effect of this chain of events at Murray Spring—and, apparently, Lehner—was that the ponding event that produced the black mat was also responsible for rapidly but gently burying and thereby preserving the remains that were identified by Haury, Haynes, and their colleagues almost 13,000 years later.

The principal cultural features identified at all three sites are bone beds and associated activity areas left behind by Clovis hunters. The smallest assemblage of animal bone was recovered from the Naco site, which yielded the incomplete remains of a single Columbian mammoth (*Mammuthus columbi*) in direct association with eight Clovis projectile points but no butchering tools or lithic debitage (flakes produced during stone tool manufacture). The larger, more species rich megafaunal assemblage recovered from Lehner is dominated by the remains of thirteen individual mammoths

FIGURE 3.7 Jaw of a dire wolf (*Canis dirus*) found on the Clovis surface in Area 3. Nearly all of the remains at Murray Springs are from extinct species of bison, mammoth, and horse, evidently important food animals. In many cases, a number of skeletal elements can be confidently attributed to specific, individual animals. Courtesy of C. Vance Haynes.

(*Mammuthus* sp.) and lesser numbers of other extinct species including camels (*Camelops* sp. and *Hemiauchenia* sp.), horse (*Equus* sp.), and tapir (*Tapirus* sp.). These remains were recovered from a single bone bed in direct association with thirteen Paleoindian projectile points (including eight Clovis points), a small assemblage of scrapers, a pebble tool, and a small quantity of lithic debitage. Two Clovis-age hearths were also excavated at Lehner, yielding charcoal for some of the first radiocarbon ages processed for Clovis following the development of radiocarbon dating by Willard Libby in 1949 (see pages 58-67).

In all, five animal kill and processing areas were identified at Murray Springs, which yielded almost 1,000 individual bone specimens attributed to the individual remains of at least four Columbian mammoth (*Mammuthus columbi*), fourteen bison (tentatively ascribed to *Bison antiquus*), one camel (*Camelops hesternus*), and two horses (tentatively ascribed to *Equus occidentalis*). [Figure 3.7] These materials were associated a diverse suite of lithic artifacts that includes eighteen Clovis points, seven bifaces, thirteen blades (but no blade cores), a host of unifacial and bifacial retouched tools, expedient flake tools, scrapers, gravers, and almost 15,000 pieces of lithic debitage (predominantly associated with bifacial technology) distributed

FIGURE 3.8 A mammoth bone "shaft straightener" *in situ* in Area 3 at Murray Springs. It was found in close association with wolf remains, a battered cobble, scatters of charcoal, and a few flakes. The tool was broken in half with one part visible on the surface and the other half buried slightly below. Courtesy of C. Vance Haynes.

among forty-eight discrete clusters across the site. At least twenty different lithic raw materials were employed in the Murray Springs stone tool industry, including nine varieties of chert, four varieties of chalcedony, and single varieties of jasper, obsidian, basalt, siltstone, limestone, and quartzite, many of which were probably widely available in the region. The few exotic, non-local raw materials in the Murray Springs assemblage—such as petrified wood—may have been obtained from as far away as 350 kilometers to the north. Among the diminutive collection of bone artifacts from the site's Clovis level is a distinctive mammoth bone "shaft straightener" that bears a strong similarity to the *batons de commandement* encountered at Old World Paleolithic sites. The occurrence of this artifact form at a Clovis site is unique to Murray Springs. [Figures 3.8 and 3.9]

The age of the Clovis deposit at Murray Springs is determined by Haynes to range between 12,870–12,700 yr BP, based on the average of what appear to be the eight most-accurate radiocarbon samples from the site's Clovis level. Determining the age of the Clovis material at Lehner was far more problematic, perhaps owing to the newness of the radiocarbon dating technique at the time of the site's investigation and issues related to the emerging techniques employed to remove sample contaminants. The initial age determination for the deeper of the Clovis hearths at Lehner, for example, ranged from 8,790–6,940 yr BP to 10,410–8300 yr BP and was considered unacceptable to Haury and his colleagues because, in Haury's words, "no aboriginal fire was kept

FIGURE 3.9 (top) The mammoth bone "shaft straightener," measuring 259 millimeters in length and crafted from a mammoth long bone, presumably from a freshly butchered animal. (bottom) Lithic artifacts from Murray Springs Area 3. From left to right: a graver, projectile point tips, flake tool fragment, blade fragment, and two Clovis projectile points. Courtesy of C. Vance Haynes.

burning that long!" The subsequent removal of contaminants from two samples recovered in the same hearth produced an age of 13,770–13,560 yr BP, which was more or less consistent with the findings of independent analyses conducted at the University of Michigan and the University of Copenhagen. Haynes's later 1974–1975 visit to Lehner recovered twelve charcoal samples from the Clovis occupation surface whose average age ranges 12,910–12,710 yr BP, which is remarkably consistent with the age obtained for Murray Springs and tends to be cited by scholars as Lehner's actual age. The Naco site has remained undated.

The Naco, Lehner, and Murray Springs sites played pivotal roles in mid-twentieth century formulations of Clovis as a Late Pleistocene lifeway that was primarily focused on the hunting of now-extinct a big game, and even scholars who are skeptical of this narrow conception of Clovis have continued to view the three sites as containing firm evidence of at least this aspect of Clovis life. As Clovis scholarship moved forward through the twentieth century and into the present one, new sites were discovered across the entire North American continent, and the Clovis archaeological database grew into its present form. During this time, the understanding of Clovis evolved into a vastly more variable lifeway that was highly adaptable to the continent's variable conditions. Some key issues in Clovis scholarship have been definitively resolved while some remain to be examined and adjudged, but the Naco, Lehner, and Murray Springs sites helped get the discussion started.

Map by David Pedler

The Shoop is a large Paleoindian campsite, dated to ca. 12,750 yr BP, located on a low hill 10 kilometers east of the Susquehanna River. Over 6,800 artifacts are known to have been collected from the 70–80 years of archaeological activity at this site.

4 SHOOP

LOCATION DAUPHIN COUNTY, PENNSYLVANIA, UNITED STATES

Coordinates restricted.

Elevation 238 meters above mean sea level.

Discovery George Gordon in 1930.

The Shoop site occupies a unique place in the archaeology of eastern North America. When the first major scholarly article reporting the site was published in 1952, Shoop was only the third Paleoindian locality that was known to exist east of the Mississippi River. Its physical setting was thought to be exemplary (but ultimately demonstrated to be anomalous), and its position in what was later established to be a continent-wide continuum of the Clovis fluted point tradition went unrecognized. In fact, the site's lithic technology was initially thought to represent a blade-based industry akin to Old World Upper Paleolithic sites, and was considered an ancestor rather than a contemporary of the Clovis complex. This assessment ultimately did not survive, but the Shoop site has nonetheless remained a key locality in eastern North American Paleoindian archaeology which may yet reveal compelling data on short-term human adaptation in the dramatically changing landscape and climate of northerly latitudes at the end of the Pleistocene.

The Shoop site is located in Dauphin County, east-central Pennsylvania, about 10 kilometers east the Borough of Halifax, 25 kilometers northeast of Harrisburg, and 150 kilometers northwest of Philadelphia. Situated within the eastern half of the massive 375,000 hectare lower Susquehanna River–Penns Creek watershed, the site lies in a small, upland valley less than 1 kilometer north of the divide between two Susquehanna tributary streams, Armstrong Creek on the north and Powell Creek on the south. Headwater streams flow within 150–500 meters of the site's boundaries in all directions but south, which is upslope. The region is also quite complex in physiographic terms, as the mapped boundaries of no

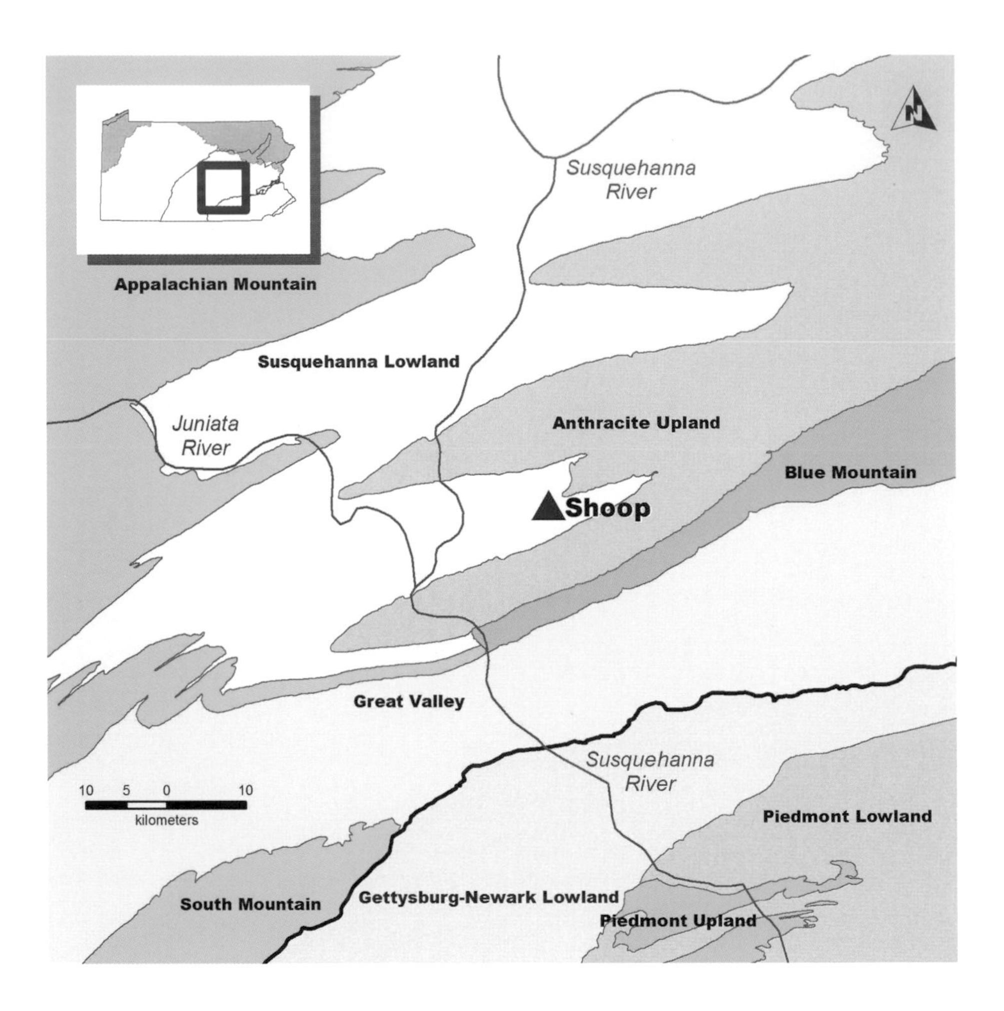

FIGURE 4.1 Map showing the boundaries of the nine physiographic sections that occur within 50 kilometers of the Shoop site. The inset shows the geographic extent of the main map, the boundaries of Pennsylvania's physiographic provinces, and the maximum southern extent of the Wisconsin glacier (in red).

fewer than nine major zones (or physiographic sections) occur within 50 kilometers of the site, and four are mapped within less than 15 kilometers. [Figure 4.1]

The maximum southern extent of the Laurentide ice sheet lay about 100 kilometers north of the Shoop site, but by about 15,000 yr BP the ice front had retreated through present-day New York and north of the St. Lawrence Valley. Following this glacial retreat, the regional vegetation transitioned from an open parkland tundra (populated by scattered stands of spruce, fir, and pine trees interspersed with grasses, sedges, heath species, and shrubs) to a boreal forest dominated by white pine and white birch. Locally, the site's surrounding slopes were covered in spruce and pine, and level areas were covered in grasses and deciduous trees such as oak. A presently buried swamp to the east of the site was populated by birch and other wetland plant species. A series of springs on the site's southeastern margins would have provided its visitors with an accessible (though not necessarily nearby) year-round supply of water.

FIGURE 4.2 Two centuries of plowing at Shoop have obliterated any Paleoindian cultural features that might have been present. The lack of charcoal from features such as hearths has precluded radiocarbon dating at the site. Based on the lithic artifact assemblage recovered, it is assumed the site is of Clovis age. Photograph by Kurt Carr.

The Shoop site covers an area of about 15 hectares in rolling terrain on a north-facing slope at elevations ranging about 230–250 meters above sea level. Relief between the site's surface and the elevation of local stream beds is about 20 meters to the east and west, and about 35 meters to the north. As the local soils for the most part are exceptionally shallow, the Shoop site's cultural materials are also quite shallowly buried and have been significantly disturbed by agricultural activity dating back to the eighteenth century, when the land was first farmed by the Shoop family. [Figure 4.2] The prevailing view of the surrounding landscape from the site is to the south, where in the absence of tall trees one could expect an unobstructed sightline through a narrow stream gorge and the Armstrong Creek Valley beyond for a distance of almost 5 kilometers. This unique vantage point apparently prompted one of the site's first professional archaeological investigators, John Witthoft of the State Museum of Pennsylvania in Harrisburg, to deem the site an

overlook locality and camp site that was established by ancient prehistoric hunters.

The Shoop site was discovered in the 1930s by avocational archaeologist George Gordon, and has been heavily collected by amateur artifact collectors and professional looters ever since, despite the best efforts of the current group of landowners. Limited (but unfortunately undocumented) professional test excavations were conducted in 1942 by Edgar Howard while he was a research associate at the University of Pennsylvania Museum of Archaeology and Anthropology in Philadelphia.

The site was subsequently investigated in 1950 by John Witthoft, who surveyed the about 8 hectares of the ground surface and recovered almost 2,000 lithic artifacts from a total of eleven artifact concentrations he

FIGURE 4.3 (above) Students from Mercyhurst and Clarion Universities search for Paleoindian artifacts on the Lesher property at Shoop. Given reports of finds in adjacent fields, it appears the area currently designated as the Shoop site is too limited and simply an artifact of access granted by other local property owners. Photograph by Kurt Carr.

FIGURE 4.4 (right) Despite the fact that the Shoop site has been known to amateur collectors for over seventy years, the boundaries of the site have never been thoroughly defined. Edgar Howard of the University Museum in Philadelphia, the first professional to visit the site in 1941, described it as an area of 20 acres (8 hectares) in plowed fields on high ground. Field work in the 1950s by John Witthoft, archaeologist at the State Museum of Pennsylvania, and in 2008 by teams sponsored by the State Museum of Pennsylvania, Mercyhurst University, and Clarion University extended the area under investigation to about 15 hectares. Courtesy of Kurt Carr.

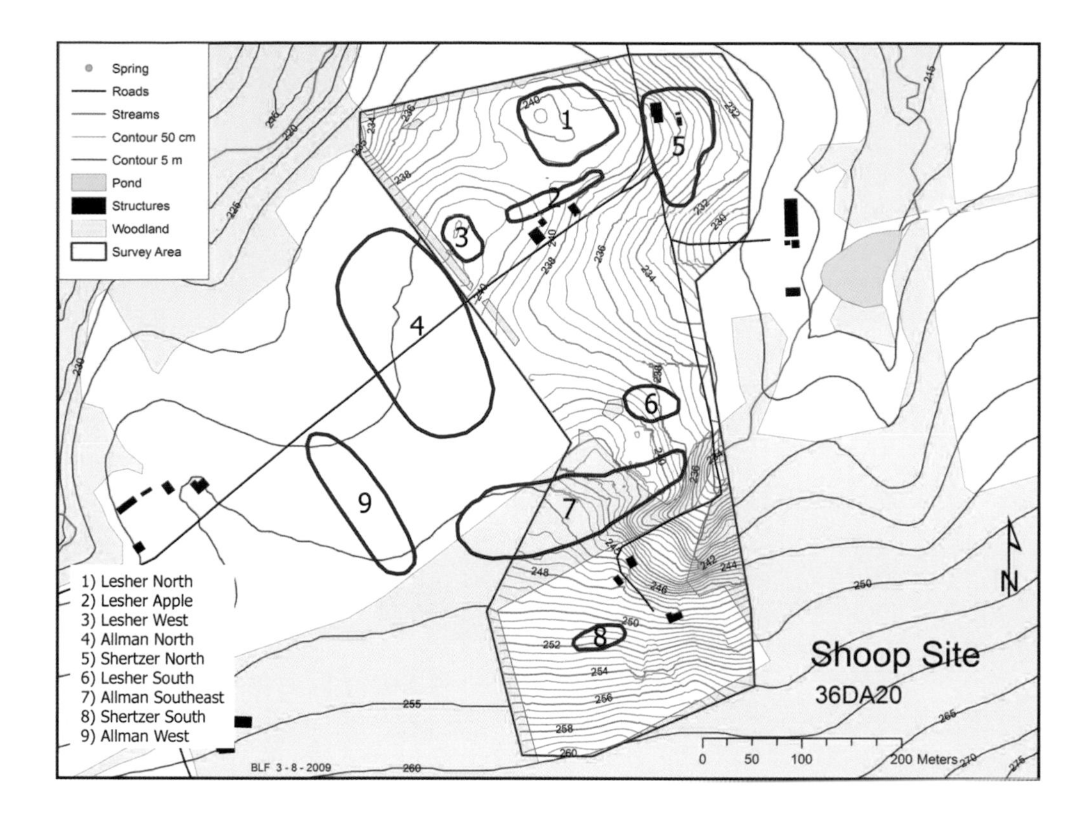

identified there. These broadly dispersed and relatively small (measuring less than 10 meters in diameter) concentrations produced 1,500 pieces of lithic debitage flakes, 400 stone tools, and fifty-three projectile points. Witthoft determined that all of these artifacts were made from Onondaga chert, derived from a formation that stretches from the lower Great Lakes to New York's Hudson Valley hundreds of kilometers to the north. This led him to postulate that these artifacts were deposited by newcomers to region who were unaware of more-locally available, high-quality lithic raw materials. Perhaps most significantly, Witthoft also characterized the site's artifacts—which attributed to the newly minted Enterline Chert Industry—as representing a blade-based lithic technology that he believed was older, and potentially much older, than Clovis and among the earliest cultural manifestations in the New World.

Several reexaminations of Witthoft's artifacts from Shoop were conducted from the mid-1950s through mid-1980s, but the site was not again examined by professional archaeologists until a joint project conducted in 2008 by the State Museum of Pennsylvania, Mercyhurst University, and Clarion University of Pennsylvania. [Figure 4.3] The project's principal field objectives included the testing and mapping of a known and relatively undisturbed artifact concentration named Lesher West (after landowner Ron Lesher), the search for subsurface cultural features (*e.g.*, fire pits, living floors, *etc.*) and material therein that would be suitable for radiocarbon dating, and the production of an accurate site map. [Figure 4.4]

The systematic surface reconnaissance of Lesher West (which had been shallowly plowed, agriculturally disked, and left exposed to rainfall for one month prior) covered about 4,000 square meters, and the area's subsurface

FIGURE 4.5 (above) Tan shaly bedrock in one of the forty-two 1 square meter excavation pits. The soil at the Shoop site is very thin. The arable soil is 20–23 centimeters deep with bedrock encountered at 30–46 centimeters. Photograph by Kurt Carr.

FIGURE 4.6 (below) Two jasper adzes from the Shertzer Shoop collection. Photograph by Kurt Carr.

deposits were examined via the excavation of seventy-two shovel tests and forty-two test pits (each measuring 1 square meter). [Figure 4.5] The relatively small total of 844 prehistoric artifacts recovered from this work were overwhelmingly composed of Onondaga chert unmodified debitage. A high frequency of minute pieces of debitage along with a low frequency of tools suggested to the investigators that this portion of the site had been heavily collected since its discovery in the 1930s.

Over 6,800 artifacts are known to have been collected from the 70–80 years of archaeological activity at the Shoop site, and more are known to exist in additional private collections. The known site assemblage is dominated by lithic debitage at about 70 percent, followed by minor constituents of various forms of utilized debitage flakes and scrapers, which together compose about 23 percent of the total. [Figure 4.6] Over 1,500 heavily used and extensively reworked tools have been collected from the site, and their large number and broad variety of forms suggest that a diverse range of activities occurred at Shoop, notably including the working of wood, bone, and antler.

Confirming and refining Witthoft's earlier assessment of the lithic raw material source for the Shoop artifacts, recent lithological analyses have determined that over 98 percent of the site's Paleoindian artifacts were made from Onondaga chert quarried at Divers Lake, located about 40 kilometers west of Buffalo, New York, and over 300 kilometers northwest of the site. More-recent assessments of Witthoft's concept of the Enterline Chert Industry by the Shoop site's investigators, however, have concluded that the lithic artifacts do not primarily reflect a blade-based technology, but rather, an admixture of both blade core and bifacial core technologies. [Figure 4.7] As only twenty fragmentary cores have been recovered from the site, however, a more-definitive resolution of this research issue seems unlikely at the present time.

The Shoop Paleoindian artifact assemblage's most distinctive form, and undoubtedly the one most sought by collectors on their visits to the site, is the fluted point. [Figure 4.8] Seventy-seven fluted points (including twenty-four complete specimens and fifty-three fragments) are known from the available artifact collections, and at least nineteen more are thought to reside in private collections. The most common type represented among the points is the Clovis form, which in the Shoop assemblage exhibits wide variety in overall shape and are both shorter and narrower at the base than those typical of early Paleoindian contexts in the Northeast. [Figure 4.9] The assemblage also includes a small number of specimens reported to resemble the Crowfield and Hardaway-Dalton types, both of which tend to occur in archaeological contexts that range from slightly later (in the case of Crowfield) to about 1,500 years later (in the case of Hardaway-Dalton) than Clovis.

In the absence of cultural features and the preserved organic material, few conclusions can be decisively drawn concerning the subsistence of Shoop's Paleoindian visitors. Most archaeologists who have analyzed the site's artifacts and its place in regional settlement systems, however, are more or less in agreement that Shoop was in some way associated with hunting of caribou and/or other migratory cervid species, probably by micro-bands of up to thirty people from the lower Great Lakes region who visited the site seasonally and for relatively short durations over the course of just a few decades sometime around 12,750 yr BP. The precise dating of site has remained problematic since the first publication on the site in 1952.

The Shoop site's 2008 principal investigator and long-standing authority on the site, Kurt Carr of the State Museum of Pennsylvania, has indicated that an area of deeper soils at the site have demonstrated the potential to contain deeply buried, undisturbed, and possibly datable Paleoindian material. [Figure 4.10] Should this prove to be the case, future studies at Shoop may well resolve the site's age and refine our understanding of precisely when the site was visited during a time of dramatic climate change and the emergence of the Holocene landscape.

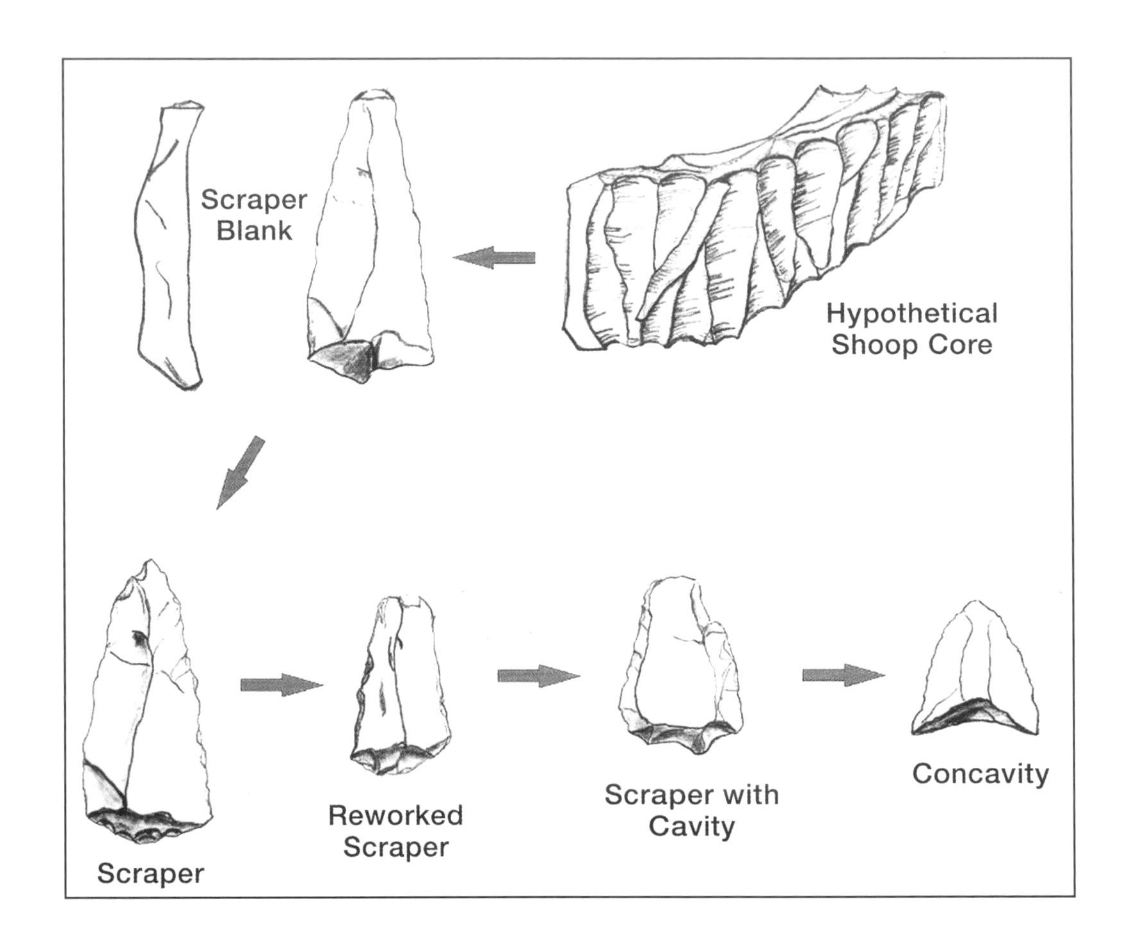

FIGURE 4.7 (above) Hypothetical Shoop lithic core and the evolution of endscrapers. Courtesy of Kurt Carr.

FIGURE 4.8 (below) Shoop fluted points from the State Museum of Pennsylvania. Photograph by Don Giles.

FIGURE 4.9 (above) Shoop fluted points from the State Museum of Pennsylvania. Photograph by Don Giles.

FIGURE 4.10 (below) Kurt Carr, archaeologist at the State Museum of Pennsylvania (left), co-author J. M. Adovasio, then Director of Mercyhurst Archaeological Institute (center), and Paul Shoop (right), whose family has farmed in the area since the late nineteenth century, inspect Shoop artifacts. Carr and Adovasio directed fieldwork at the site for three months in 2008. Courtesy of Kurt Carr.

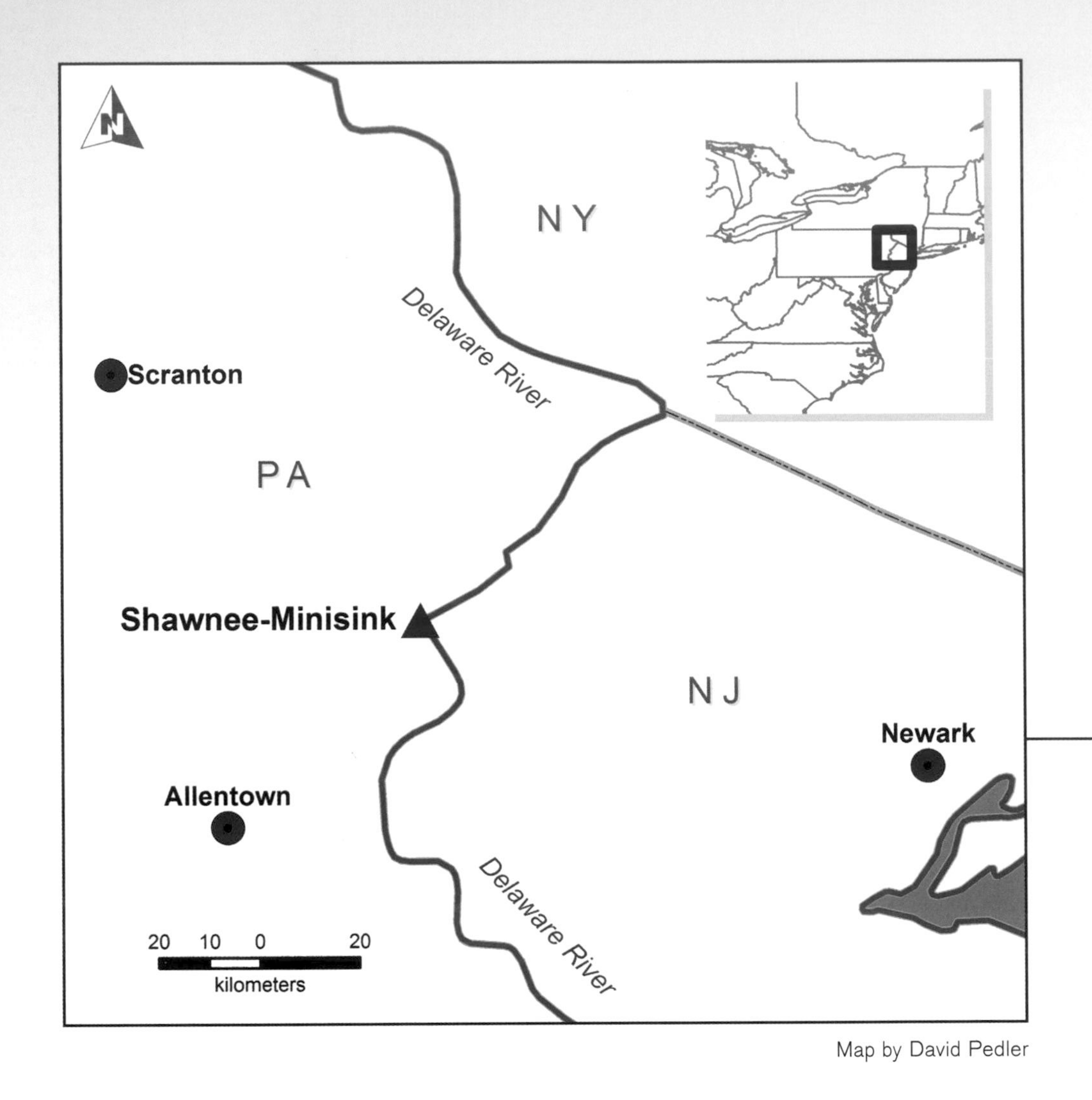

Map by David Pedler

Shawnee-Minisink is located near the confluence of the Delaware River and Brodhead Creek in Smithfield Township, Monroe County, Pennsylvania. Only 10 percent of this approximately 4,000 square meter site has been excavated.

5 SHAWNEE-MINISINK

LOCATION MONROE COUNTY, PENNSYLVANIA, UNITED STATES

Coordinates restricted.

Elevation 113 meters above mean sea level.

Discovery Donald Kline in 1972.

The Shawnee-Minisink site is considered to be among the best-dated and most-intact archaeological localities in eastern North America. It is also quite rare among the region's Clovis-age sites, which for the most part are shallowly buried localities that have been discovered as a consequence of plowing and other ground surface disturbances. Shawnee-Minisink, on the other hand, is a deeply stratified site whose unique depositional circumstances have provided well-preserved archaeological materials from intact deposits for reliable radiocarbon dating.

Shawnee-Minisink is located in Monroe County, northeastern Pennsylvania, about 5 kilometers east of the Borough of Stroudsburg, 50 kilometers northeast of Allentown, and 100 kilometers northwest of New York City. The site lies within the upper Delaware River valley in a geographic region that historically has been known as the Minisink, a name that was bestowed upon it by Dutch settlers after (or at least in association with) the Munsee peoples who inhabited the area prior to European contact. Set in the Blue Mountain section of the Ridge and Valley physiographic province, the region is characterized by low linear mountain ridges with intervening shallow valleys. The site's position on the Delaware River, just above the Delaware Water Gap about midway between the glaciated Appalachian Plateau to the north and the Great Valley to the south, places Shawnee-Minisink in a key location on what regionally is perhaps the most efficient, natural access route connecting these two important ecological zones. [Figure 5.1]

In the waning years of the late Pleistocene the upper Delaware Valley was covered by the Laurentide ice sheet,

FIGURE 5.1 View from Shawnee-Minisink looking south down the Delaware River. Photograph by Joseph A. M. Gingerich.

which achieved its maximum southern extent about 20 kilometers south of the Shawnee-Minisink site. Having already melted from the surrounding uplands, this glacier began its retreat north through the valley about 15,000 yr BP, leaving deep deposits of gravel (in some locations measuring 60–80 meters thick), sand, clay, and silt in its wake. Also included in this sediment are possible deposits of glacier-associated, wind-borne layers of fine sediment known as "loess" that apparently are associated with the site's Paleoindian material. The Delaware River, which presently flows about 100 meters to the east of the site and may have been closer during Paleoindian times, appears to have been at its present position and level for at least the past 6,000 years. The site occupies an ancient riverbank deposit about 6–7 meters above the present level of the Delaware River. The site is currently surrounded by woodlands and low-density residential development.

The vegetation present in the region following the glacial retreat appears to have been composed of an open parkland tundra populated by scattered stands of spruce, fir, and pine trees interspersed with grasses, sedges, heath species, and shrubs. Around the time of the site's Paleoindian occupation, the region is thought to have been transitioning from pre-boreal forest conditions to a true boreal forest dominated by white pine and white birch. This setting would have provided the site's Paleoindian visitors with abundant plant and animal resources, complemented by additional resources from the adjacent Delaware River.

Though known to local artifact collectors for years prior, Shawnee-Minisink was formally discovered and registered by avocational archaeologist Donald Kline in the winter of 1972 while exploring an eroded portion of the lower first terrace of the Delaware River. Kline noted the presence of the higher second terrace at the site, paying particular attention to that terrace's southern exposure and its protection by nearby hills to the north. In March of that year, Kline excavated about 6 square meters of the site in two test pits spaced 100 meters apart, and recovered Woodland and Archaic materials from a depth of about 60 centimeters below the ground surface. As other sites on the same landform elsewhere in the Delaware River Valley were known to be deeply stratified, that summer Kline decided to excavate much deeper at Shawnee-Minisink and ultimately recovered lithic debitage flakes and two lithic scrapers at depths ranging about 2.5–3 meters below ground surface. The additional excavation of about 17 square meters in 1973 recovered over 5,000 artifacts (including twenty-seven scrapers) and associated charcoal which produced a radiocarbon age of 13,060–11,600 yr BP.

Archaeologists William Gardner of Catholic University and Charles W. McNett, Jr., of American University were alerted to the findings and declared the site's deeper material (particularly the scrapers) to be ascribable to the Paleoindian period, despite the lack of diagnostic projectile points. The decisive radiocarbon age determination of Shawnee-Minisink's deep antiquity led to the formation of the Upper Delaware Valley Early Man Project and the inception of an intensive four-year excavation project, all under the direction of McNett, and sponsored by American University. McNett's 1974–1977 excavations removed about 380 square meters of the site deposits and recovered over 55,000 artifacts ascribing to the Woodland, Late Archaic, Early Archaic, and Paleoindian periods. In 2003–2006, additional excavations at the site were initiated by Donald Kline with the help of professional archaeologist Joseph Gingerich. [Figures 5.2 and 5.3] Under their direction, students from Temple University excavated an additional 40 square meters from the site, recovering over 30,000 additional artifacts from the site's Paleoindian level. Only about 10 percent of this 4,000 square meter site has been excavated since its discovery. [Figure 5.4]

The archaeological investigations at Shawnee-Minisink have identified thirteen major strata (and as many as twenty-eight discrete soil horizons), predominantly composed of river-deposited sand, which extend to a layer of glacially deposited outwash gravel that is encountered almost 5 meters below the modern ground surface. The four major cultural zones are restricted to the upper 3 meters of this stratigraphic sequence. [Figure 5.5] The Woodland period artifacts represent all phases of that cultural episode, though all intervals but the Late Woodland occupation(s) are rather sparsely represented. [Figure 5.6] The recovered Late Woodland materials include abundant lithic artifacts, pottery, and a double line of post molds outlining a 20 meter long structure apparently representing a longhouse.

The Woodland remains are underlain by a spectacular array and quantity of Archaic period materials ranging from the earliest of Archaic times (probably representing the transition from the preceding Paleoindian period) through a lightly represented Middle Archaic level and again heavily represented Late Archaic deposit. Unfortunately, no radiocarbon dates are available from the site's Archaic levels, but the large number of recovered diagnostic projectile points (including Kline, Hope Stemmed, Abbott, Brodhead Side-Notched, LeCroy, Kanawha, Brewerton, Kittatinny, Lackawaxen, Lamoka, and Orient Fishtail forms) decisively affirm their temporal and cultural ascription. [Figures 5.7 and 5.8]

The Paleoindian artifacts from all years of investigation at the site include both bifacial and unifacial implements, modified and utilized flakes, "rough-stone" implements, and debitage, along with a suite of more formalized tools such as endscrapers, bifaces, cores, hammerstones, and two Clovis projectile points. The artifacts are primarily made from a locally available black chert, with minor constituents of other chert, jasper, and argillite lithic materials from further afield. [Figure 5.9]

McNett's 1974–1977 excavations also identified at least ten cultural features in the site's Paleoindian levels, mostly clusters of lithic materials (some of which were quite dense), as well as a few ephemeral remnants of apparent hearths. Two additional hearths and several clusters of lithic material were identified during the subsequent 2003–2006 investigations. More recently processed AMS radiocarbon dates of charred seed (hawthorn and plum) samples initially recovered by McNett's research team have produced early age determinations ranging from about 13,010–12,770 yr BP to 12,830–12,700 yr BP. These Paleoindian materials are reported by McNett to have been recovered from the upper reaches of higher elevation loess deposits at the site and were covered by a layer of red clay and a 1 meter thick layer of river-deposited culturally sterile sand. The continuous cover provided by both of these

FIGURE 5.2 (left) The first projectile point excavated from the Clovis level at Shawnee-Minisink in the 1970s. Photograph by James Di Loreto.

FIGURE 5.3 (below) *In situ* artifacts uncovered in 2006. Photograph by Joseph A.M. Gingerich.

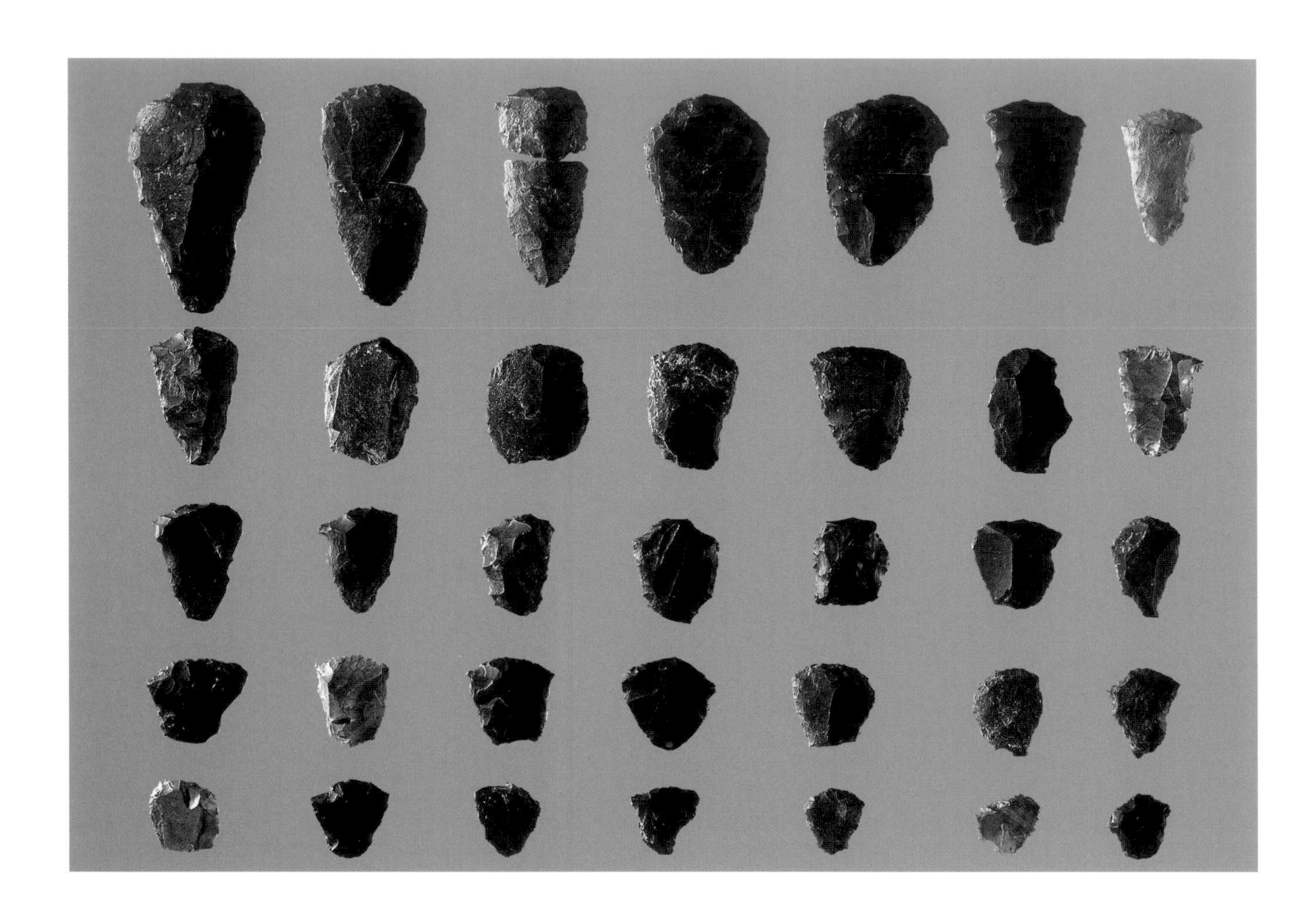

FIGURE 5.4 (top) Selection of the over 200 endscrapers that have been recovered at Shawnee-Minisink during the 2003–2006 field seasons. Most show use-wear indicating various types of hide scraping. Photograph by Joseph A.M. Gingerich.

FIGURE 5.5 (bottom) One of the 2003 excavation units. Since 1974 over 85,000 Woodland, Late Archaic, Early Archaic, and Paleoindian period artifacts have been recovered from Shawnee-Minisink. Photograph by Joseph A.M. Gingerich.

FIGURE 5.6 (top) Incised pebble from the Clovis deposits at Shawnee-Minisink. Incised stones have been recovered from the Clovis deposits at Gault (see pages 116-129) and El Fin del Mundo (see pages 130-139). Photograph by Joseph A. M. Gingerich.

FIGURE 5.7 (bottom) Dense cluster of artifacts uncovered during the 2003–2006 field seasons. This cluster has been interpreted as a clean-up or dump pile, and is Feature 6 shown in the Figure 5.8 plan map. Photograph by Joseph A. M. Gingerich.

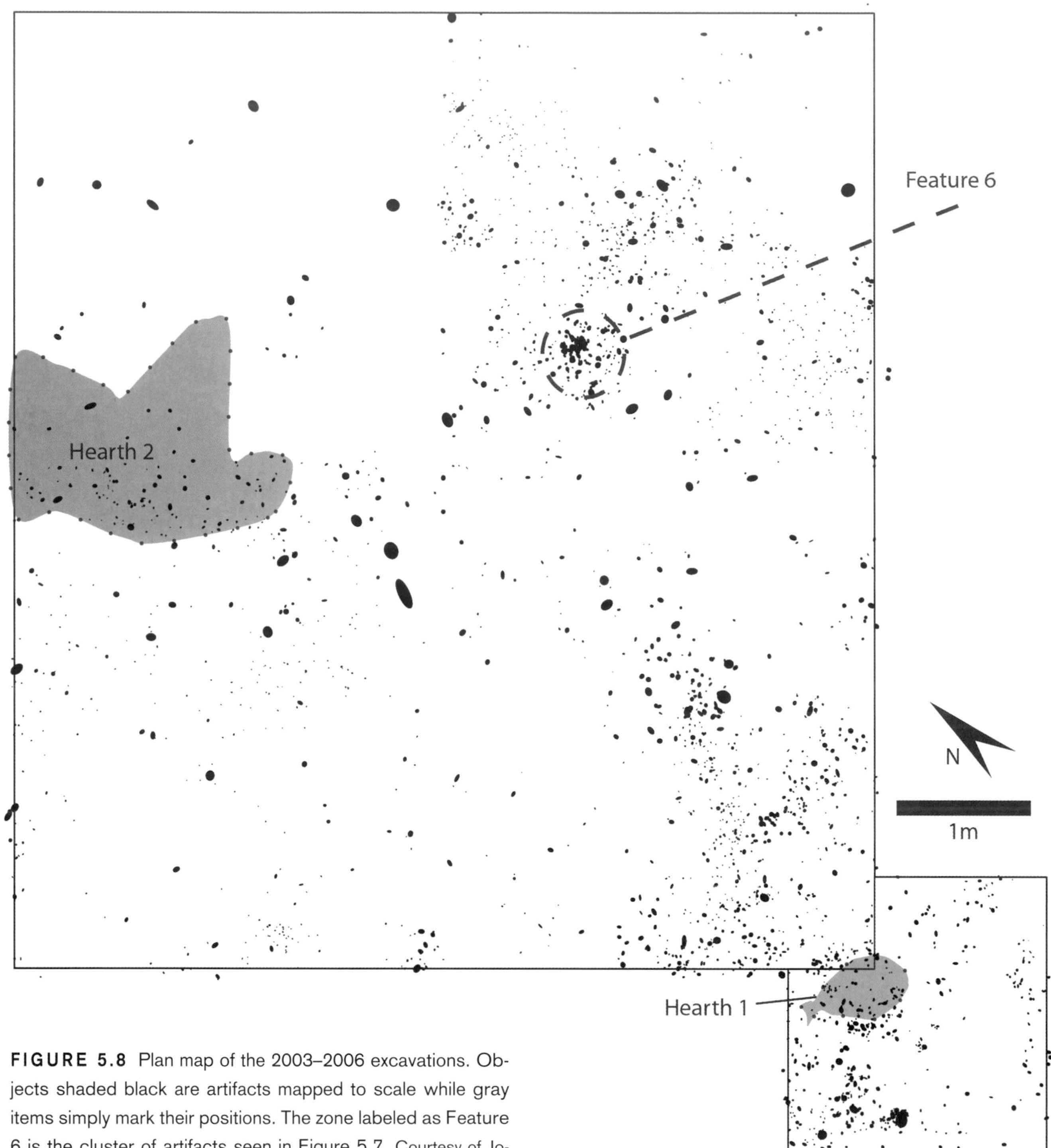

FIGURE 5.8 Plan map of the 2003–2006 excavations. Objects shaded black are artifacts mapped to scale while gray items simply mark their positions. The zone labeled as Feature 6 is the cluster of artifacts seen in Figure 5.7. Courtesy of Joseph A. M. Gingerich.

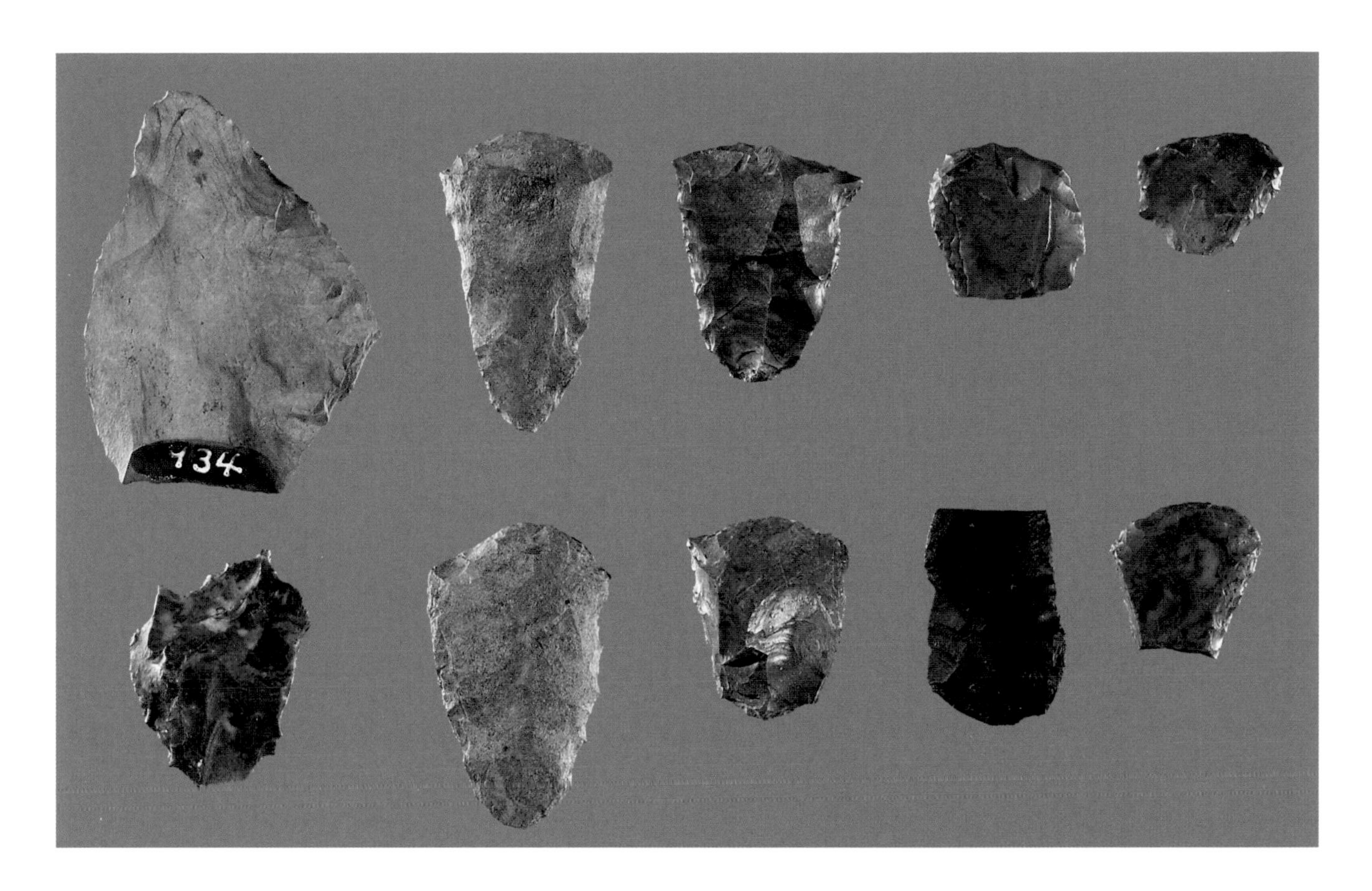

FIGURE 5.9 Tools made from exotic raw materials like these make up less than 20 percent of the formal tool assemblage (scrapers, gravers, retouched flakes, *etc.*) and less than 10 percent of the entire assemblage (flakes, debitage, *etc.*) recovered from Shawnee-Minisink. Photograph by Joseph A. M. Gingerich.

layers essentially precluded the possibility of disturbance from animal burrowing or other natural processes.

The most distinctive artifacts recovered from the site to date are the two Clovis points, one recovered in 1975 and the other in 2006. The former and larger (about 79 millimeters long) of the two exhibits the distinctive single flutes on both faces that extend about one quarter of the point's length and terminate with hinge fractures. [Figure 5.10] The latter and smaller (about 39 millimeters long) specimen is similarly configured and exhibits flutes which appear to extend almost halfway to its midpoint. The larger specimen is made of Onondaga chert, which occurs from the lower Great Lakes through central New York to the upper Hudson Valley, and the smaller is made of a heavily weathered, white-spotted gray chert whose source is believed to be within 80 kilometers of the site. The latter point was recovered in direct association with a hearth dated to 12,980–12,720 yr BP.

Plant remains from the Paleoindian level at Shawnee-Minisink firmly belie the notion that Clovis-age peoples were primarily focused on the relentless hunting of late Pleistocene megafauna. Instead, the data recovered from the site suggest that their generalized tool kit was also

FIGURE 5.10 One of the two Clovis points excavated in 2006, dated to 12,980–12,720 yr BP. Photograph by Joseph A. M. Gingerich.

FIGURE 5.11 Hawthorn seeds recovered from hearths during excavations in 2003–2006 dated to 12,980–12,720 yr BP. Hawthorn and other plants found in smaller quantities likely represent opportunistic gathering within the immediate vicinity of the site. Recent analysis of endscrapers, which show high frequency of hide scraping, supports the hypothesis that hawthorns and other fruits with high caloric return and low processing costs were likely supplements to meat. Photograph by Joseph A. M. Gingerich.

employed in a variety of foraging tasks. Numerous carbonized seeds from Paleoindian hearths at the site, for example, indicate the collection of a variety of plant foods (such as grape, hawthorn plum, hackberry, and blackberry, among others) and charred fish bones indicate the taking of fish from the nearby river and stream. [Figure 5.11] It is highly unlikely, however, that the Shawnee-Minisink Paleoindian peoples were exclusively generalized foragers. Almost certainly, their diet would have relied to at least some significant degree upon the taking of game animals, despite the apparent lack of evidence in the site deposits.

The prevailing view of the Clovis occupation at the Shawnee-Minisink site is that it represents a short-term, autumn campsite that served as a resource-processing locality. The most recently interpreted evidence from the site suggests that it probably saw multiple visits throughout Clovis times and, based on features that appear to represent site maintenance in the form of the "cleaning up" of lithic debris, those visits might also have been of longer duration than had previously been thought. In view of other Clovis-age localities discovered in riverine settings in northeastern North America, there remains the intriguing possibility that Shawnee-Minisink instead represents a base camp whose extensive unexcavated portions may well provide a much fuller picture of early Paleoindian life far from what has long been regarded as the Clovis heartland.

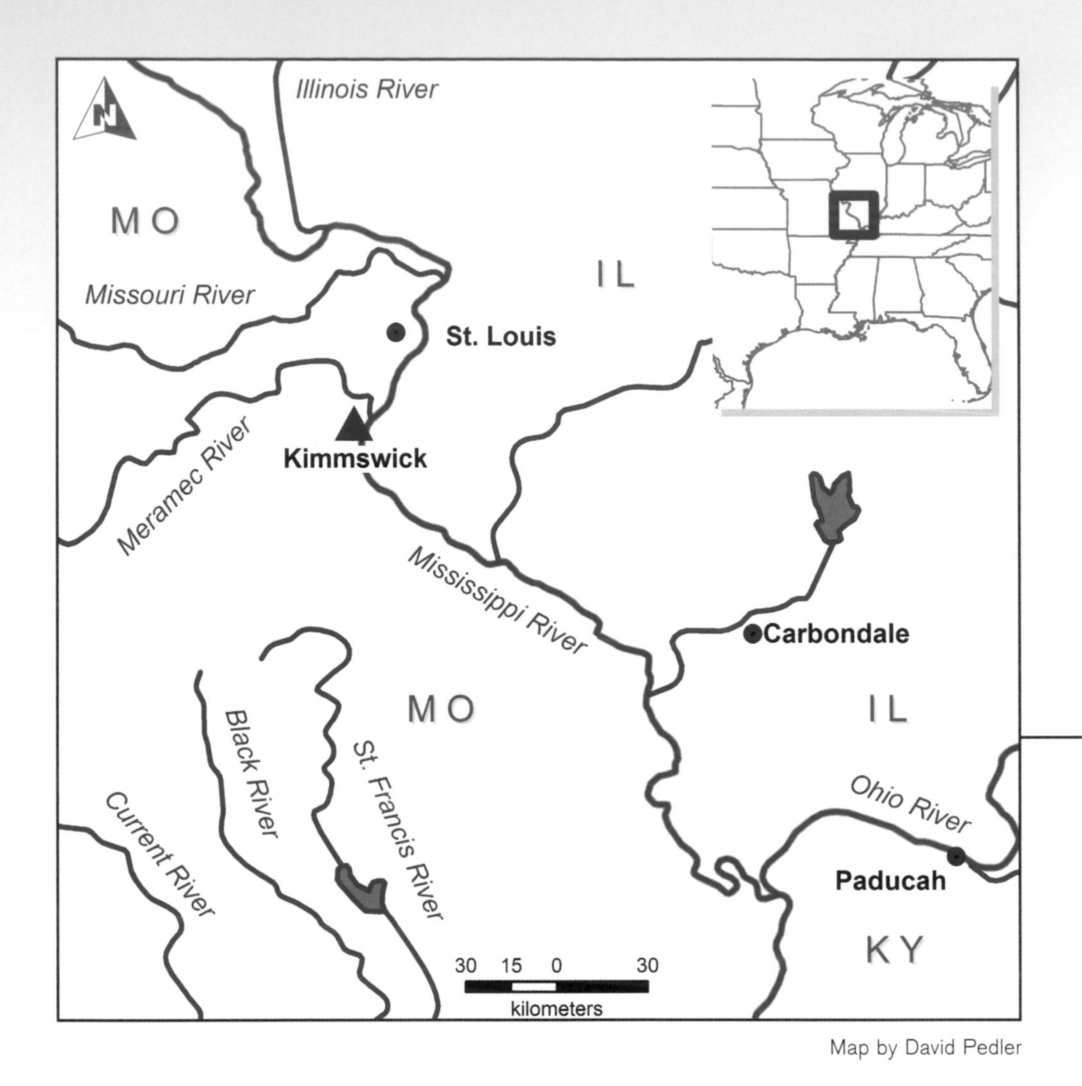

Map by David Pedler

Kimmswick may represent a Clovis mastodon kill and processing site with limited occupation. The animal remains found there suggest that Clovis subsistence, in part, depended on a wide assortment of game species (everything from fish and turtles to terrestrial species of all sizes) rather than a sole concentration on big game.

6 KIMMSWICK BONE BED

LOCATION JEFFERSON COUNTY, MISSOURI, UNITED STATES

Coordinates 38°22'43.67"N, 90°23'5.38"W.

Elevation 127 meters above mean sea level.

Discovery Pierre Chouteau in 1790s.

As the number of Clovis finds grew through the latter half of the twentieth century, archaeologists came to expect that if the remains of extinct megafauna (*i.e.*, large animals) were found at a Clovis site, they most frequently would be those of mammoth or, in environments like the eastern woodlands where mammoths were rare, the remains of mastodon instead. Though both species are frequently encountered in paleontological contexts, however, the archaeological recovery of mastodon remains in direct association with artifacts is quite rare. Kimmswick is unusual among Clovis sites in that it is one of the very few North American localities which has preserved mastodon remains (as opposed to mammoth) in a credible archaeological context firmly indicating that some form of human exploitation such as hunting—or at least butchering—was directly responsible for their presence.

Included within the 1.7 square kilometer Mastodon State Historic Site, the Kimmswick site (also known as the Kimmswick Bone Bed Site) is located in Jefferson County, Missouri, immediately west of its namesake town of Kimmswick, about 32 kilometers southwest of St. Louis, 125 kilometers northwest of Carbondale, Illinois, and 210 kilometers northwest of Paducah, Kentucky. The site lies within the upper Mississippi Valley, 2.5 kilometers west of the Mississippi River and 3.6 kilometers southwest of that river's confluence with the Meramec River. The 200 kilometer long stretch of the Mississippi Valley in the general vicinity of the Kimmswick site is notable as a region of confluence for four of North America's major drainage basins (those of the upper Mississippi, Missouri, Ohio, and Tennessee Rivers), which collectively drain a total area of about

FIGURE 6.1 In his 1841 *Description of the Missourium, or Missouri Leviathan*, "Dr." Albert C. Koch claimed the "gigantic skeleton" he had unearthed at Kimmswick had webbed feet and other anatomical features that distinguished it from a terrestrial mastodon. It was, he concluded, nothing other than the biblical Leviathan described in the Old Testament Book of Job, chapter forty-one. Courtesy of the Missouri Historical Society.

2.4 million square kilometers or almost 30 percent of the contiguous United States landmass.

The Laurentide ice sheet appears to have reached its maximum extent about 350 kilometers east of Kimmswick, but by the time of the site's occupation the glacier's southern margin lay over 800 kilometers to the north, having receded into the present-day upper Great Lakes basin. As the glacier retreated, massive amounts of sediment were deposited in the upper Mississippi Valley and those sediments can be found in the lower stratigraphic levels at Kimmswick. Several shallow basins or ponds appear to have formed in these lower deposits at the site. Throughout the Late Pleistocene, the Kimmswick region was probably situated in mixed deciduous (oak and hickory) forest mixed with prairie grasses on the adjacent bottomlands, and lay well to the south of more northerly boreal forest, forest-tundra, and tundra vegetation.

Kimmswick occupies a small terrace elevated about 127 meters above sea level at the confluence of Rock and Black Creeks, about 5 meters above creek level and more than 10 meters above the local elevation of the Mississippi River. Kimmswick's long axis is roughly orientated southwest to northeast, and the portion of the site examined since its discovery encompasses about 2,300 square meters. The site is bounded on the north by a 20 meter high limestone bluff, but its setting otherwise affords unobstructed views of the Mississippi River to the southeast and the Rock Creek valley to the south and west. Kimmswick formed in Pleistocene-age alluvial (*i.e.*, water-deposited) overbank stream deposits and colluvial (*i.e.*, downslope) deposits from the adjacent bluff. The nearby Rock Creek valley has been filled in with a later Holocene-age silts, sands, and gravels.

The Kimmswick site is thought to have been discovered by French creole fur trader and St. Louis co-founder Pierre Chouteau in the 1790s. It was first noted by Scottish botanist John Bradbury during the first of three field trips that would ultimately be chronicled in his book *Travels in the Interior of America in the Years 1809, 1810, and 1811*. The earliest known excavation of the site was conducted in 1839 by "Dr." Albert C. Koch, an apparently uncredentialed charlatan and popular showman who recovered the skeletal remains of what he dubbed the Missouri Leviathan—actually, American mastodon (*Mammut americanum*)—for display at his St. Louis Museum, where for twenty-five cents one could view assorted curiosities from the natural and cultural worlds. [Figure 6.1] The site had more or less lapsed into obscurity until its rediscovery by St. Louis businessman and inventor Charles W. Beehler in 1893. Beehler began his excavations in 1897 and by 1900 had constructed an on-site museum for the viewing of his finds, ultimately attracting the attention of various regional and national scholarly institutions with his claims for the association of human archaeological remains with the site's well-known mastodon remains. This early interest in the site culminated in a 1902 visit from the Smithsonian Institution archaeologists William Henry Holmes (see page 22) and Gerard Fowke, who after excavating a 3–4 meter deep trench both declared that any cultural remains present at the site had been recently introduced and none were in direct association with Kimmswick's fossil remains.

The Kimmswick site was revisited by archaeologist Robert M. Adams of the St. Louis Academy of Science in 1940–1942 under the auspices of the Works Progress Administration. The modern era of professional investigation began in the late 1970s under the sponsorship of the Missouri Department of Natural Resources, which had purchased the site and surrounding tract of land for incorporation into the Mastodon State Historic Site. [Figures 6.2 and 6.3] Conducted in 1979–1980 and 1984 under the directorship of Illinois State Museum paleontologist Russell W. Graham, this project provided the first definitive evidence of the site's human affiliation through the discovery of a Clovis projectile point in direct association with mastodon bones. Graham and his research team carefully excavated 52 square meters along with nine machine-excavated exploratory trenches. [Figure 6.4]

Seven discrete strata were identified in the Kimmswick site sediments, two of which (designated Strata C_1 and C_3 by the excavators) yielded Clovis-age artifacts and Late Pleistocene animal remains. [Figure 6.5] These Clovis levels are overlain by what appears to be an Early Archaic period cultural horizon (Stratum D) which in turn is capped by a relatively thin, organic-rich zone of disturbed surface deposits. The Clovis-age deposits contained a small series of sediment-filled, ponded basins containing numerous remains of extinct and extant large mammals such as mastodon (*Mammut americanum*), long-nosed peccary (*Mylohyus nasutus*), and Harlan's ground sloth (known variously as *Glossotherium harlani* or *Paramylodon harlani*), as well as numerous species of smaller vertebrates (both extinct and extant) and artifacts. [Figure 6.6] Although no definitive cultural features were identified within the site's Clovis deposits, several clusters of bone and/or artifacts suggested the presence of buried cultural features, including one possible hearth surrounded by Clovis artifacts and the remains of Harlan's ground

FIGURE 6.2 (left) In 1900, Charles Beehler initiated excavations at Kimmswick, housing the fossils and artifacts he unearthed in a small museum. It is not known what became of most of his collection. Only four Clovis points from the site are currently at the Field Museum of Natural History, one of which is shown here. Courtesy of the Field Museum.

FIGURE 6.3 (below) Excavations at Kimmswick in 1980 under the direction of Illinois State Museum paleontologist Russell Graham. The area was once swampy and the mastodons that visited may have become trapped in the mud, which made them easy targets for Paleondian hunters and later helped preserved their bones. Photograph by Russell Graham.

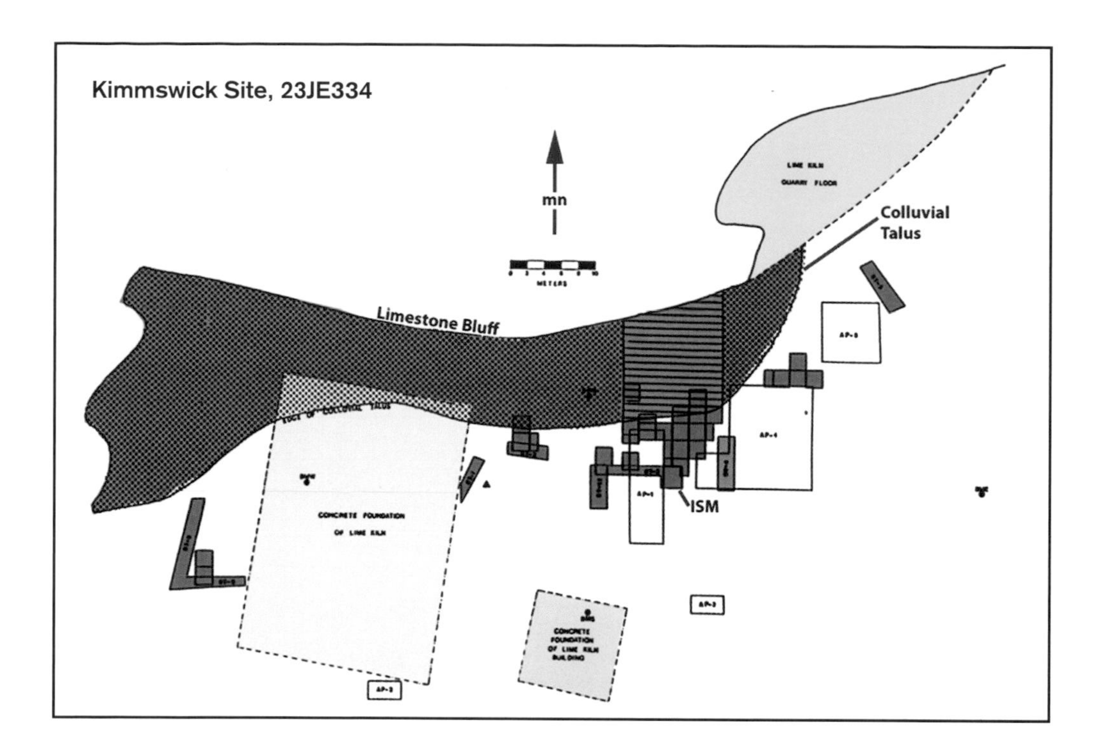

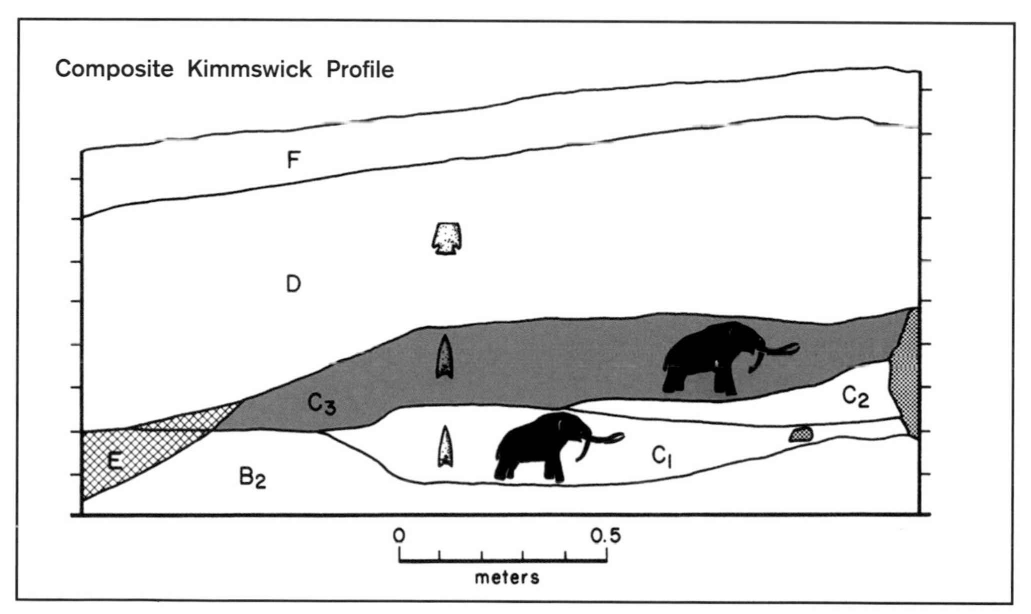

FIGURE 6.4 (above) Historical compilation of excavations at Kimmswick. The blue rectangles represent the 1979–1980 Illinois State Museum excavations, most of which were hand-excavated 2 meter by 2 meter squares. The larger blue rectangles represent machine-excavated trenches. Courtesy of Marvin Kay.

FIGURE 6.5 (below) Stratigraphic profile of deposits at Kimmswick. From top to bottom: (F) disturbed organic matter; (D) brown clayey silt of tan colluvium with debitage; (C_3) olive-green silty clay of upper pond deposits with Clovis points and mastodon fossils; (C_2) brown, clayey, silty gravel of upper colluvial gravel containing no fossils or artifacts; (C_1) bluish-gray silty clay of lower pond deposits containing Clovis points and mastodon fossils; (E) B horizon (zone of accumulation) developed in B_1 and C_3; (B_2,) colluvial limestone gravels. Courtesy of Russell Graham.

FIGURE 6.6 A plaster-jacketed mastodon rib with a Clovis point *in situ* below and to the right, unearthed in 1980 from Zone C_3. Though both species are of the same taxonomic order (Proboscideans), mastodons are a distinct species from mammoths (which are more closely related to modern elephants). Mastodons were slightly smaller than mammoths with shorter legs and lower, flatter heads. While both were herbivores, mastodons had cone-shaped cusps that show they tended to eat softer vegetation such as twigs and leaves. Typical of grazing animals, mammoth teeth bore a series of low ridges that were effective for grinding vegetation like grasses. Photograph by Russell Graham.

sloth. Unfortunately, no radiocarbon dates are available for Kimmswick, but its artifacts, context, and the species composition of its associated faunal remains permit the site's firm ascription to the 13,300–12,800 yr BP Clovis horizon.

Only a small number of lithic tools were recovered from Kimmswick, though the small debitage flakes in the assemblage number in the hundreds. [Figure 6.7] The eight formal tools known from the site are predominantly of Clovis age and include a "knife-like" implement and resharpened Clovis point, both of which were recovered in the early 1900s and are currently housed at the Field Museum of Natural History in Chicago. The remaining tools recovered from the Illinois State Museum excavations include two Clovis points from Stratum C_3, a bifacial tool fragment from C_3, a utilized flake from C_3,

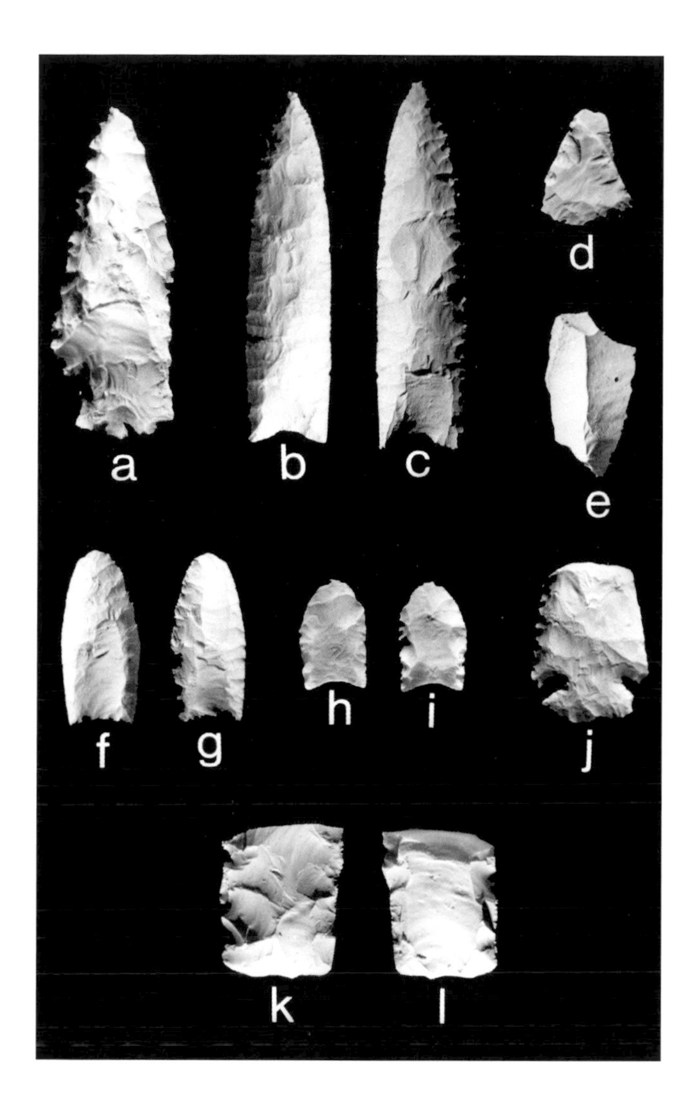

FIGURE 6.7 Kimmswick Bone Bed stone tools: (a) knife-like implement found in the early 1900s, presumably in association with mastodon bones; (b–c) obverse and reverse sides of Clovis point from Zone C_3; (d) bifacial tool fragment from Zone C_3; (e) utilized flake from Zone C_3; (f–g) obverse and reverse sides of resharpened Clovis projectile point also found in the early 1900s, presumably in association with mastodon bones; (h–i) obverse and reverse sides of Clovis projectile point from Zone C_3; (j) base of St. Charles projectile point from Zone D; (k–l) obverse and reverse sides of fluted preform with reverse hinge fracture from Zone C_1. Photograph by Russell Graham.

a fluted tool preform from C_3, and the basal fragment of an Early Archaic age St. Charles point from Stratum D. At least four lithic raw materials are represented among the stone tool assemblage, but all derive from sources within a 25 kilometer radius of the site and are thus considered local. [Figure 6.8]

The most distinctive artifact forms from a reliable stratigraphic context at the site are the two Clovis points recovered by Graham and his colleagues. [Figure 6.9] The artifacts lay *in situ* within 1.25 meters of each other and at virtually the same level, being separated only by a vertical distance of less than 1.5 centimeters. One of them lay in horizontal orientation among the disarticulated foot bones of an adult mastodon, and the other was identified beneath (but in contact with) a large mastodon bone at a tip-down angle of 34° from the stratum's horizontal plane. [Figure 6.10] Both points display damage to their tips, and the possibility exists that the latter artifact was embedded in the animal's flesh. [Figure 6.11] One of the points, Specimen K-L22-32, is thought by the Kimmswick research team to resemble Clovis specimens recovered from Blackwater Draw Locality No. 1 (see pages 46-57) and the Naco site (see pages 58-67) in southeastern Arizona, located 1,250 kilometers and 2,000 kilometers southwest of Kimmswick, respectively.

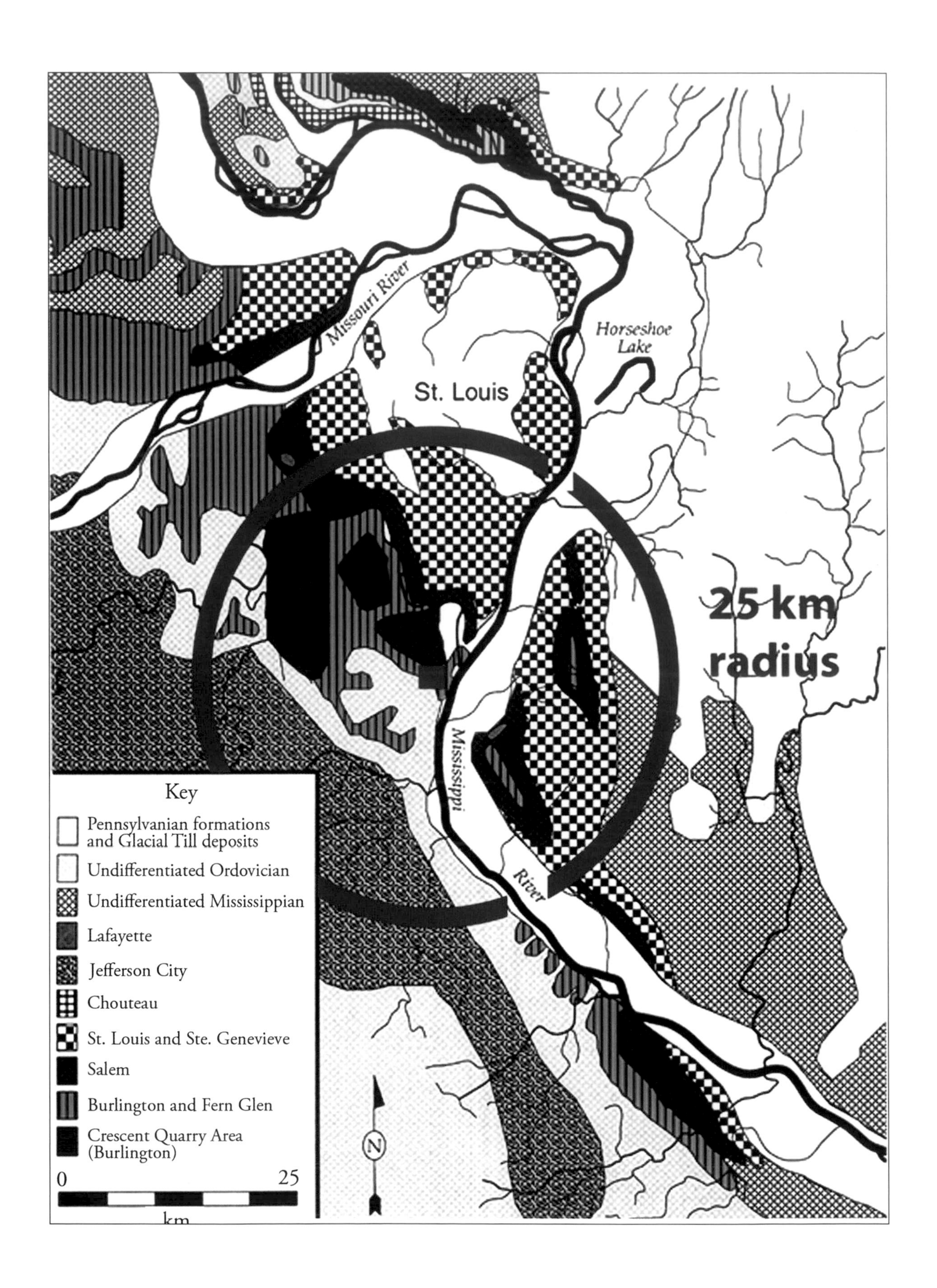
Missouri River
Horseshoe Lake
St. Louis
25 km radius
Mississippi River
Key
Pennsylvanian formations and Glacial Till deposits
Undifferentiated Ordovician
Undifferentiated Mississippian
Lafayette
Jefferson City
Chouteau
St. Louis and Ste. Genevieve
Salem
Burlington and Fern Glen
Crescent Quarry Area (Burlington)
0
25
km
N

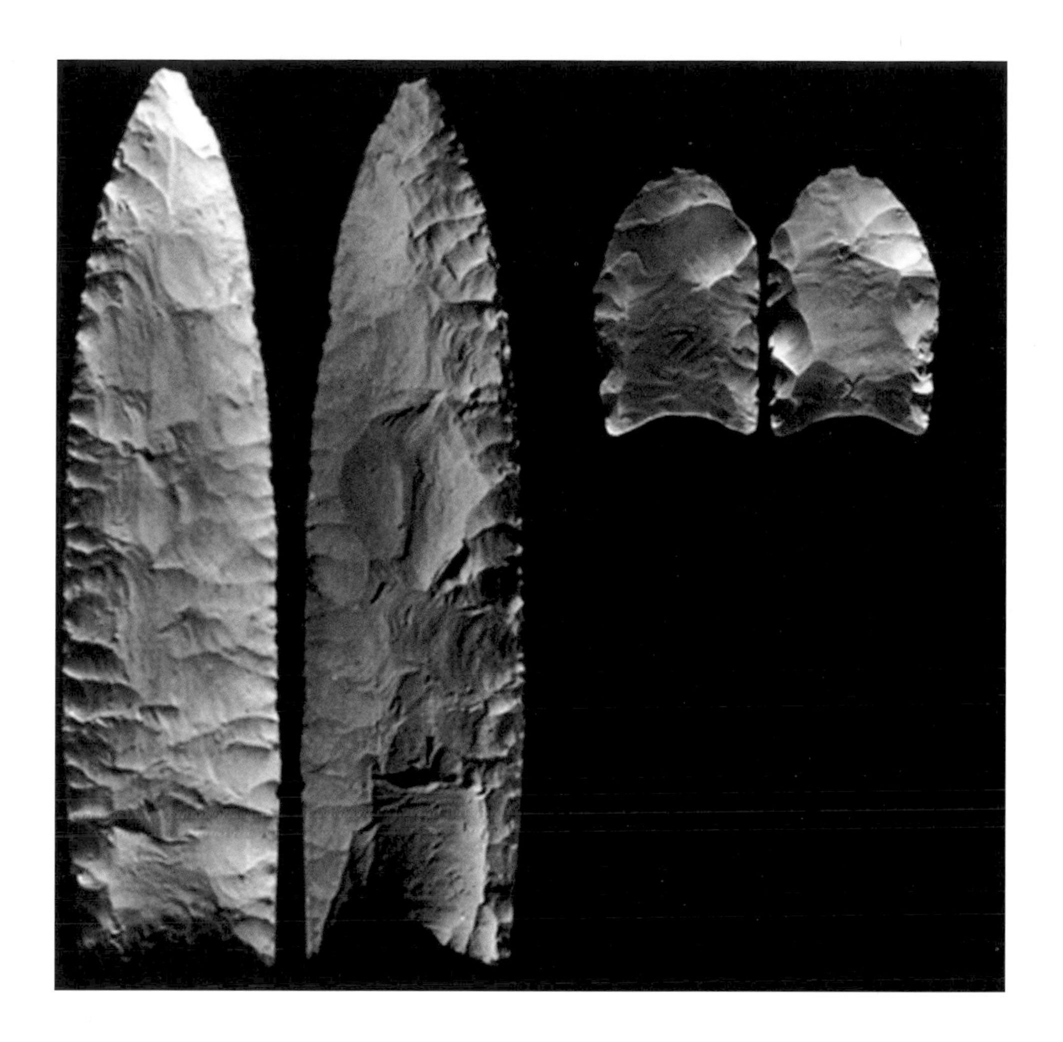

FIGURE 6.8 (left) Likely sources of the lithic raw materials from which the Clovis artifacts were manufactured at Kimmswick (red square). Courtesy of Marvin Kay, Michael Ross, and the Illinois State Museum.

FIGURE 6.9 (above) Kimmswick Clovis points recovered by the Illinois State Museum in 1980 from Zone C_3, dated to about 13,000 yr BP. Courtesy of Marvin Kay, Michael Roos, and the Illinois State Museum.

FIGURE 6.10 (above) The first Clovis point (in white circle) recovered during the Illinois State Museum excavation in 1979. This and other Clovis points found in direct association with mastodon bones were the first solid evidence that Paleoindians coexisted with and hunted this large prey. Photograph by Russell Graham.

FIGURE 6.11 (right) Photomicrographs of use-wear (lower right) and schematic of use-wear patterns (upper right), indicating of use on Clovis point (far left) from the 1979 excavation at Kimmswick. Courtesy of Russell Graham.

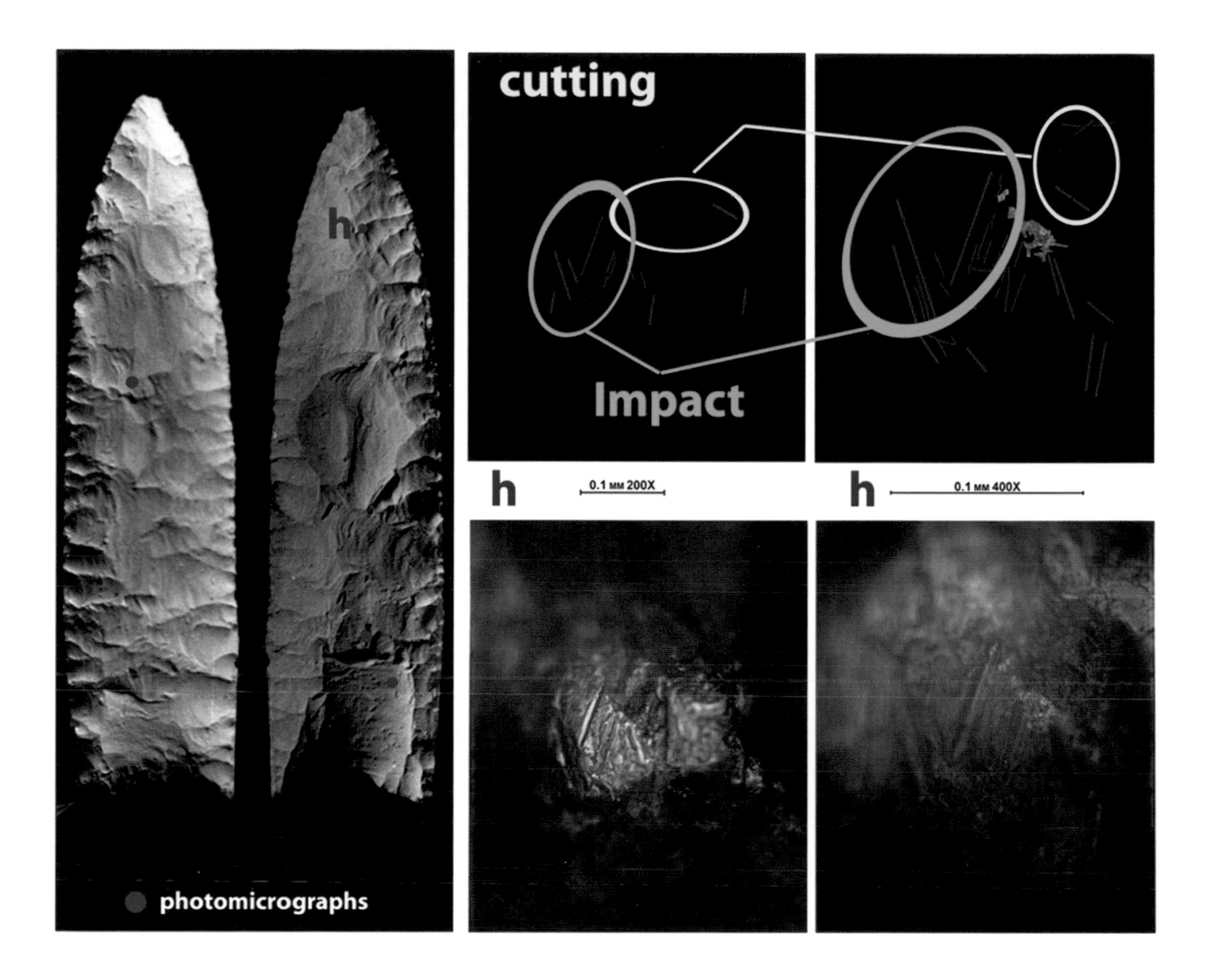

The artifacts from Kimmswick, though relatively few in number, and the closely associated faunal remains suggest that human activity at the site involved far more than the simple predation and processing of mastodons. The presence of diverse animal species apart from mastodon—including white-tailed deer, fish, amphibians, reptiles such as turtles, birds, and numerous small mammals—also suggests that the Clovis lifeway was availed a broader spectrum of dietary choices than the one presumed by a tight focus on big game. Moreover, the Kimmswick site demonstrates that its Clovis visitors were readily adaptable to environmental conditions which were quite different from those faced by Clovis hunters hundreds of kilometers to the south and west.

Map by David Pedler

Bonfire Shelter, the oldest mass bison-kill site in the New World, lies in a deeply entrenched canyon 1 kilometer above its mouth on the Rio Grande near Langtry, Texas. This large rockshelter, hidden behind a massive roof fall, holds three bone deposits and perhaps two occupational levels, each stratum separated from the others by sterile cave fill.

7 BONFIRE SHELTER

LOCATION VAL VERDE COUNTY, TEXAS, UNITED STATES

Coordinates 29°48'56.34"N, 101°32'56.99"W.

Elevation 430 meters above mean sea level.

Discovery Michael Collins in 1958.

Bonfire Shelter is a Late Pleistocene (Folsom) through late Holocene age bison jump archaeological site that is the oldest and southernmost locality of its kind. The heartland for sites known as bison jumps, defined here as the final destinations for large numbers of bison that were herded by communal groups of prehistoric hunters along established routes or "drive lanes" over steep cliffs, lies more than 2,000 kilometers to the north in central Montana and southern Alberta. Moreover, the nearest bison jump to this western Texas site is about 1,200 kilometers to the north in north-central Colorado, which makes Bonfire Shelter unique if not anomalous among such localities. It has been proposed that the site's Folsom visitors migrated to the region from the High Plains, over 200 kilometers to the north.

Bonfire Shelter is located in Val Verde County, western Texas, about 50 kilometers west of Pecos River and less than 1 kilometer north of the Mexican border near the historic railroad town of Langtry. The shelter is situated on the Stockton Plateau, which is considered to be a western extension of the much larger Edwards Plateau, and thus lies in a transitional zone between the desert scrub and desert grasslands of the Trans Pecos on the west and live oak-mesquite savannah that is more typical of the Edwards Plateau on the east. At the time of the site's earliest human occupation, the region's vegetation appears to have been a mosaic of woodlands, pinyon parklands, and scrub grasslands.

The present-day vegetation on the canyon floor near the site is sparse and limited to hardy, heat- and flood-tolerant species like creosote bush and mesquite, while the greater region is characterized by mid-height grasslands sparsely covered with drought-deciduous and conifer shrubs and small trees, short to mid height grasses, and oak savanna. Today, species such as white-tailed deer, peccary, ring-tailed cat, fox, skunk, jackrabbit, endangered wild cats (ocelot and jaguarundi), quail, turkey, dove, a variety of rodents, and over thirty species of snakes (some poisonous) and lizards are found in the area.

The site lies within the east wall of Mile Canyon about 1 kilometer from the canyon's southern opening on the Rio Grande and about 300 meters from this box canyon's northern end, where one encounters a canyon wall and an abrupt 20 meter tall cliff. The canyon is

FIGURE 7.1 (above) Bonfire Shelter is situated on the east side of the canyon about 18 meters above the valley floor and 20 meters below the canyon rim. The entrance to the rock-shelter, which can be seen from the opposite rim of the canyon, is not visible from the canyon floor due a massive fan-shaped jumble of talus and large limestone boulders sloping down from the lip of the shelter. Photograph by David Meltzer.

FIGURE 7.2 (right) Sometime before the shelter was occupied by humans, a huge portion of the canyon's cliff face collapsed into a pile of huge angular blocks, partially enclosing the entrance. Photograph by Jack Skiles.

relatively narrow, ranging about 50–100 meters in width, and is about 90 meters wide in the vicinity of the site. The canyon's steep walls attain a maximum height of about 60 meters.

Bonfire Shelter's northwest-facing opening is about 85 meters wide and presently sits about 18 meters above the present-day canyon floor. [Figure 7.1] When the shelter was formed by fluvial (*i.e.*, river derived) erosion sometime in the Late Pleistocene well before the site's human occupation, it is presumed to have been situated on the valley floor. Over time, continued erosion appears to have detached and deposited a number of very large, boulder-sized limestone blocks at the shelter's opening, effectively forming a wall that impeded the flow of water into the site. Somewhat later, after the canyon was cut even deeper, a massive fan-shaped jumble of talus deposits accumulated below the site. This barrier provided protection from the elements and made the site virtually invisible from the canyon floor. Behind this limestone barrier, the shelter's downward-sloping floor measures about 10–20 meters in depth and its roof measures about 4–10 meters in height. [Figures 7.2 and 7.3] The canyon rim above the shelter's opening reaches a height of about 20 meters above the shelter floor, and its edge is distinguished at the site's southern end by the presence of a steeply sloping V-shaped cleft under which there has accumulated a large quantity of bison bone on the talus cone.

Though Bonfire Shelter had been cursorily examined in 1936 by archaeologists from the Witte Museum in San Antonio during a survey of Mile Canyon, the site was first identified as a potential archaeological locality by archaeologist Michael Collins in 1958, when he was a high school student. Collins's single shallow test pit identified a dense but relatively thin layer of very friable burned bone from what appeared to be large animals, and he was able to extract an intact mandible that was later identified as that of a modern bison (*Bison bison*).

The site's landowners subsequently contacted representatives of the Texas Archeological Salvage Project at the University of Texas, who were conducting archeological research associated with the imminent flooding of the Amistad Reservoir, situated on the Rio Grande at its confluence with the Devils River about 60 kilometers

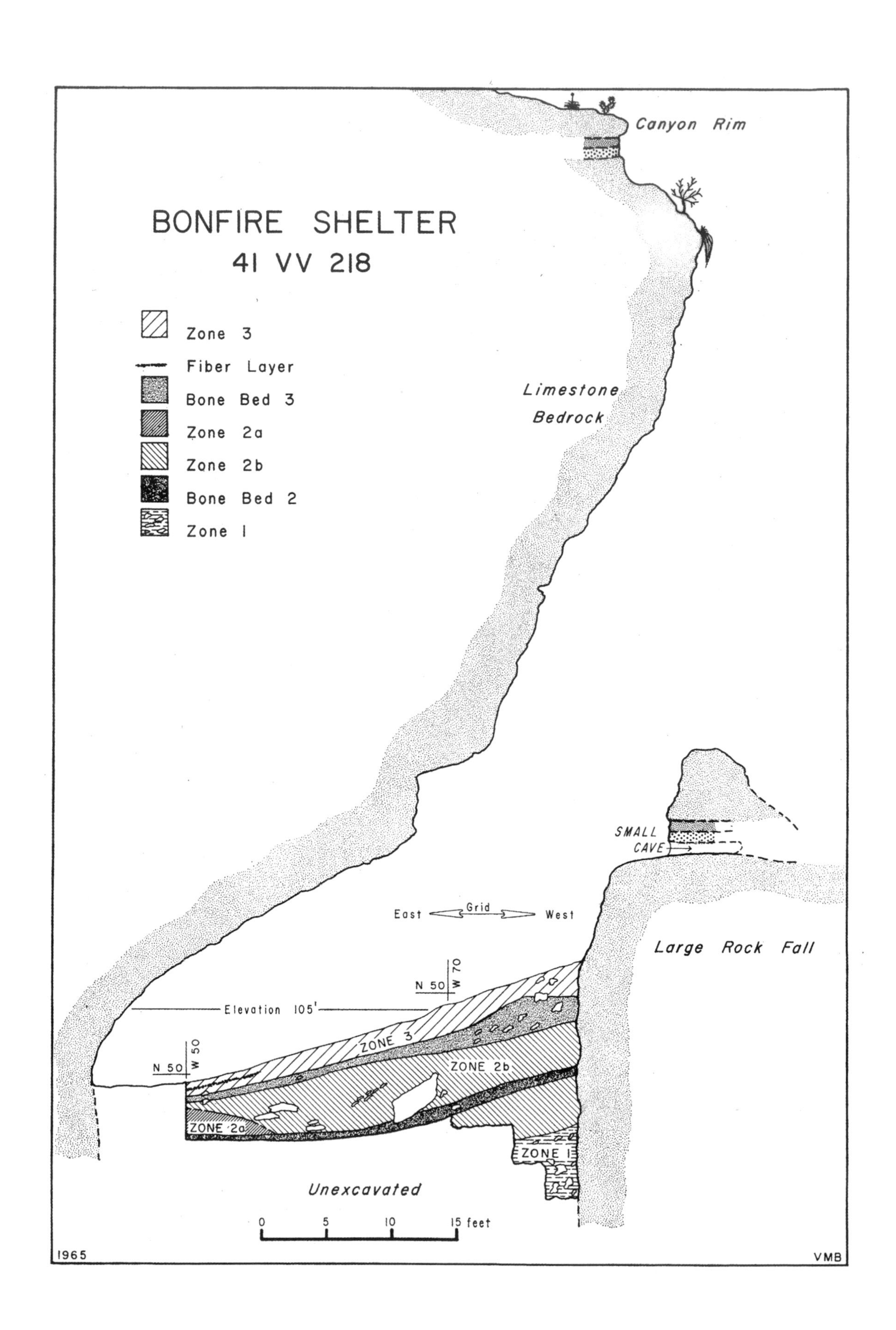

BONFIRE SHELTER
41 VV 218
Zone 3
Fiber Layer
Bone Bed 3
Zone 2a
Zone 2b
Bone Bed 2
Zone 1
Canyon Rim
Limestone Bedrock
SMALL CAVE
East
Grid
West
Large Rock Fall
N 50
W 70
Elevation 105'
ZONE 3
N 50
W 50
ZONE 2b
ZONE 2a
ZONE 1
Unexcavated
0
5
10
15 feet
1965
VMB

FIGURE 7.3 (left) Bonfire Shelter, a spacious semi-circular vault that is 85 meters wide and 10–20 meters deep, was formed by stream-undercutting of the canyon's cliffs. The floor of the shelter slopes up from the back wall toward the front. The shelter's 4–10 meter high ceiling curves smoothly upward and outward from the rear. From *Bonfire Shelter: A Stratified Bison Kill Site, Val Verde County, Texas* by David S. Dibble and Dessamae Lorrain.

FIGURE 7.4 (below) The first major excavation at the shelter began on 24 September 1963 and continued until 27 February 1964 under the supervision of University of Texas archaeologist David Dibble. During the five months of fieldwork at the shelter, fifteen pits were excavated almost exclusively within the southern third of the site and covering about 200 square meters. From *Bonfire Shelter: A Stratified Bison Kill Site, Val Verde County, Texas* by David S. Dibble and Dessamae Lorrain.

southeast of the site. In the fall of 1962, a small survey team under the direction of Mark Parsons returned to Bonfire Shelter, excavating a test pit adjacent to Collins's earlier excavation. This test excavation relocated the previously identified layer of burned bison bone and in apparent association recovered a burned, Late Archaic period Montell dart point dating to about 3000–2000 yr BP.

The first major excavation at Bonfire Shelter began on 24 September 1963 and continued until 27 February 27 1964 under the supervision of David Dibble, again under the auspices of the University of Texas. Dibble's crew excavated a total of fifteen excavation units that were heavily concentrated in the southern and south-central portions of the site, covering a total excavated area of about 200 square meters. [Figure 7.4] After a twenty-year hiatus, Dibble returned to the site, this time as principal investigator of a field crew directed

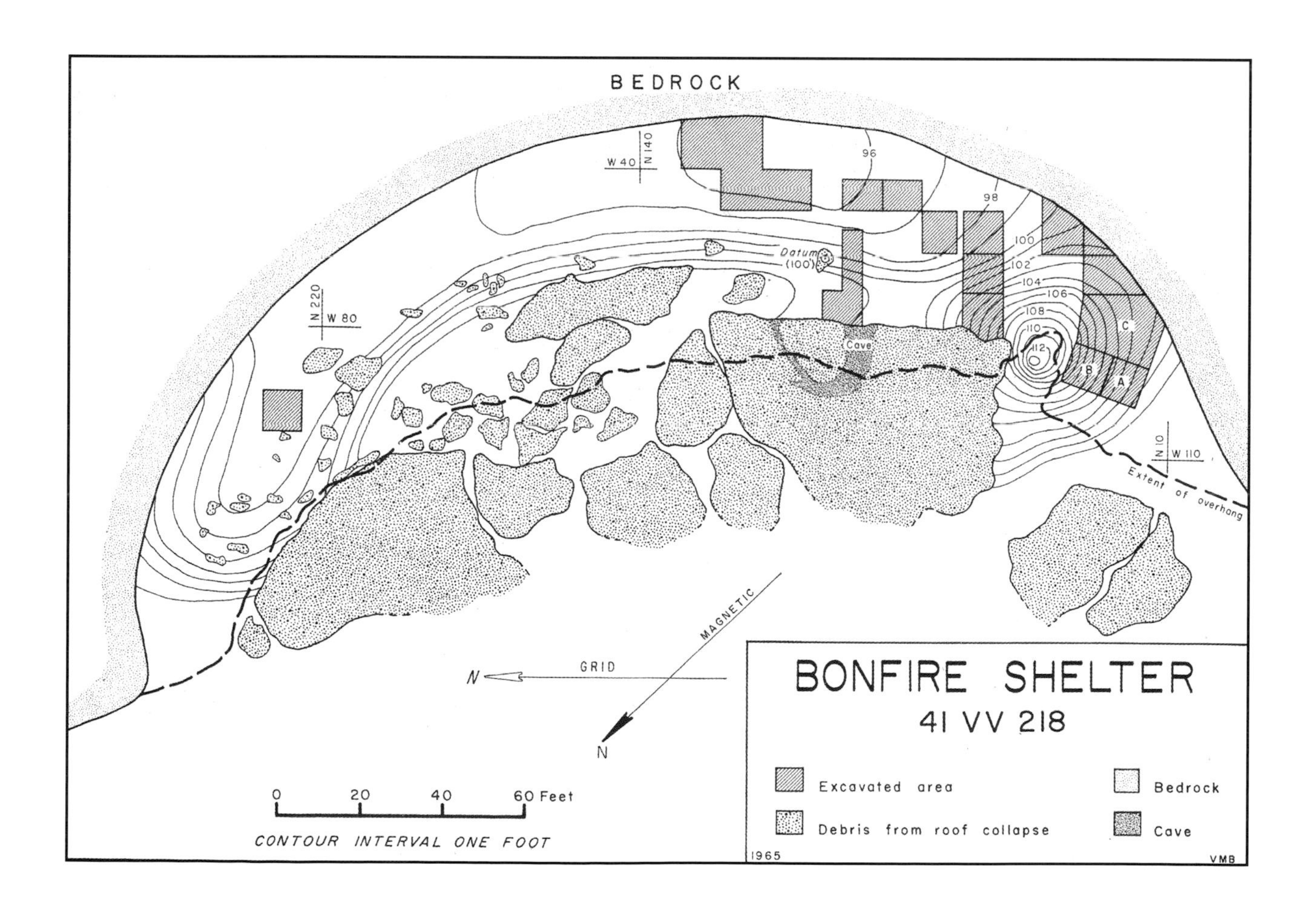

Zone
3b
Fiber Layer
Zone
3a
Bone
Bed
3
Zone
2
c
Bone
b
Bed
a
2
Zone
1
Elev.
102'
101'
100'
99'
98'
97'

FIGURE 7.5 (left) Section of the north wall. Bonfire Shelter contains three principal stratigraphic deposits. Bone Bed 1 dates to 16,060–13,370 yr BP (Paleoindian) and contains bones of extinct Pleistocene mammalian species, including horse (*Equus francisci*), camel (*Camelops hesternus*), antelope (*Capromeryx* sp.) and mammoth (*Mammuthus* sp.). Bone Bed 2 dates to 12,430–11,330 yr BP (Paleoindian) and contains bones of extinct bison (*Bison antiquus*). Bone Bed 3 dates 2725–3210 yr BP (Late Archaic) and contains bones of modern bison (*Bison bison*). Roy Little points to Bone Bed 2 in profile near talus accumulation. Photo by David Dibble.

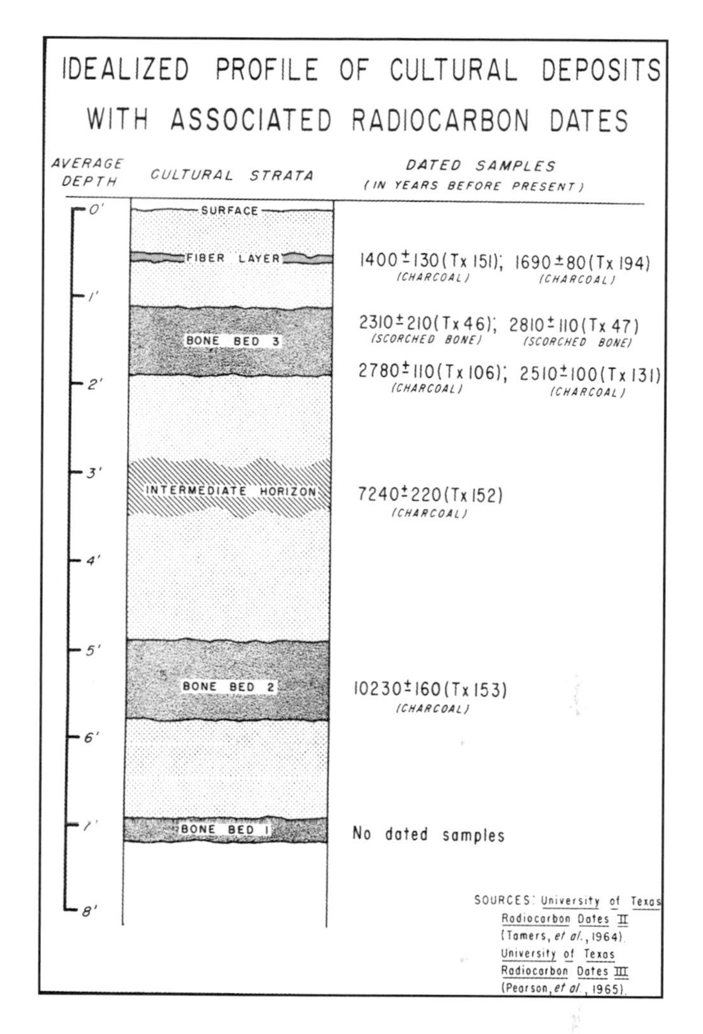

FIGURE 7.6 (right) The natural and cultural deposits within Bonfire Shelter were excavated to a maximum depth of almost 6 meters below the modern floor of the shelter. The depth of the oldest, universally accepted archaeological deposit ranges from 1.3 meters (in the front) to about 2.6 meters (in the back) below the shelter floor's original surface. From *Bonfire Shelter: A Stratified Bison Kill Site, Val Verde County, Texas* by David S. Dibble and Dessamae Lorrain.

by Solveig Turpin, conducting additional work from January to April 1984 and once again for a final time in April 1984. Including all archaeological work at the site, a total area of about 225 square meters was excavated at Bonfire Shelter.

The ten months of investigation at Bonfire Shelter identified three discrete bone beds within about 3 meters, vertically, of the site deposits. [Figures 7.5 and 7.6] Uppermost Bone Bed 3 (BB3), which corresponds to the thin deposit of burned bison bone identified by Parsons in 1962, ranges 30–60 centimeters in thickness and dates to as early as 3210–2730 yr BP (based on hearth charcoal). Cultural associations with BB3 include a lithic artifact assemblage composed of Late Archaic projectile points (Castroville and Montell), biface fragments, scrapers, hammerstones, debitage flakes, and two hearths. [Figure 7.7] Dibble estimated that the remains in the BB3 deposit represented at least 800 individual animals.

Underlying BB3 by a depth of about 1 meter, the excavators identified a second, continuous 45 centimeter thick layer of larger disarticulated bone, designated BB2 (or, as defined in 1983, Strata A–C) that was ultimately identified as the remains of extinct forms of bison (either *Bison antiquus* or *Bison occidentalis)* and thought to represent the remains of at least twenty-seven individual animals. [Figures 7.8, 7.9, 7.10, and 7.11] Dibble and his colleagues concluded that BB2 was composed

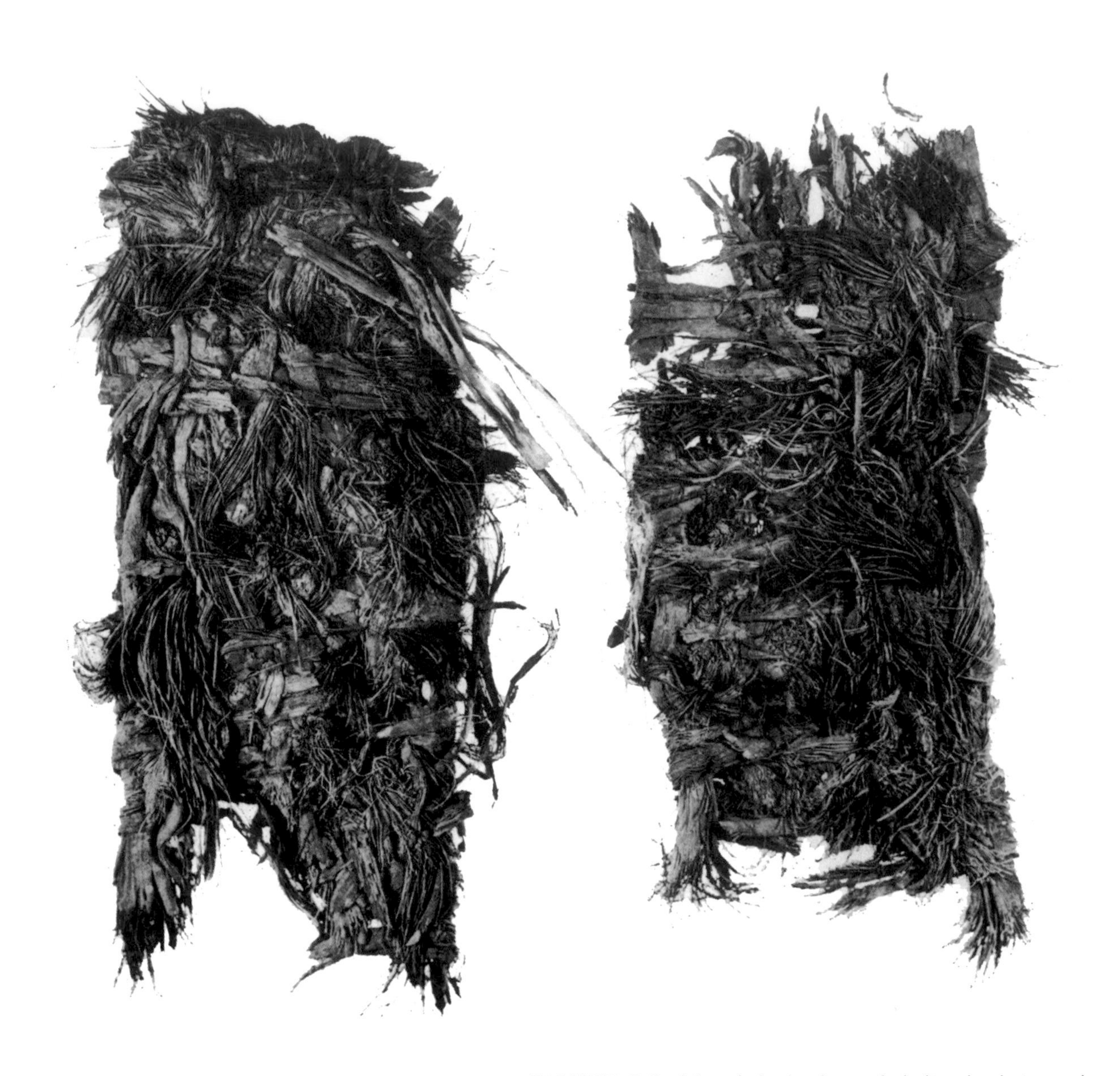

FIGURE 7.7 (above) At the front of shelter, basketry and Sandal fragments were recovered from a thin layer within Zone 3. Photograph by David Dibble.

FIGURE 7.8 (right, top) Zone 2 contained the remains of two hearths. A single piece of charcoal from one of the hearths was radiocarbon dated to 12,430–11,330 yr BP. Photograph by David Dibble. Courtesy of Texas Beyond History.net, Texas Archeological Research Laboratory, University of Texas-Austin.

FIGURE 7.9 (right, bottom) Burned Montell dart point as found in Bone Bed 3. Photograph by Jack Skiles.

FIGURE 7.10 (left) Typical Late Archaic Castroville and Montell points from Bone Bed 3. Photograph by David Dibble.

FIGURE 7.11 (below) Crude biface and flake scrapers from Bone Bed 2. Photograph by David Dibble.

of three discrete horizons or jump events. Associated cultural materials included five points (four Plainview and one Folsom), five point fragments, two bifaces, five scrapers, three worked flakes, seventeen debitage flakes, and remains of a single hearth. A charcoal sample from a hearth in the upper reaches of BB2 (Stratum A) was radiocarbon dated to 12,430–11,330 yr BP. [Figure 7.12]

Underlying BB2 at depths ranging between just over 1 meter to 3.7 meters, the excavators identified a third and final bone bed, designated BB1 (or, as defined in 1983, Strata E–H), which was populated by the remains of extinct species including mammoth (*Mammuthus* sp.), horse (*Equus francisci*), bison (*Bison antiquus*), camel (*Camelops hesternus*), and antelope (*Capromeryx* sp.). [Figure 7.13] Initially, based on its stratigraphic position below BB2, Dibble hazarded a guess that it BB1 was several millennia older and deposited in the latter part of the last Ice Age. Later work at the site confirmed the radiocarbon age of BB1 to be 16,060–13,370 yr BP, based on charcoal from the middle of this layer. Although no definitive proof of a human presence has been identified in the BB1 deposit, several bones recovered from the 1983–1984 excavation display putative cut marks and spiral fractures suggesting butchering. In the absence of bona fide artifacts, however, the deposit's archaeological validity remains conjectural.

Dibble and his colleagues maintained that the modern bison bones in the BB3 deposits and the extinct bison bones in the BB2 deposits resulted from multiple jumps of human-stampeded bison that were driven down though the cliff's cleft and over its edge, landing maimed (or worse) on the talus cone below—a distance estimated by Dibble to have been about 23 meters when BB2 was deposited. [Figure 7.14] To conclude

FIGURE 7.12 In contrast to Bone Bed 3, the bones in Bone Bed 2 were fully disarticulated and sorted. Here the bones are seen in spoke-like arrangement around a limestone anvil or butcher block. From *Mammalian Faunal and Cultural Remains in the Late Pleistocene Deposits of Bonfire Shelter, 41VV218, Southwest Texas* by Leland C. Bement.

FIGURE 7.13 *Bison antiquus* horn core from Bone Bed 2. The bones of Bone Bed 2 were presumed to have been butchered within the shelter. The first step was to sever and separate the major parts of the animal. After sorting, the skulls were bashed apart, presumably to extract the brains. The mandibles were split apart and separated from the maxilla, probably to free the tongue. The front and rear legs were cut apart at the joints, and the meat was carefully cut from each bone. Following this, the long bones were systematically broken open, no doubt to extract the fat-rich marrow. Photograph by David Dibble.

FIGURE 7.14 Aerial photograph of Mile Canyon. The black arrow points to the spot where bison were allegedly driven over the cliff above Bonfire Shelter. Photograph by David Dibble. Courtesy of Texas Beyond History, Texas Archeological Research Laboratory, University of Texas-Austin.

otherwise would have meant that the bison either were stampeded directly into the shelter or were killed on the canyon floor and dragged almost 20 meters upslope to the site for processing. Dibble, his colleagues, and other scholars who have analyzed data from the site agree that both of these latter scenarios seem unlikely.

While no one questions the radiocarbon age of the bone beds at Bonfire Shelter, nor the archaeological validity of the upper two bison deposits, some questions have remained about the precise human role in how the site came to be deposited. In 2005 and 2007, a group of archaeologists from Southern Methodist University, the University of Arizona, and Colorado State University raised doubts about the status of the Bonfire's BB2 as a bison jump. Noting that the bison bones recovered from BB2 were dominated by limb and axial (*i.e.*, trunk and vertebrae) skeletal elements, they claim the assemblage is the opposite of what one would expect from a kill site. Drawing support of their view from an earlier critique by anthropologist Lewis Binford on the site's status as a bison jump, these archaeologists instead suggested that the BB2 more closely resembles a butchering or processing area and that the kill site probably lay elsewhere.

In the absence of new data from the site, it is probably unlikely that these doubts will be definitively or even satisfactorily resolved. If the BB2 deposit at Bonfire Shelter is indeed a bison jump kill site, its position virtually on the Mexican border and its placement in time make it, in the words of the site's 2005–2007 investigators, "an event isolated by almost 1,800 kilometers and by nearly 4,300 years from anything like it." In any case, however, there is firm evidence that Folsom-age peoples were processing bison at Bonfire Shelter as early as about 12,400 yr BP.

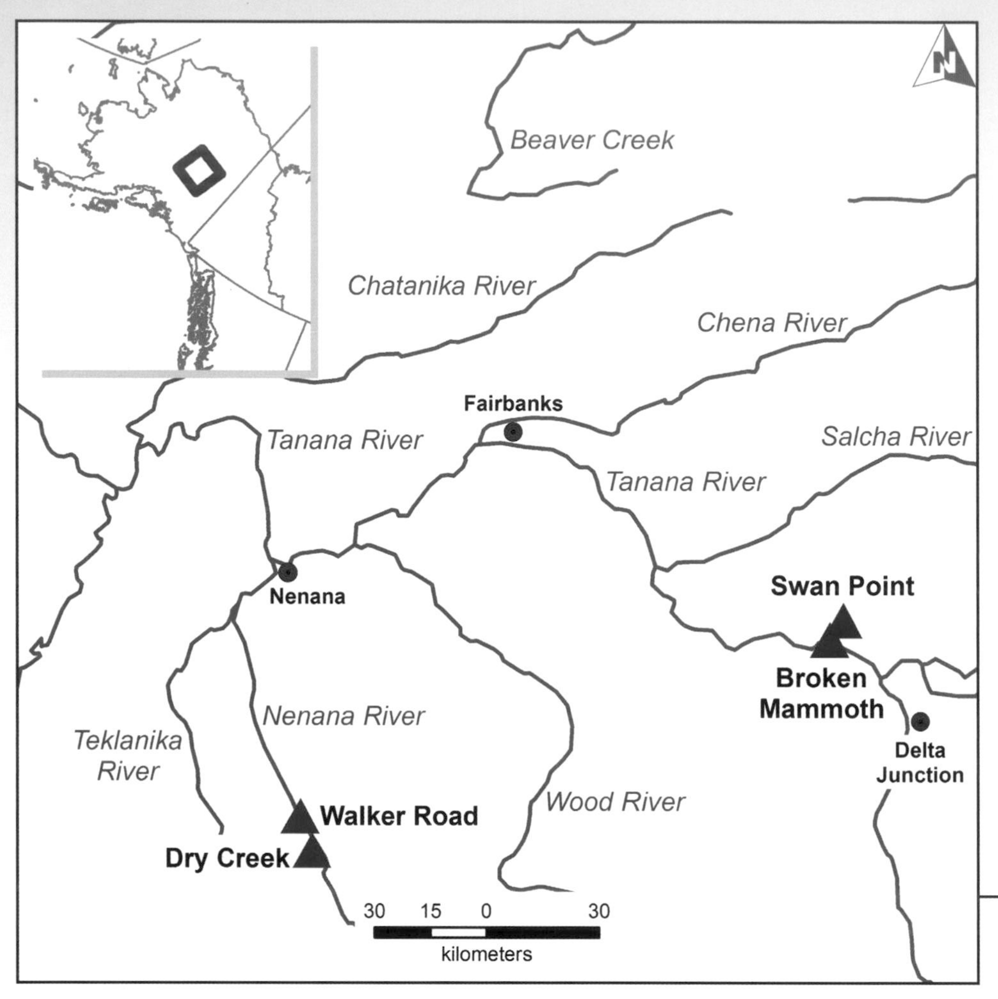

Map by David Pedler

The Nenana and Tanana River valleys in central Alaska were at the eastern extent of a vast, mostly ice-free landmass that extended from Siberia at the end of the Pleistocene and later. The valleys apparently experienced repeated visits by Paleoindian hunter-gatherers from about 14,500 yr BP until 1600 yr BP. Broken Mammoth, Dry Creek, Swan Point, and Walker Road are among the dozens of Late Pleistocene and Early Holocene archaeological sites that have been identified in these valleys.

8 CENTRAL ALASKA (BROKEN MAMMOTH, DRY CREEK, SWAN POINT, AND WALKER ROAD)

LOCATION FAIRBANKS NORTH STAR BOROUGH, ALASKA, UNITED STATES

Coordinates Broken Mammoth, 64°15'38.23"N, 146° 6'35.46"W; Dry Creek, 63°52'53.40"N, 149° 2'21.24"W; Swan Point, 64°18'7.31"N, 146° 1'58.02"W; Walker Road, 63°58'0.00"N, 149° 5'0.00"W.

Elevation above mean sea level: Broken Mammoth, 305 meters; Dry Creek, 470 meters; Swan Point, 320 meters; Walker Road, 430 meters.

Discovery Broken Mammoth by Charles Holmes in 1989; Dry Creek by Charles Holmes in 1973; Swan Point by Tom Dilley and Richard VanderHoek in 1991; Walker Road by John Hoffecker and S.M. Wilson 1980.

Central Alaska has proven to be one of the most dynamic (if not simply one of the harshest) environments for archaeological research in the New World over the past twenty years. Site discovery in the region has virtually exploded, and research questions that were presumed to have been answered decades ago are once again open to active inquiry. The earliest, most thoroughly investigated and well-known Late Pleistocene archaeological sites in Alaska are those of the central Nenana and Tanana River valleys. We have chosen four localities as being representative of this time frame, though as many as ten additional sites of this period are presently known in the region.

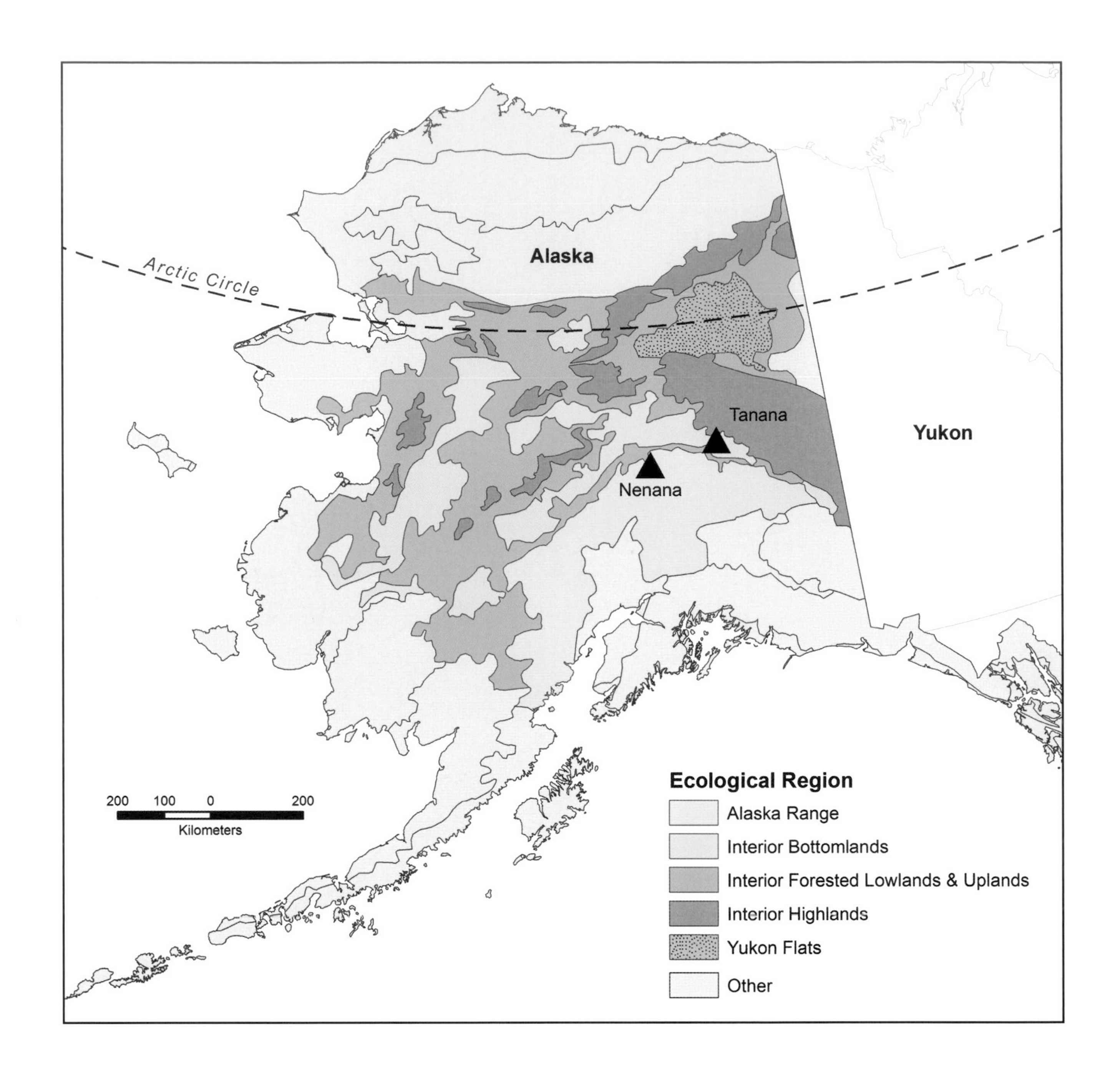

The sites of the central Nenana and Tanana River valleys are located about 120 kilometers southwest and 100 kilometers southeast of Fairbanks, Alaska, respectively. Anchorage and the Gulf of Alaska lie 300–400 kilometers to the south. The region is bounded on the south by the steep, very high peaks of the Alaska Range and on the north by the low, rounded mountains of the Interior Highlands. The tributary Nenana River watershed is the smaller of the two, draining a 10,100 square kilometer portion of the much larger 115,500 square kilometer Tanana River watershed. Together, these watersheds define the south-central margin of the vast Yukon River basin, which drains a total area of about 850,000 square kilometers by the time it reaches the Bering Sea on Alaska's west coast. The fourth largest watershed in North America, the Yukon River basin covers an area that is larger than the Northeast region of the United States. [Figure 8.1]

FIGURE 8.1 (left) The sites of the Nenana River valley (*i.e.*, Dry Creek and Walker Road) lie within a zone of rounded hills and rolling lowlands that are covered in a complex mosaic of bog, scrub vegetation, and forest. Today, the area's rugged terrain is used primarily for fishing and hunting with relatively light human settlement. The sites of the Tanana River valley (*i.e.*, Swan Point and Broken Mammoth) lie within a flat zone dotted by numerous lakes and covered in forest, scrub, and wetland vegetation. Also valued as a place for fishing and hunting, this area contains the Alaskan interior's densest human settlement. Its rivers, which provide critical food sources and transportation routes necessary for human survival and trade, have attracted Native American and Euro-American peoples for millennia. Map by David Pedler.

The Dry Creek and Walker Road sites in the Nenana River valley straddle the boundary between the Interior Forested Lowlands and Uplands ecoregion and the Alaska Range ecoregion. This zone's rolling lowlands are covered in a complex mosaic of bog and scrub vegetation, spruce forest, and broadleaf forest composed principally of balsam poplar and quaking aspen. The Swan Point and Broken Mammoth sites in the Tanana River valley lie within the Interior Bottomland ecoregion, an area dotted with numerous oxbow lakes (*i.e.*, U-shaped waterbodies formed from cutoff stream meanders) and small shallow lakes that have formed in depressions created by melting permafrost. The bottomland's regional vegetation is dominated by spruce and hardwood forest, scrub thickets, and wetland plant species.

The Dry Creek and Walker Road sites are located 10 kilometers from each other in Denali Borough near the town of Healy, the borough's seat, in the foothills of the Alaska Range. [Figures 8.2 and 8.3] The Dry Creek site occupies a southeast-facing bluff on the north bank of Dry Creek at an elevation of 470 meters above sea level, about 3 kilometers west of the Nenana River. The site was discovered in May 1973 by Charles E. Holmes, formerly of the Alaska Department of Natural Resources' Office of History and Archaeology. The subsequent work at the site in 1974 and 1976 excavated a total area of 347 square meters and recovered over 35,000 lithic artifacts along with a small assemblage of faunal remains. Dry Creek was the first deeply stratified archaeological site in central Alaska to be dated to the Late Pleistocene. In 2011, Dry Creek was revisited for limited test excavations (10 square meters) by a team of archaeologists led by Texas A&M University archaeologist Kelly Graf.

The Walker Road site was discovered in August 1980 by University of Colorado archaeologist John Hoffecker along an unnamed creek on the east side of the Nenana River at an elevation of 430 meters above sea level. The site was investigated in 1985–1990 by several teams of archaeologists who excavated a total area of 200 square meters and recovered about 5,000 lithic artifacts. [Figure 8.4] At both sites, the Late Pleistocene components are overlain by more-recent cultural materials.

The oldest stratigraphic levels at both sites are composed of glacier-associated, wind-borne layers of fine sediment (known as loess) that were deposited to depths of up to 1–1.5 meters below the modern ground surface and contain numerous discrete activity areas and hearths. The lithic artifact assemblages from the sites' lowest levels are dominated by flakes and flake fragments, with relatively low percentages of tools that include projectile points, bifaces, scrapers, blades, and flake tools. These levels have been ascribed to an entity known as the Nenana complex based on the unique suite of tools recovered, the absence of microblades, and the presence of distinctive small, teardrop-shaped bifaces known as Chindadn projectile points. [Figure 8.5] The Nenana complex was formally proposed by John Hoffecker and University of Alaska archaeologist William R. Powers in a 1989 synthesis of archaeological research in the middle Nenana River valley.

The Nenana complex material at the Dry Creek site was initially dated to 13,130–12,770 yr BP based on charcoal obtained from the site's Loess 2 level, which makes it older than the Folsom site (see pages 36-45) and only slightly younger than the Clovis occupation at Blackwater Draw (see pages 46-57). Graf's subsequent work at Dry Creek, however, identified three hearths in the site's lower levels that date to as early as 13,485–13,305 yr BP, which is marginally—but still, clearly—older than Clovis. [Figure 8.6] Three AMS

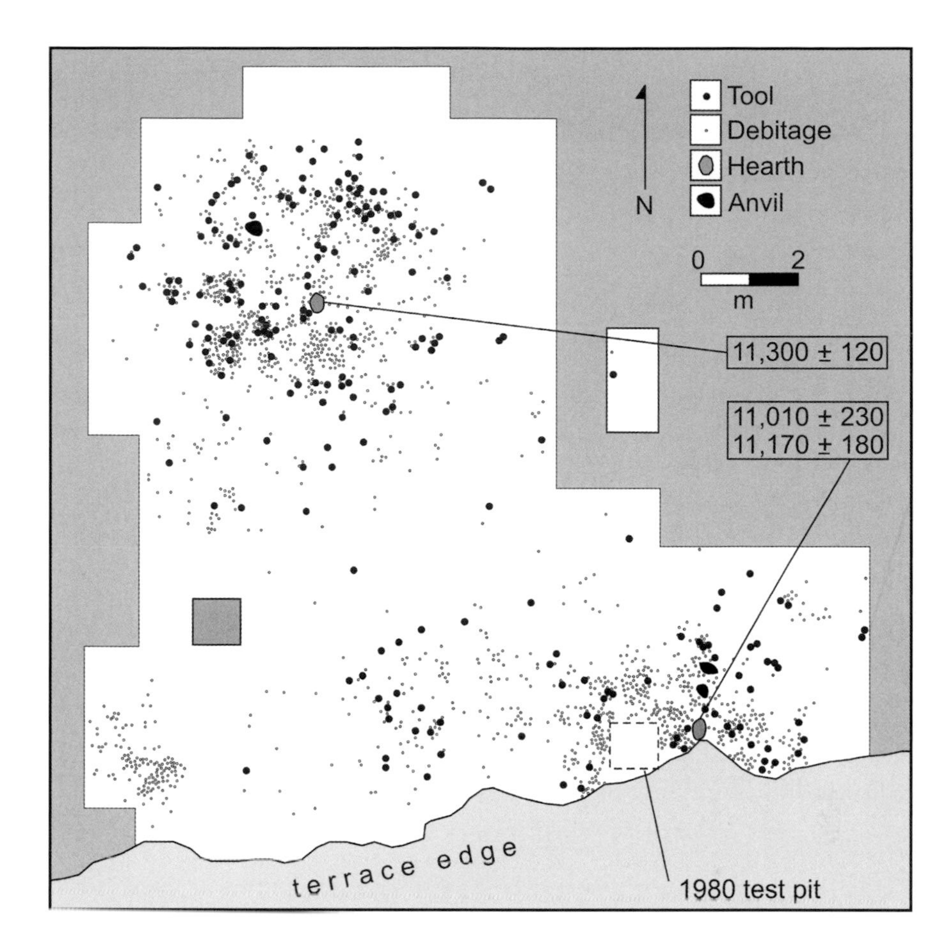

FIGURE 8.2 (left, above) Aerial view of Dry Creek, one the most important archaeological sites in Beringia. Two separate cultural layers, called Components 1 and 2, the latter dated to as early as 13,485–13,305 yr BP, have been identified. Multiple types of stone tools have been found at the site, as have remains of processing a variety of animals. The site, which its earliest dates mark as pre-Clovis, is valuable for the insight it yields into the critical transitional period at the end of the most recent Ice Age. Courtesy of Kelly Graf.

FIGURE 8.3 (left, below) An aerial view of the Walker Road archaeological site, which is located on a bluff, looking south over the Nenana River valley. The earliest deposits at this site date to between 13,420–12,900 yr BP and 13,360–12,520 yr BP, and contain four separate activity areas where tool manufacture and repair evidently took place. Tools such as scrapers, blades, and wedges were found, along with cores and debitage flakes. Hearths were found in two of the activity areas and flakes of red ochre were found in each of the hearths. No identifiable animal remains were discovered in excavation but small fragments of charred bone were found in the hearth areas. The large quantity of lithic remains at this site (almost 5,000 pieces including 218 tools) has provided archaeologists with a great deal of information about the tool types and manufacturing techniques of the Nenana people. Courtesy of Ted Goebel.

FIGURE 8.4 (above) Walker Road site map. Walker Road is located a south facing bluff 60 meters about the Nenana River valley floodplain. While Chindadn points have been recovered at Walker Road, the majority of tools are retouched/utilized flakes and blades, endscrapers, side scrapers, and cobble tools. Retouched tools and blades make up nearly 50 percent of the entire tool assemblage. Courtesy of Ted Goebel.

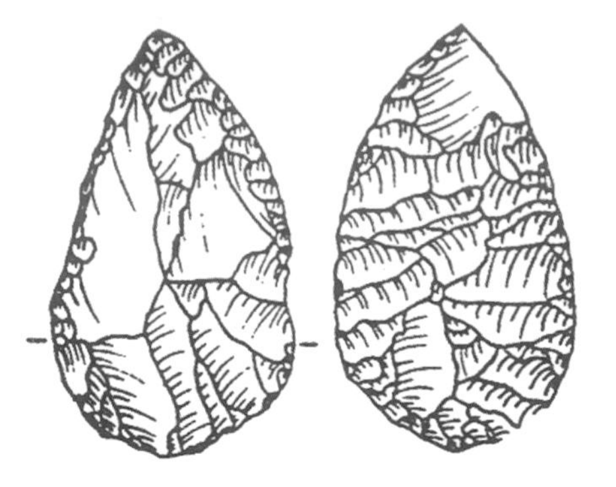

FIGURE 8.5 (left) A Chindadn point from Walker Road. The lithic technology of the first inhabitants of the Nenana River valley was characterized by finely crafted, small teardrop shaped Chindadn points. New dating of the oldest levels at Walker Road suggests the site was occupied as early as 13,485–13,305 yr BP, which expands the duration of Nenana Complex lithic technology to about 2,000 years or twice as long as was previously thought. Courtesy of Ted Goebel

FIGURE 8.6 (below) Stratigraphic profiles from the 2011 excavations at Dry Creek. Profiles of blocks A and B represent the eastern walls, while the profile of block C represents the western wall. Excavations followed standard procedures. Site sediments were removed by hand troweling. All excavated sediment was dry-screened through 1/8-inch mesh. Artifacts, bones, and charcoal samples recovered *in situ* were recorded with three-point grid charts. Reproduced by permission of the Society for American Archaeology from *American Antiquity* "Dry Creek Revisited" vol. 4, no. 1, 2015.

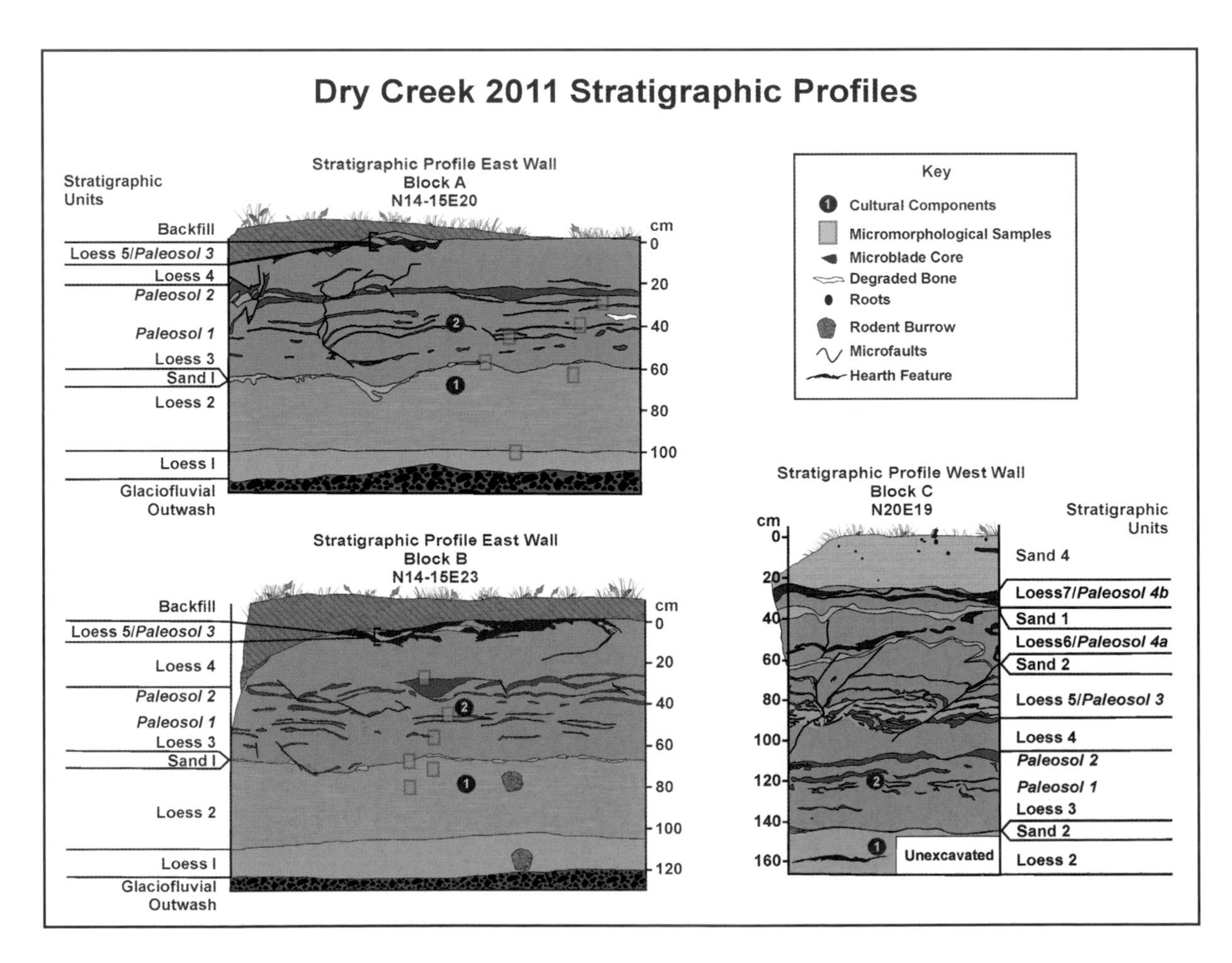

radiocarbon ages based on charcoal from Hearth 1 in the Walker Road site's Nenana complex level range from as early as 13,420–12,900 yr BP to a slightly later date of 13,360–12,520 yr BP. (A fourth conventional [*i.e.*, non-AMS] radiocarbon date obtained from the same hearth has been rejected as being 600–700 years too old.) The presence of a second, later occupation at the Dry Creek site has played a crucial role in interpreting the Nenana River valley's cultural stratigraphy and its relationship to that of the Tanana River valley and central Alaska. This later material, which ascribes to an entity known as the Denali complex, ranges in age from 12,730–12,380 yr BP to 11,840–11,280 yr BP. (Interestingly, Graf's more-recent test excavations at Dry Creek appear to indicate a much younger age of 11,060–13,590 yr BP for this site level.) Unlike the Nenana complex, Denali lithic artifact assemblages include distinctive wedge-shaped microblade cores, microblades, multifaceted burins, lanceolate projectile points, and knives. Denali complex sites are found elsewhere in the Nenana and Tanana River valleys, and the signature presence of microblades continues well into the subsequent archaeological complexes and traditions of the Holocene.

The Tanana River valley Broken Mammoth and Swan Point sites are spaced 10 kilometers apart in the watershed of Tanana River tributary Shaw Creek, 100 kilometers southeast of Fairbanks. The Broken Mammoth site is located near the confluence of Shaw Creek and the Tanana River at an elevation of 305 meters above sea level and 25 meters above the level of the Tanana River. The site was discovered in 1989 by Charles Holmes, whose attention was drawn by the presence of mammoth ivory fragments and other animal remains eroding from the bluff there. Episodic fieldwork between 1990 and 2010 excavated a total area of 408 square meters and recovered thousands of artifacts and well-preserved animal remains.

The stratigraphy at the Broken Mammoth site is composed of four major layers of glacially derived, windborne sand and loess whose lower levels are interspersed with three thin layers representing episodes of soil development and plant growth (called paleosols). [Figure 8.7] The oldest level in the site deposit (labeled Cultural Zone 4 or CZ4) displayed at least nine discrete clusters of faunal remains, three lithic manufacturing areas focused on the reduction of quartz, and eight hearths. The CZ4 level has also yielded 800 lithic tools—largely comprised of scrapers, retouched flakes, cobble tools, and cores—and a small number of mammoth ivory artifacts. Flakes from bifacial reduction were recovered from CZ4, but the excavators did not encounter bifacial tools or any other diagnostic artifacts. The faunal remains from CZ4 include large to medium-sized mammals—mostly elk (*Cervus canadensis*) with lesser representations of bison (*Bison priscus*), moose (*Alces alces*), Arctic fox (*Alopex lagopus*), and river otter (*Lutra canadensis*)—as well as a large array of waterfowl and smaller mammals. The CZ4 level the Broken Mammoth site is overlain by CZ3, which is about 1,000 years younger and yielded triangular bifaces that appear to be variants of the Chindadn points of the Nenana complex. Unlike the Nenana complex sites, however, CZ3 at the Broken Mammoth site produced microblades.

The Swan Point site lies on the northern edge of the Shaw Creek Flats on a knoll that is 320 meters above sea level and 25 meters above the surrounding lowlands, which are presently dissected by meandering streams and dotted with small lakes, ponds, and wetlands. [Figure 8.8] The north bank of the Tanana River lies about 6 kilometers to the southwest. The site was identified in August 1991 by archaeologists Tom Dilley and Richard VanderHoek (Alaska's current State Archaeologist), working under the direction of Charles Holmes. The initial work at the site began in 1992–1992 and continues to the present day. As of 2013, 65 square meters have been excavated there.

The stratigraphy at the Swan Point site is broadly similar to that identified at the Broken Mammoth site, as it is also composed of several layers of wind-deposited sand and loess that contain three paleosol zones. [Figure 8.9] Four cultural levels (*i.e.*, Cultural Zones 1–4) spanning the Late Pleistocene through Holocene epochs were identified at the site. The oldest of these zones, CZ4, yielded mammoth ivory artifacts, megafaunal remains (including horse and mammoth teeth, a

FIGURE 8.7 (left, top) Archaeological excavations at Broken Mammoth in the Tenana River valley. The Broken Mammoth site was occupied at least three separate times in its history, the last occupation occurring approximately 2500 yr BP. The oldest occupation of the site occurred between 13,560–13,120 yr BP and 13,420–13,120 BP. Well-defined hearths and a variety of scrapers were recovered from the site's deepest deposits. Courtesy of Richard VanderHoek, Alaska Office of History and Archaeology.

FIGURE 8.8 (left, bottom) An aerial view of the Swan Point archaeological site, which is located in the central Tanana River valley, an area that was unglaciated during the end of the Last Glacial Maximum (20,000–19,000 yr BP). Swan Point has been occupied at least five times since ca. 14,500 yr B.P with evidence of charcoal that has been dated to approximately 14,000 B.P. Swan Point is the oldest known site in the Tanana River valley. Courtesy of Charles Holmes.

FIGURE 8.9 (below) Middle Taiga period artifacts from Swan Point dated to ca. 6,000–3,000 yr BP: a–b, notched point forms; c–e, lanceolate biface bases; f–g, multiplatform tabular microblade cores; h, large unifacial side scraper; i, large, thin biface; j–l, transverse flake burins. Courtesy of Charles Holmes.

FIGURE 8.10 A microblade core and core preparation flake from the oldest cultural level Swan Point, which ranges in age from 14,490–14,050 yr BP to 14,030–13,760 yr BP. Other artifacts found at this level include worked mammoth tusk fragments, blades, burins, red ochre, pebble hammers, and quartz hammer tools and choppers. The microblades found at this zone are significant as they are the oldest securely dated microblades in eastern Bergingia and may represent initial human exploration of central Alaska's unfamiliar landscapes. Courtesy of Charles Holmes.

large mammoth tusk, and numerous ivory fragments), avian remains, three large hearths that appear to have been fueled using bone, and a relatively small but unique suite of lithic artifacts. These artifacts demonstrate a microblade tool production technique that is unrepresented elsewhere in the New World and resembles that employed at Diuktai complex sites in eastern Siberia. [Figure 8.10] Bifacial technology is present, but limited. Cultural Zone 4 at Swan Point, like CZ4 at Broken Mammoth, is overlain by a CZ3 level that produced a variety of bifaces and projectile points, at least some which suggest affiliation with bifaces of the Nenana complex. [Figure 8.11] And again like Broken Mammoth—but no other Nenana complex site—Swan Point's CZ3 lithic artifact assemblage includes microblades, which puts to rest the notion that the Nenana complex lacked a microblade industry.

The CZ3 zone at the Broken Mammoth site produced a number of radiocarbon ages that might best be represented by an AMS radiocarbon date of 12,390–11,810 yr BP on charcoal from a hearth in the lower reaches of the zone. This age is more or less consonant with the age of charcoal from two hearths in the CZ3 level at the Swan Point site, which range in age from 12,660–12,510 yr BP to 11,820–11,290 yr BP. These ages are at least 1,000 years younger than what one otherwise might presume to be their related contemporaries in the Nenana River valley.

The lowermost CZ4 zones at the two Tanana River valley sites, which have no known analogues in the Nenana River valley, range in age from as old as 14,490–14,050 yr BP to 14,030–13,760 yr BP at the Swan Point site to the more-recent age of 13,560–13,120 yr BP to 13,420–13,120 BP at the Broken Mammoth site. [Figure 8.12] These rather conservative age estimates—which are based exclusively on charcoal from hearths at both Tanana River valley sites—indicate that the CZ4 zones are at least contemporary with Clovis (as at Broken Mammoth) if not almost 1,000 years older (as at Swan Point). These ages have had dramatic impacts on the archaeological understanding of central Alaska. (The Swan Point site has not been included with the pre-Clovis sites discussed in this book only because the investigations there are not yet fully published.)

When Powers and Hoffecker published their findings and proposed the concept of the Nenana complex in 1989, some scholars came to view it as Alaska's earliest cultural horizon. It was further speculated that the Nenana complex was directly related to the Clovis tradition of the continental United States, and that its general character was distinctly North American in comparison to the northeast Asian aspect of the Denali complex. The archaeological discoveries

FIGURE 8.11 Lithics recovered from the oldest cultural zones at Swan Point. The lithic technology of these artifacts is reminiscent of the Diuktai technique of microblade production. (see pages 14-15) There is strong evidence that hearths in this zone were fueled by burning bones. Dates on charcoal, horse and mammoth teeth, and burnt hearth reside have a provisional age of ca. 14,000 yr BP. Courtesy of Charles Holmes.

Microblade cores and refit "ski" spalls (platform rejuvenation flakes)

Transverse/polyhedral burins

Microblade core biface preform

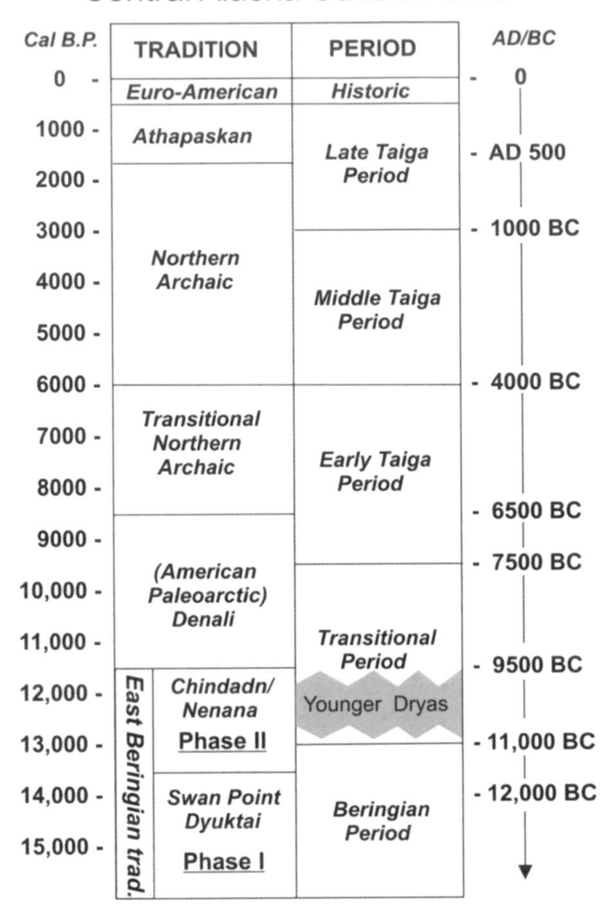

FIGURE 7.12 This chart shows the cultural traditions and periods of Central Alaska over the past 16,000 years. Traditions are identified on the basis of long-persisting, apparently related patterns of cultural remains (artifact types, domestic structures, etc.) over a specific geographic region, whereas periods are considered as intervals of time through which those traditions occur over a broader area. The sites of the Nenana and Tanana River valley fall within the Eastern Beringian Tradition, whose constituent phases spanned the so-called Beringian and Transitional periods. The Younger Dryas (shown in the early Transitional period) was a global climatic event that saw a sharp decline in the northern hemisphere's temperature at the end of the Pleistocene. Courtesy of Charles Holmes.

since then, however—and especially those at the Broken Mammoth and Swan Point sites—have occasioned a fundamental recasting of this earlier scenario.

It now appears that rather than being a precursor cultural phenomenon, the Nenana complex represents a unique adaptation that occurred over a 2000-year (or perhaps even longer) interval against the background of an otherwise continuous microblade technology with a decidedly trans-Beringian cast. The reasons for this adaptation are far from clear, though they could be related to terminal Pleistocene climate change, large mammal extinctions, adaptation to unique site habitats, site function, the season of site occupation, and/or even the introduction of discrete cultural groups. While some scholars have continued to view the Nenana complex as archaeologically valid, others like Charles Holmes have questioned the utility of maintaining the designation, instead proposing that it is best considered as a brief, post-Diuktai phase of a greater East Beringian tradition rather than a stand-alone cultural complex. [Figure 8.13]

Whatever the ultimate disposition of Nenana, the Late Pleistocene lifeway of central Alaska has been demonstrated to be remarkably complex, adaptable, and diverse. These early Alaskans engaged in specialized big-game at some localities and broad-spectrum foraging focused on small game, fish, and (probably) plant resources at others. Some of their sites were occupied only briefly, while others were more elaborately developed for animal butchery, hide processing and craft, tool manufacture, and the burning of numerous hearths. Future research will determine whether these wide-ranging endeavors were pursued within a single cultural system.

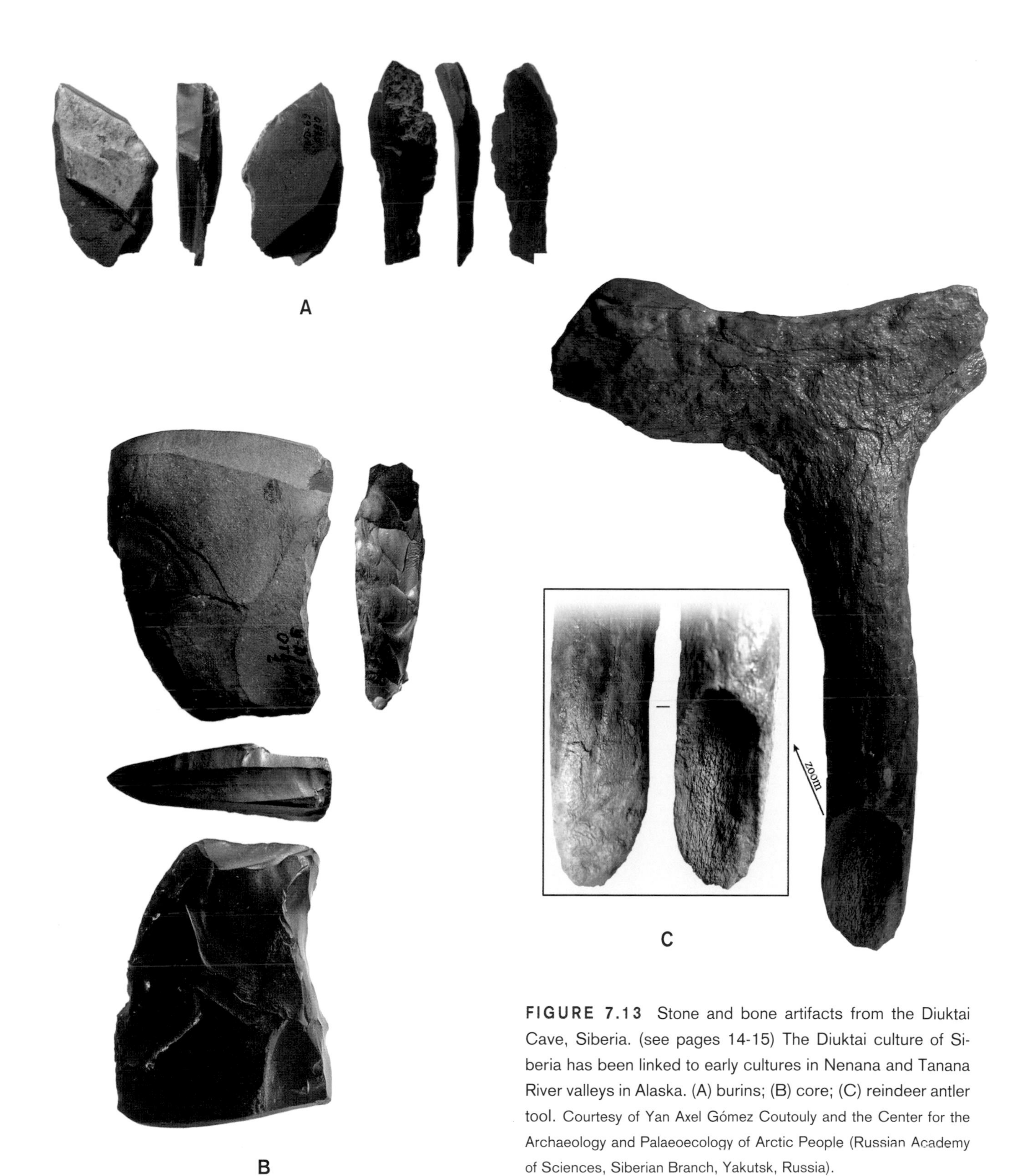

FIGURE 7.13 Stone and bone artifacts from the Diuktai Cave, Siberia. (see pages 14-15) The Diuktai culture of Siberia has been linked to early cultures in Nenana and Tanana River valleys in Alaska. (A) burins; (B) core; (C) reindeer antler tool. Courtesy of Yan Axel Gómez Coutouly and the Center for the Archaeology and Palaeoecology of Arctic People (Russian Academy of Sciences, Siberian Branch, Yakutsk, Russia).

Map by David Pedler

El Fin del Mundo is only one of four Clovis sites found south of the Mexican border and the southernmost of all known Clovis localities. Archaeological excavations at this site have expanded the known geographic and temporal range of the Clovis culture.

9 EL FIN DEL MUNDO

LOCATION SONORA, MEXICO

Coordinates 29°45'31.04"N, 111°40'35.89"W.

Elevation 625 meters above mean sea level.

Discovery Argonaut Archaeological Research Fund (University of Arizona) and Instituto Nacional de Antropología e Historia survey crew in 2007.

El Fin del Mundo is an important—and potentially quite crucial—Clovis site that has been only very recently reported. It is the southernmost North American Clovis site discovered to date, and one of that cultural horizon's earliest securely and directly dated sites. El Fin del Mundo is also a rare example of a locality that preserves both Pleistocene megafaunal remains and associated Clovis camps, and is the only Clovis site discovered to date that contains the remains of a proboscidean species other than mammoth or mastodon. The investigations at this very large site are ongoing, and the forthcoming results have the potential to profoundly influence our understanding of Clovis and the peopling of the New World.

The El Fin del Mundo site is located in the southeastern reaches of the Sonoran Desert in the Mexican state of Sonora, about 100 kilometers northwest of Hermosillo, the state capital. The Gulf of California lies about 90 kilometers to the west of the site, and the Clovis mammoth kill sites of the San Pedro Valley in southeastern Arizona (see pages 58-67) lie about 240 kilometers to the northeast. The site lies within a chain of volcanic hills that occupy a broad valley between two low mountain ranges in the Sierra Madre de Occidental foothills, which is typical of this basin and range landscape. [Figure 9.1] Local drainage is controlled by the Arroyo Bacoachi, a normally dry network of erosional channels whose watershed encompasses almost 11,000 square kilometers and ranges in elevation from 1,300 meters in the east to 5 meters above sea level near the Gulf of California. The Bacoachi reaches it southern terminus in the lowlands at the 34 square kilometer Playa San Bartolo, a dry playa lake situated about 20 kilometers from the Gulf of California.

FIGURE 9.1 El Fin del Mundo lies in the Sonoran Desert within a basin surrounded by volcanic hills. According to one of its principal researchers, Vance Holliday, an anthropologist at the University of Arizona, the site got its name (Spanish for "The End of the World") because to reach it requires a bone-jarring three-hour drive from the nearest paved road. Courtesy Henry Wallace, *Archaeology Southwest Magazine*.

The portion of the Sonoran Desert surrounding El Fin del Mundo, known as the Sonoran Plain, is presently covered in scrub and open forest that is dominated by foothill palo verde, ironwood, and mesquite with localized occurrences of elephant tree, organ pipe cactus, and tree morning glory. Although relatively little is known about the region's Late Pleistocene environment, the available data indicate that pinyon–juniper woodland with isolated examples of Joshua tree and extensive grasslands dominated the landscape. Sometime around 13,000 yr BP, however, the onset of hotter and dryer climatic conditions lead to the decline of the region's forest communities. The pinyon–juniper forest appears to have been completely eradicated by about 10,000 yr BP, and the region's contemporary vegetation became established as early as 8000 yr BP.

El Fin del Mundo is composed of at least twenty-five discrete zones (termed "localities" by the excavators) that are spread over an area of about 1,800 meters north–south by a maximum distance of 700 meters east–west.

These localities include a buried, stratified bone bed (Locality 1) that contains artifacts associated with the remains of Pleistocene mammals, several upland Clovis camp areas within 500–1,000 meters of Locality 1, and two lithic quarries. At least six of the Clovis camps appear to be centered on Locality 1, which was exposed in a west–east trending arroyo that has cut through the edge of a large bajada. (A *bajada*, which literally translates as a "descent" or "slope," is a broad complex of water-transported rock and debris that collects at the base of uplands.) Locality 1 was discovered atop an erosional "island" created by the flow of water through the arroyo, which removed surrounding sediments that might have provided a more complete picture of its relationship to the other site localities.

Alerted to the site by a local rancher in 2007, the El Fin de Mundo research team discovered Clovis artifacts and extinct animal remains eroding from the arroyo at what was ultimately defined as Locality 1. [Figures 9.2 and 9.3] Excavations and surveys were carried out between 2007 and 2012 under the direction of University of Arizona archaeologist Vance Holliday, Guadalupe Sánchez of the Instituto de Geología–Universidad Nacional Autónoma de México, and Joaquín Arroyo-Cabrales of the Mexican Instituto Nacional de Antropología e Historia. The excavation of Locality 1 encompassed about 56 square meters (14 meters by 4 meters) and proceeded to a depth of at least 2.6 meters. The key association identified during this work, and the primary focus of this chapter, is that between Clovis artifacts and a bone bed containing the remains of two gomphotheres (*Cuvieronius* sp.), a large, extinct ancestral elephant species that distinctively bore four tusks—two on the mandible, and two on the maxilla. Gomphothere remains have been identified in the southern United States and are common from Mexico to southern South America.

The excavators identified three strata at El Fin del Mundo. [Figure 9.4] Lowermost Stratum 2 measures up to 3 meters thick and is composed of a pebble-filled sandy clay that becomes finer in its upper reaches. In areas of the site beyond the arroyo where it has not been eroded, Stratum 2 is capped with Late Pleistocene through Holocene age spring-deposited sediments. Within the arroyo around Locality 1, it appears that overlying Stratum 3 was deposited into a basin which had formed within Stratum 2 sometime prior to the site's deposition. Stratum 3 is a pebble-filled sandy clay that becomes finer in its upper reaches, and is judged by the excavators to have been deposited by variably discharged, spring-fed water. The gomphothere bone bed was partially buried in the upper 15 centimeters of this level, and produced a radiocarbon age of 13,490–13,270 yr BP. [Figure 9.5] Uppermost Stratum 4 is a thick layer of diatomaceous (*i.e.,* containing fossilized remains of algae) clay and silt with a pure layer of diatomite at is base. The composition of Stratum 4 indicates that standing water conditions prevailed at the site for some time, beginning soon after the deposition of the bone bed. Hence, the excavators have concluded that in Late Pleistocene times, the site deposit lay in a wet, perhaps marshy basin that was at least 32 meters wide.

The complete exposure of the 40 square meter bone bed at Locality 1 revealed two concentrations of large mammal bones in its upper reaches. [Figures 9.6 and 9.7] Bone concentration 1 contained the vertebrae, long bones, a complete pelvis, and various lower limb bones of a gomphothere that is probably at the younger limit of its assigned 13–24 year age range. Bone concentration 2, the remains of 0–12 year old gomphothere, contained long bones, cranial fragments (including a mandible and teeth), vertebrae, ribs, pelvic bones, and a scapula. That none of the exposed bones were identified in anatomical position led the excavators to consider it doubtful that these animals met a natural death. Instead, they infer human agency in the apparently selective arrangement of bones in two discrete piles amid lithic artifacts of indisputable human manufacture. Extensive weathering of the gomphothere bone surfaces, unfortunately, has so far precluded the identification of direct human modification (*e.g.*, cut marks, breakage for marrow extraction, *etc.*).

Thirty-one lithic artifacts were recovered from within and immediately around Locality 1, including

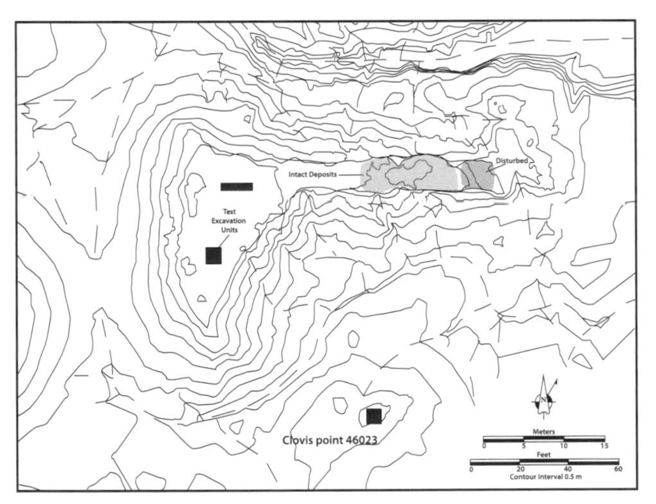

FIGURE 9.2 (above) Survey and excavation at El Fin del Mundo, which took place between 2007 and 2012, revealed a Clovis camp and the sites where two extinct ancestral elephants, gomphotheres (*Cuvieronius* sp.), had been butchered. Courtesy Guadalupe Sánchez.

FIGURE 9.3 (below) Topographic map of the Locality 1 and surrounding area showing the main excavation area (pink), disturbed area (gray), and three test units (blue) located on the west end of the locality and the area where a Clovis point was found. (See Figures 9.9 and 9.10.) Courtesy Vance Holliday.

FIGURE 9.4 Stratigraphy at Locality 1 showing layers (Strata 2, 3, 4), the layer of diatomite (fossilized algae [d]) at the base of Stratum 4, and the materials used for dating (rectangles represent charcoal, the circle represents land snail shell). The calibrated ages of the radiocarbon dates shown here are as follows: 8870 = 10,190–9740 yr BP; 9715 = 11,250–11,070 yr BP; 11,550 = 13,490–13,270 yr BP. Rectangles = charcoal; circle = snail shell. Courtesy Guadalupe Sánchez.

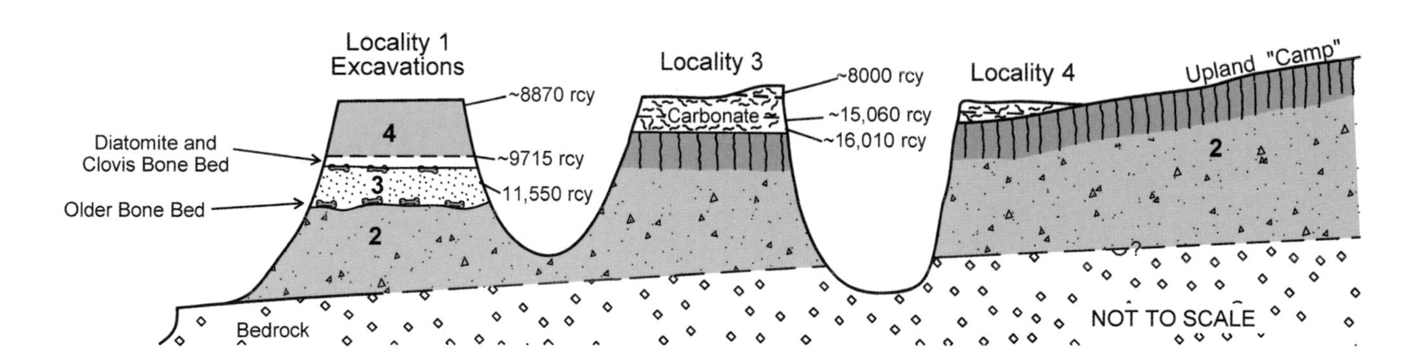

FIGURE 9.5 Schematic cross section of Localities 1, 3, and 4 and the Upland camp where Clovis materials were found. The calibrated ages of the radiocarbon dates shown for Locality 1 are as follows: 8870 = 10,190–9740 yr BP; 9715 = 11,250–11,070 yr BP; 11,550 = 13,490–13,270 yr BP. The calibrated ages for non-cultural Locality 3 range about 19,300–8900 yr BP. Courtesy Vance Holliday.

FIGURE 9.6 (above) In 2007 researchers at El Fin del Mundo found disarticulated skeletal remains initially thought to be mammoth or mastodon. Courtesy Guadalupe Sánchez.

FIGURE 9.7 (below) The numerous cusps on the molars of this mandible identified it as a gomphothere. (see Figure 9.6) Courtesy Guadalupe Sánchez.

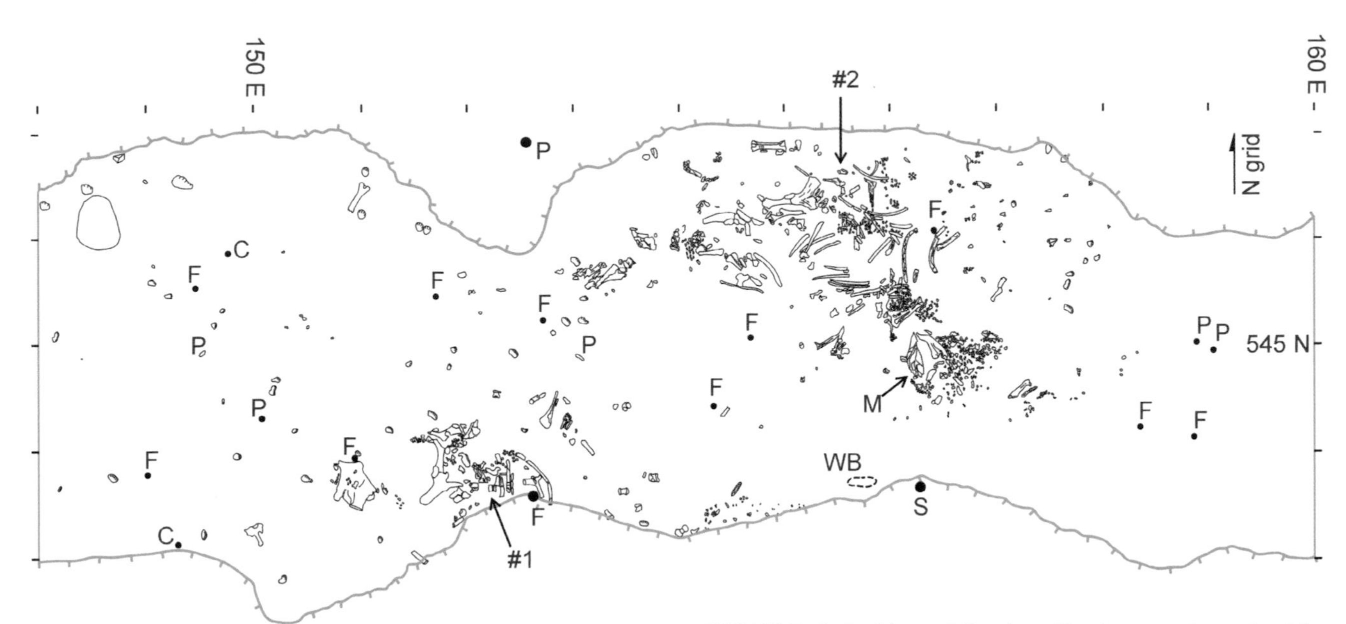

FIGURE 9.8 Map of the Locality 1 upper bone bed in, showing bone concentrations 1 (#1) and 2 (#2), and highlighting the mandible (M). Also shown are key archaeological finds recovered *in situ*: (C) charcoal, (F) flake, (P) projectile point, and (WB) worked bone with incised V. Courtesy Vance Holliday.

seven projectile points, two biface fragments, one scraper, and twenty-one retouched and bifacial thinning flakes. All of the points are well-made, fluted bifaces and at least four of them were identified *in situ* in direct association with the bone bed. [Figure 9.8] One of the points, a distal (*i.e.*, pointed end) fragment, appears to have been detached from the rest of the artifact via an impact fracture. The remaining three points were recovered from nearby and/or disturbed contexts. Other El Fin del Mundo localities have also been quite productive, particularly the upland Clovis camp localities surrounding Locality 1, which yielded a total of thirteen complete and fragmentary Clovis points, with five examples from Locality 5 alone. These reaches of the site also yielded a variety of point preforms, end scrapers, large blades, blade cores, and core tablets (*i.e.*, a large flake struck from the top a core to permit the continued removal of blades).

Several varieties of mostly local, lithic raw material are represented in the El Fin del Mundo point assemblage, predominantly including chert from nearby channel deposits and rhyolite from the quarry at Locality 2, followed by lesser quantities of obsidian, clear quartz from an outcrop 5 kilometers away, and quartzite. Of the two dominant raw materials, the preference for chert projectiles is precisely even between Locality 1 and the upland areas, but rhyolite is represented in the upland by a factor of 5:1 over Locality 1. Whatever the reason for this difference, the excavators are nonetheless confident that the site's upland camp and bone bed localities are directly related, based on the common occurrence of lithic raw material and distinctive similarities in projectile point styles and manufacturing techniques.

The most distinctive artifacts from El Fin del Mundo are undoubtedly its numerous, finely crafted Clovis fluted points. [Figures 9.9 and 9.10] Though the points are of broadly varying size, with the complete (*i.e.*, non-fragmentary) examples ranging in length from as small as 37.1 millimeters to as large as 95.3 millimeters, all of them fit comfortably within the accepted range of

FIGURE 9.9 A crystal quartz Clovis point recovered from Locality 1. Courtesy Guadalupe Sánchez.

FIGURE 9.10 Clovis points found at El Fin del Mundo Locality 1, confirming that Clovis culture was even more widespread than originally thought. Courtesy Guadalupe Sánchez.

variation recognized for Clovis points. The bone bed also included two bone ornaments, which were recovered from an area between bone concentrations 1 and 2. Both artifacts are spherical, measuring 9.16 millimeters and 11.85 millimeters in diameter, and the smaller of the two bears a single V-shaped incision that occupies much of its face.

El Fin del Mundo's excavators claim that the site provides strong evidence for a short-term, contemporaneous presence of Clovis hunters and gomphotheres, based upon the discovery of discrete disarticulated bone piles in the same directly dated deposit as the Clovis artifacts, one of which evinces an apparent impact fracture that one would expect from hunting. Skeptics like archaeologist David Meltzer, on the other hand, are not yet quite convinced. Hardly a stranger to ancient kill sites (see pages 36-45), Meltzer points out that in the absence of clear, verifiable evidence in the form of human-induced cut marks on bone, bone breakage, and skeletal manipulation, the case for a "predator–prey relationship" at El Fin del Mundo is based on circumstantial evidence. The ultimate resolution of any doubts about the site will depend on the results of ongoing analyses and any future work conducted on the upland Clovis camp sites.

In the meantime, El Fin del Mundo shows considerable potential to significantly contribute to scholarly inquiry into the origins of Clovis. As discussed in Part 1 (see pages 24-25), it had long been all but concluded that Clovis hunters entered the Western Hemisphere from Asia only a few hundred years before 13,300 yr BP and rapidly dispersed throughout the newly discovered land. While the archaeological data are still far from sufficient to demonstrate a southerly origin for Clovis, southern sites like El Fin del Mundo at the very least suggest a far more complicated Late Pleistocene peopling scenario than that which prevailed for decades.

Rock art at Pedra Furada, Brazil.

DISPUTED PRE-CLOVIS SITES

In some respects, the mid-twentieth century archaeological search for the First Americans bore more than just a passing similarity to the quest for Glacial Man of nearly 100 years prior (see pages 21-25). By this time, the early Clovis discoveries had percolated through the professional archaeological community and the continent-wide distribution of Clovis as North America's earliest cultural horizon was regularly being confirmed, with confidence. At the same time, like their Glacial Man era predecessors, dozens of localities throughout the length and breadth of the New World were forwarded—and ultimately rejected—as being older than Clovis. (In some cases these claims bordered on preposterousness by proposing site ages well in excess of 40,000 yr BP, which effectively would have placed them in the chronological range of the Neanderthals.) Unlike the earlier Glacial Man debate, however, this time around there was at least one Aleš Hrdlička on the faculty of every major university anthropology department and several Hrdličkas on the editorial boards of major archaeological journals. Most of these sites did not survive, as they failed to meet the criteria that were firmly in place at the time of the Folsom and Clovis discoveries, in that they did not: (1) contain artifacts of indisputable human manufacture or fossil bones that were unmistakably human within (2) a site whose geological layers could be unambiguously interpreted along with (3) appropriate chronological controls such as direct association with the remains of Ice Age plants and animals and/or multiple archaeometric determinations. The following entries describe five of the most prominent disputed pre-Clovis age sites.

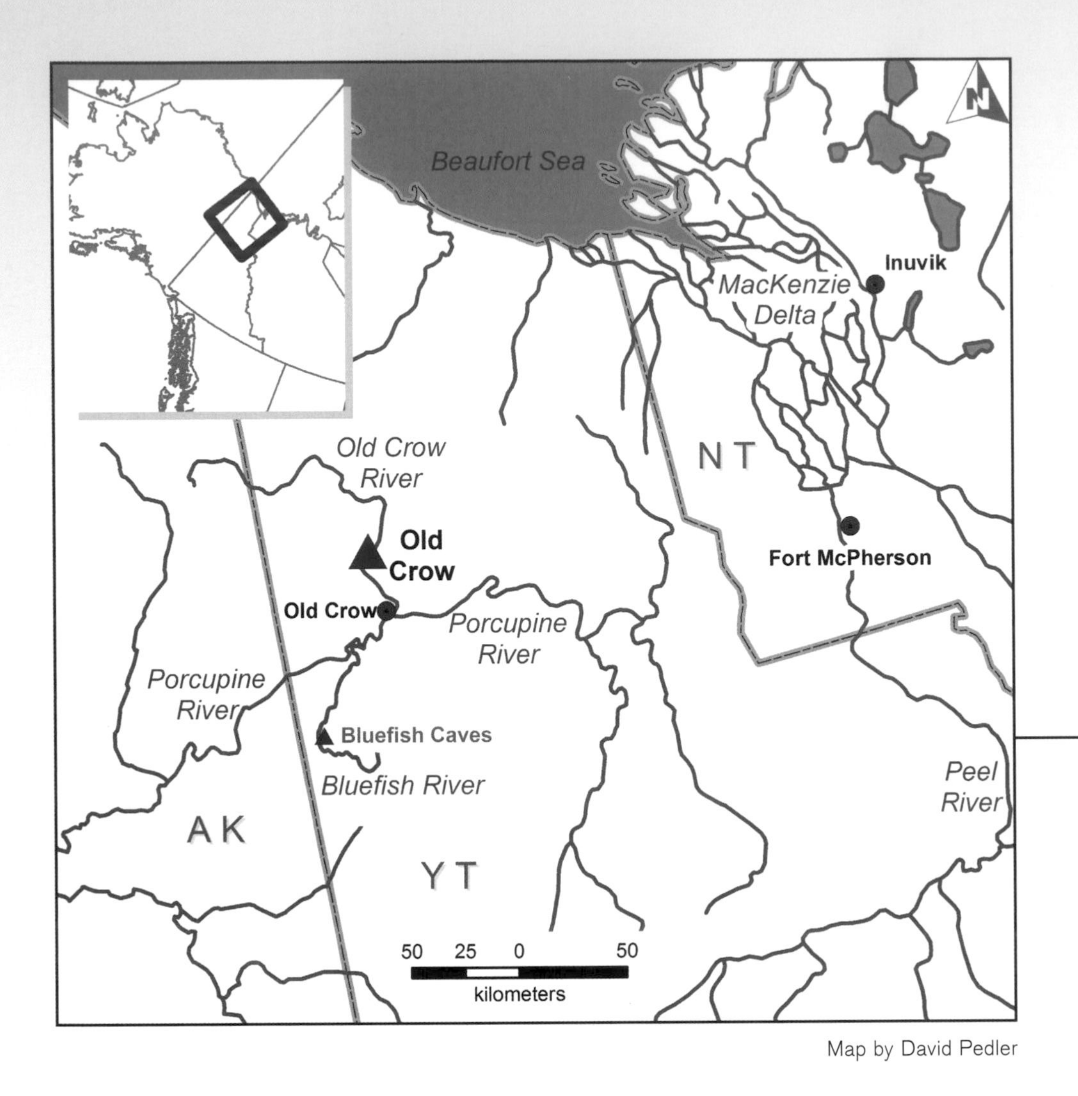

Map by David Pedler

The Old Crow archaeological sites have been the subject of controversy for over fifty years. Fossil bones recovered there have been fractured in ways that some archaeologists attribute butchering by Ice Age peoples upwards of 150,000 years ago. Others who have examined these fossils attribute the factures to natural processes and question such deep antiquity anywhere in the Western Hemisphere.

10 OLD CROW

LOCATION OLD CROW, YUKON, CANADA

Coordinates 67°50'49.86"N, 139°51'18.80"W.

Elevation 265 meters above mean sea level.

Discovery Charles Richard Harington and Peter Lord in 1966.

Old Crow has been a controversial archaeological locality for fifty years. At least one of the region's investigators has argued for a human presence in this portion of the Canadian Arctic Circle by 150,000 yr BP, despite widespread agreement among scholars that humans were not even present in the Asian far north until 120,000 years later. More-conservative age estimates have been proposed, but even those pre-date recognized New World archaeological sites by at least 10,000 years.

The Old Crow archaeological sites are located in Canada's Yukon Territory about 830 kilometers northwest of Whitehorse, 900 kilometers northeast of Anchorage, Alaska, and 160 kilometers south of the Yukon's shoreline on the Arctic Ocean's Beaufort Sea. These sites lie on the east and west banks of the Old Crow River around its confluences with tributary Johnson and Schaeffer Creeks, in a low broad valley that is underlain by continuous permafrost. [Figure 10.1] The 14,600 square kilometer Old Crow River basin flows into the much larger (118,000 square kilometer) Porcupine River basin about 30 kilometers southwest of the site at the town of Old Crow, the sites' namesake. Together, these drainage basins define the northeastern margin of the vast Yukon River basin (see pages 116-129).

The lowland terrain of the Old Crow River basin is surrounded by the relatively low and gently sloping Old Crow, Keele, and Dave Lord Mountain ranges on the north and south, and the substantially higher and steeper Richardson Mountains on the east. The present-day Old Crow River meanders through the complex network of lakes, small ponds, wetlands, and minor tributary streams that dominate the terrain of

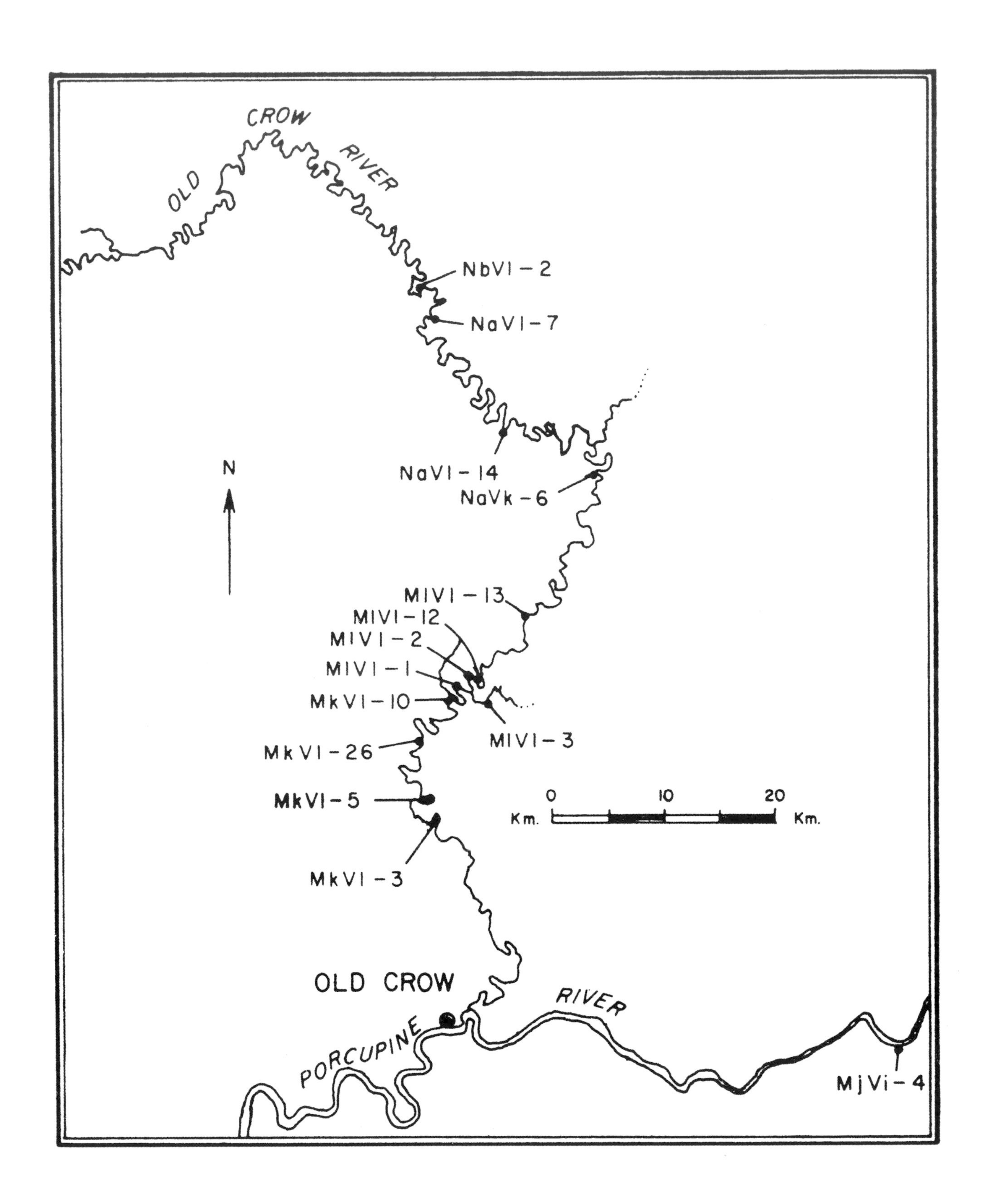
OLD CROW RIVER
NbVI-2
NaVI-7
NaVI-14
NaVk-6
N
MIVI-13
MIVI-12
MIVI-2
MIVI-1
MkVI-10
MIVI-3
MkVI-26
MkVI-5
0
10
20
Km.
Km.
MkVI-3
OLD CROW
RIVER
PORCUPINE
MjVi-4

FIGURE 10.1 At least fourteen archaeological sites have been identified near the confluence of the Old Crow and Porcupine Rivers. Locality 14N (NaVi-14) was the first to be identified in 1966 by guide Peter Lord and Canadian Museum of Nature archaeologist Charles Richard Harington. More than sixty Ice Age mammal species have been found in the Old Crow area, including many not found anywhere else in Canada. The fossils are dominated by the iconic megafauna of the Ice Age: woolly mammoth (*Mammuthus primigenius*), steppe bison (*Bison priscus*), horse (*Equus* sp.), and lion (*Panthera* sp.). Courtesy of Richard Morlan.

its 75 kilometer wide valley. The regional vegetation is transitional between boreal forest and tundra, and as such is dominated by sedges with sparse zones of shrubs including willow, birch, and black spruce interspersed with moss. As with the rest of the Yukon, the Old Crow River basin has retained extensive, near-natural ecosystems and its present-day wildlife closely resembles that of over a thousand years ago.

Prior to the Last Glacial Maximum (26,500–19,000 yr BP), when the Laurentide ice sheet reached its maximum extent, the Old Crow and Porcupine Rivers flowed east to join the Mackenzie River and ultimately the Beaufort Sea. As the glacier expanded westward and impeded the flow of the Mackenzie River, a 13,000 square kilometer waterbody known as Glacial Lake Old Crow formed in the present-day Old Crow River basin, effectively closing the area to human settlement. When the Porcupine River finally breached the divide of the west-flowing Yukon River around 15,000 yr BP, the region's entire drainage system was rearranged to flow west and Glacial Lake Old Crow drained rapidly. This catastrophic event and earlier glacial episodes left behind thick deposits whose continued erosion through deep stream-cutting—to depths as great as 40 meters—has exposed an impressive array and diversity of Pleistocene animal remains in the basin's stream banks. [Figure 10.2] Internationally renowned as a paleontological locality, the Old Crow region has produced numerous specimens such as flatheaded peccary (*Platygonus compressus*), giant beaver (*Castoroides ohioensis*), giant ground sloth (*Megalonyx jeffersonii*), horse (*Equus* sp.), lion (*Panthera* sp.), mastodon (*Mammut americanum*), steppe bison (*Bison priscus*), steppe mammoth (*Mammuthus trogontherii*), camel (*Camelops hesternus*), and woolly mammoth (*Mammuthus primigenius*), among many others.

The Old Crow sites number in the dozens and have been investigated since the mid-1960s. [Figure 10.3] The first and most famous discovery was made in 1966 at Old Crow locality 14N by paleontologist Charles Richard Harington and his guide Peter Lord, who found a bone tool that was fashioned from a caribou (*Rangifer tarandus*) tibia bone in association with Pleistocene animal remains. The two men presented their find to archaeologist William Irving of the National Museum of Man (now the Canadian Museum of History), who was conducting excavations at the nearby Klo-kut prehistoric site on the Porcupine River. Irving pronounced the artifact to be authentic, but upon visiting the place of its discovery judged the site to be a secondary deposit (*i.e.*, disturbed by subsequent natural or human agency) of relatively recent age.

Sporadic field research conducted through the early 1970s in the Old Crow area drew the interest of numerous Quaternary research specialists, particularly following the announcement of a radiocarbon age that calibrates to 31,620–30,360 yr BP for the bone implement recovered by Harington in 1967. [Figure 10.4] (Though secondarily deposited, the artifact itself genuinely appeared to be ancient.) By 1975, this growing interest had culminated in the establishment of two major multidisciplinary research enterprises known as the Yukon Refugium Project, headed by Canadian Museum of History archaeologist Richard Morlan, and the Northern Yukon Research Programme, headed by Irving and based at the University of Toronto. [Figure 10.5] The field work of these two entities continued into the mid-1980s, followed by a research and analysis phase that continued well into the 1990s. [Figure 10.6]

Both Morlan and Irving generally agreed upon the stratigraphy in the lower reaches of the Old Crow River basin, interpreting it as being composed of a 20–30 meter thick layer of ancient lake bed deposits (Unit 1 and 2) overlain by a 5 meter thick layer of glacially

FIGURE 10.2 (above) Bluffs and banks along the Old Crow River are the richest source of Ice Age fossils in Canada. Every spring, thousands of fossil bones erode from the bluffs and are deposited along river banks or in river shallows. Photograph by Alberto Reyes.

FIGURE 10.3 (below) The cranium of a steppe Bison (*Bison priscus*) at Old Crow Locality 11, an incredibly rich fossil site. Dated to about 12 000 yr BP, it is among the most-recent known steppe bison fossils from the region. Traditional stories of the region's Vuntut Gwitchin people tell of encountering the fossil bones of huge mammoths and giant beavers along the Old Crow River. Photograph by Richard Harington.

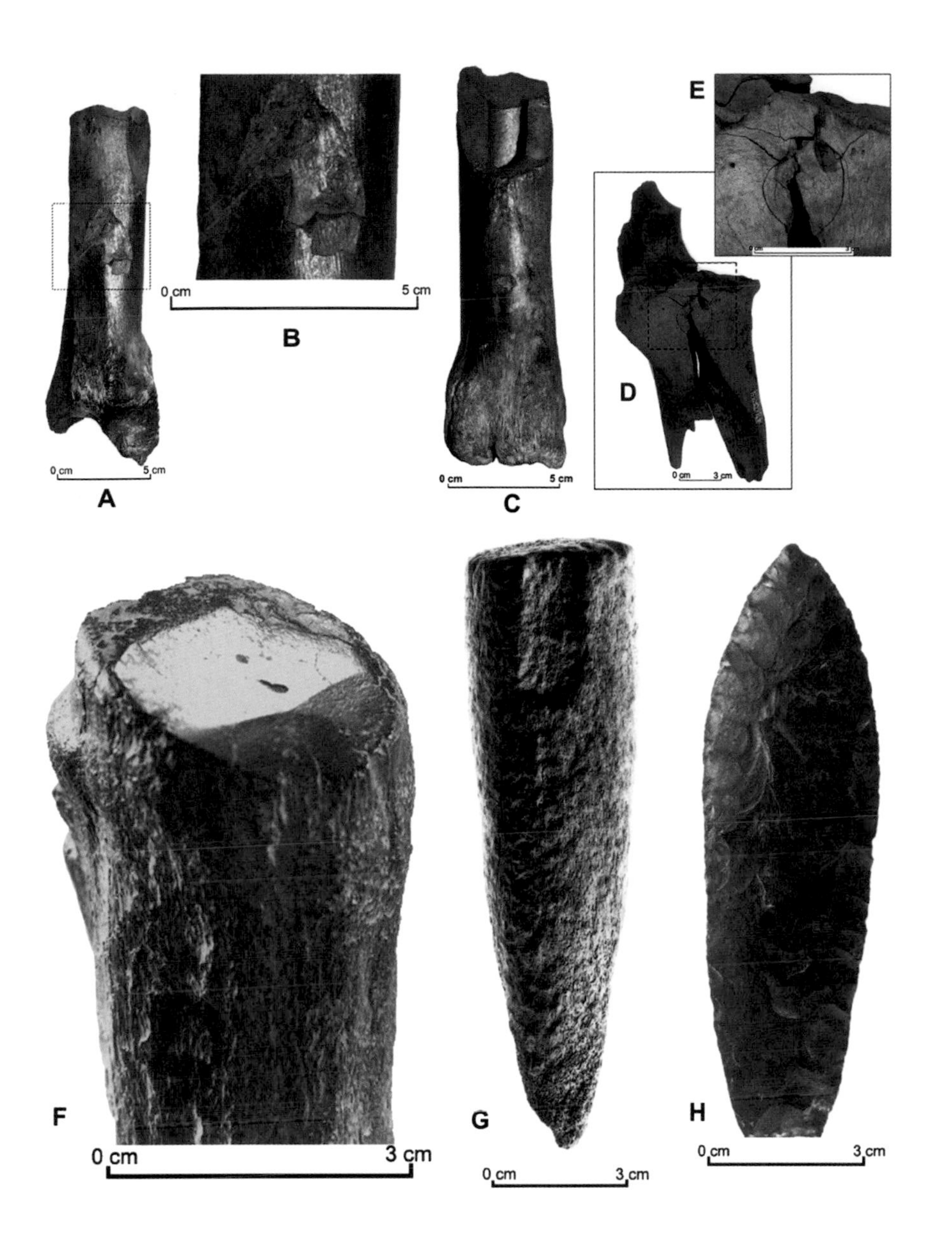

FIGURE 10.4 Examples of putative evidence for human modification of Late Pleistocene bone from the Yukon Territory, and a chert biface from Old Crow basin. (A) Distal half of a left tibia of an adult steppe Bison (*Bison priscus*), showing a large flake scar (top right) and a large batter mark with several subparallel drag (from a glancing blow?) or cutmarks adjacent to it (see arrow). (B) Detailed view of large batter mark and marks to the left of it (see upper arrow on bone). (C) View of opposite side of (A) showing particularly the spiral fracture evidently produced by a heavy blow near the middle of the shaft that removed a large flake from the opposite side. Presumably the bone was broken to obtain marrow. Radiocarbon dated to 45,840–41,420 yr BP. (D) Anterior view of proximal end of a steppe Bison (*Bison priscus*) right elbow, showing a ring crack just below the proximal articular surface of the radius perhaps made by a person wielding a cobble-size hammerstone trying to expose the marrow. The focused blow resulted in a spiral fracture extending through both the radius and the ulna. (E) Detailed view of the ring crack. Radiocarbon dated to 36,990–32,700 yr BP. (F) Caribou (*Rangifer tarandus*) antler with a polished head from Old Crow evidently resulting from use as a pestle, or as a tool for preparing hides, by humans. Radiocarbon dated to 30,390–27,700 yr BP. (G) Modified caribou (*Rangifer tarandus*) antler, interpreted as a flintknapper's punch. Radiocarbon dated to 13,430–13,030 yr BP. (H) Gray chert biface, presumably used as a spear tip from Old Crow, presumably of late glacial age about 12,000 yr BP. Courtesy of Richard Harington.

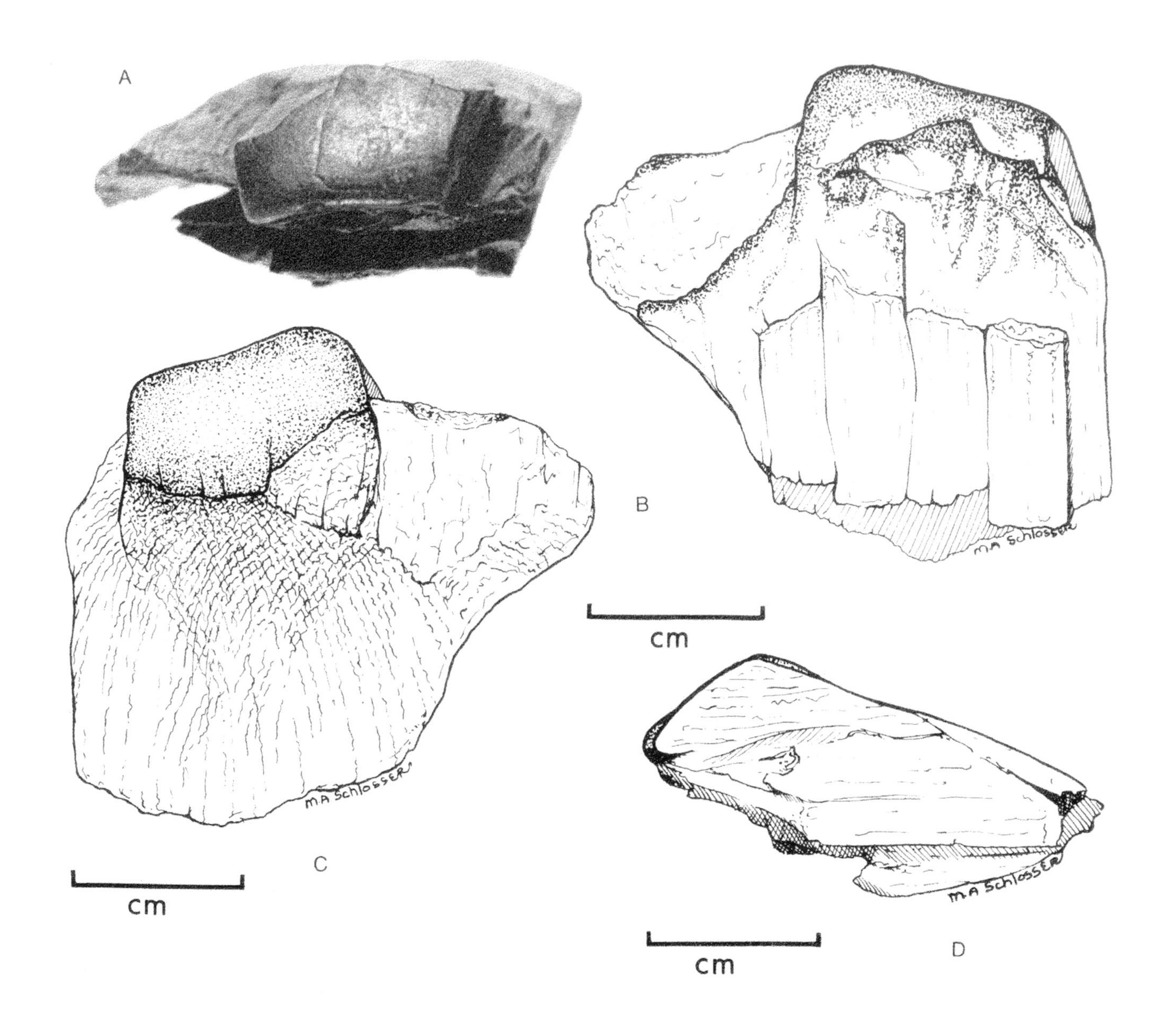

FIGURE 10.5 (above) A battered fragment of a Wooly Mammoth *(Mammuthus primigenius)* tusk from Old Crow (A and D), bearing two highly polished surfaces (B and C) dated to over 120,000 yr BP. Archaeologists from the University of Toronto contended that no known natural process could have produced them. They concluded "only a human, with hands warm enough for careful, forceful manipulation, would seem to have been able to effect the highly localized pair of polished facets." From A. V. Jopling, W. N. Irving, and B. F. Beebe "Stratigraphic, Sedimentological and Faunal Evidence for the Occurrence of Pre-Sangamonian Artefacts in Northern Yukon" Courtesy of the Arctic Institute of North America.

FIGURE 10.6 (below) A Wooly Mammoth *(Mammuthus primigenius)* limb bone collected by Richard Morlan showing evidence that it was fractured when fresh and then shaped by human hands into a tool, dated to in excess of 40,000 yr BP. Courtesy of Richard Morlan.

deposited silt and clay associated with the inundation of Glacial Lake Old Crow and the 2 meter thick layer of Holocene material that represents the modern ground surface. [Figure 10.7] Both excavators also distinguished short-lived ground surfaces within Units 1 and 2, upon which the bone artifacts were deposited before being covered by thick lake bed sediment deposits. Their views on the antiquity of humans in the basin, on the other hand, diverged radically, with Irving believing humans were present by at least 150,000 yr BP and Morlan hypothesizing a more modest 30,000–25,000 yr BP human appearance in the region.

The most distinctive—and controversial—artifacts from Old Crow are those fashioned from bone and antler via processes (*i.e.*, cutting, fracturing, flaking, and polishing) that many scholars have claimed can occur naturally, especially in glacial permafrost environments where materials can be transported great distances and under great pressure by ice, water, and sediment. [Figure 10.8] But even if one accepts the Old Crow bone artifacts as bona fide products of human manufacture—as some authorities have done—there have remained what are, in all probability, insurmountable problems relating to archaeological context at the sites. Not one ancient bone artifact from Old Crow has been recovered from a valid, undisturbed archaeological site, nor have any cultural features (*e.g.*, fire pits, living floors, *etc.*) been identified in the Old Crow River basin's deepest, most-ancient deposits. With the data presently at hand, there is no way to demonstrate with certainty that the putatively ancient bone artifacts and Pleistocene animals ever occupied the same point in time and space together.

The peril of accepting the validity of archaeological materials from secondary deposits was very well illustrated by Morlan's subsequent work with a small sample of bone tools and animal remains several years after the completion of the Old Crow field work. Using refined sample processing and AMS radiocarbon dating, Morlan and his colleagues discovered that the animal remains—which included objects that were interpreted to be human-made flakes struck from mammoth bone—were indeed ancient and consistent with the ages of the deposits from which they were recovered. The caribou antler and bone tools, however, proved to be quite recent. Ironically, the tibia bone flesher that inspired the Old Crow research actually dated to 1550–960 yr BP, at least 30,000 years younger than its originally presumed age. [Figures 10.9 and 10.10]

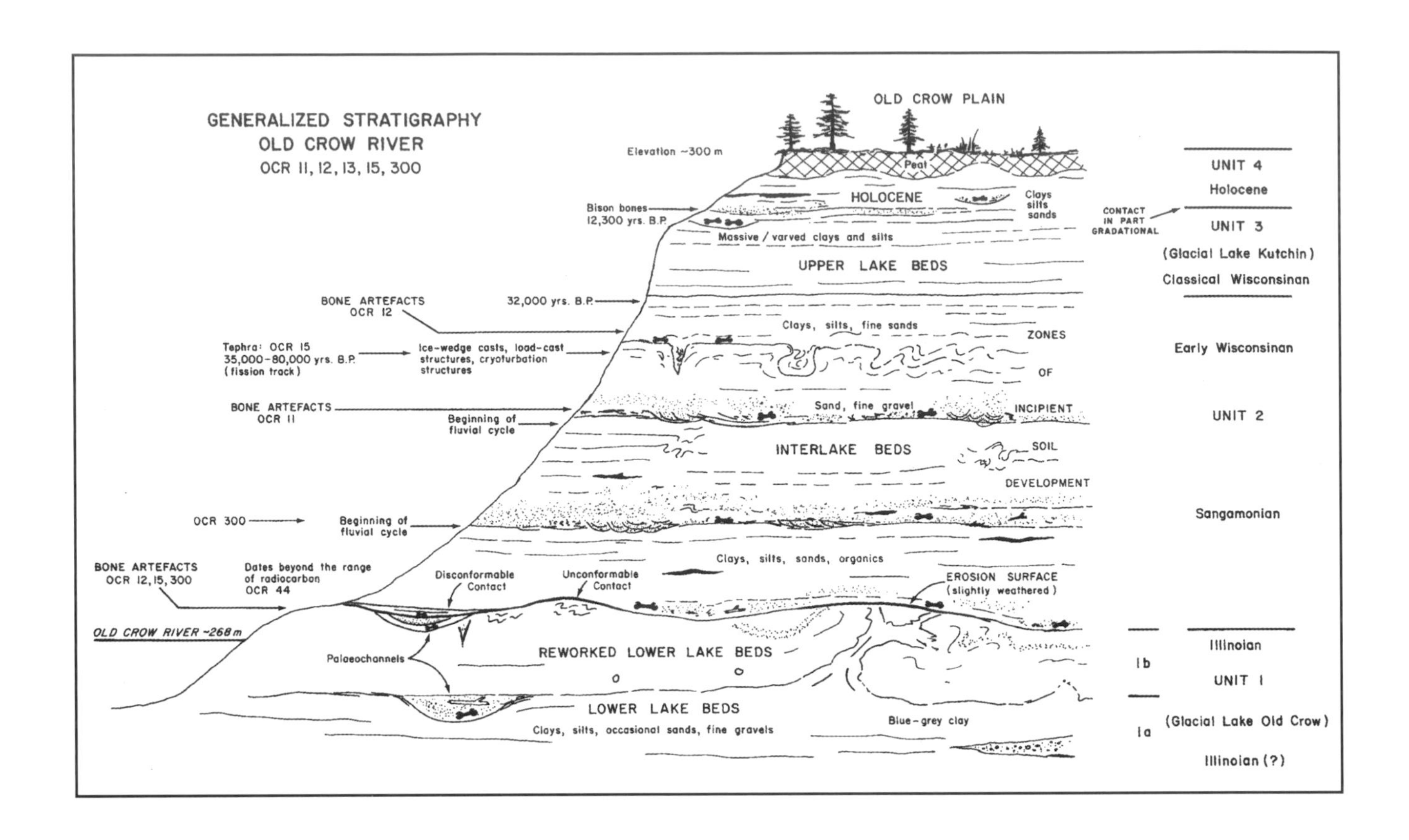
GENERALIZED STRATIGRAPHY
OLD CROW RIVER
OCR 11, 12, 13, 15, 300
OLD CROW PLAIN
Elevation ~300 m
Peat
HOLOCENE
Clays silts sands
Bison bones 12,300 yrs. B.P.
Massive / varved clays and silts
UPPER LAKE BEDS
CONTACT IN PART GRADATIONAL
UNIT 4
Holocene
UNIT 3
(Glacial Lake Kutchin)
Classical Wisconsinan
BONE ARTEFACTS OCR 12
32,000 yrs. B.P.
Clays, silts, fine sands
ZONES
Tephra: OCR 15 35,000-80,000 yrs. B.P. (fission track)
Ice-wedge casts, load-cast structures, cryoturbation structures
OF
Early Wisconsinan
BONE ARTEFACTS OCR 11
Beginning of fluvial cycle
Sand, fine gravel
INCIPIENT
UNIT 2
INTERLAKE BEDS
SOIL
DEVELOPMENT
OCR 300
Beginning of fluvial cycle
Sangamonian
Clays, silts, sands, organics
BONE ARTEFACTS OCR 12, 15, 300
Dates beyond the range of radiocarbon OCR 44
Disconformable Contact
Unconformable Contact
EROSION SURFACE (slightly weathered)
OLD CROW RIVER ~268 m
Illinoian
Palaeochannels
REWORKED LOWER LAKE BEDS
Ib
UNIT I
LOWER LAKE BEDS
Clays, silts, occasional sands, fine gravels
Blue-grey clay
Ia
(Glacial Lake Old Crow)
Illinoian (?)

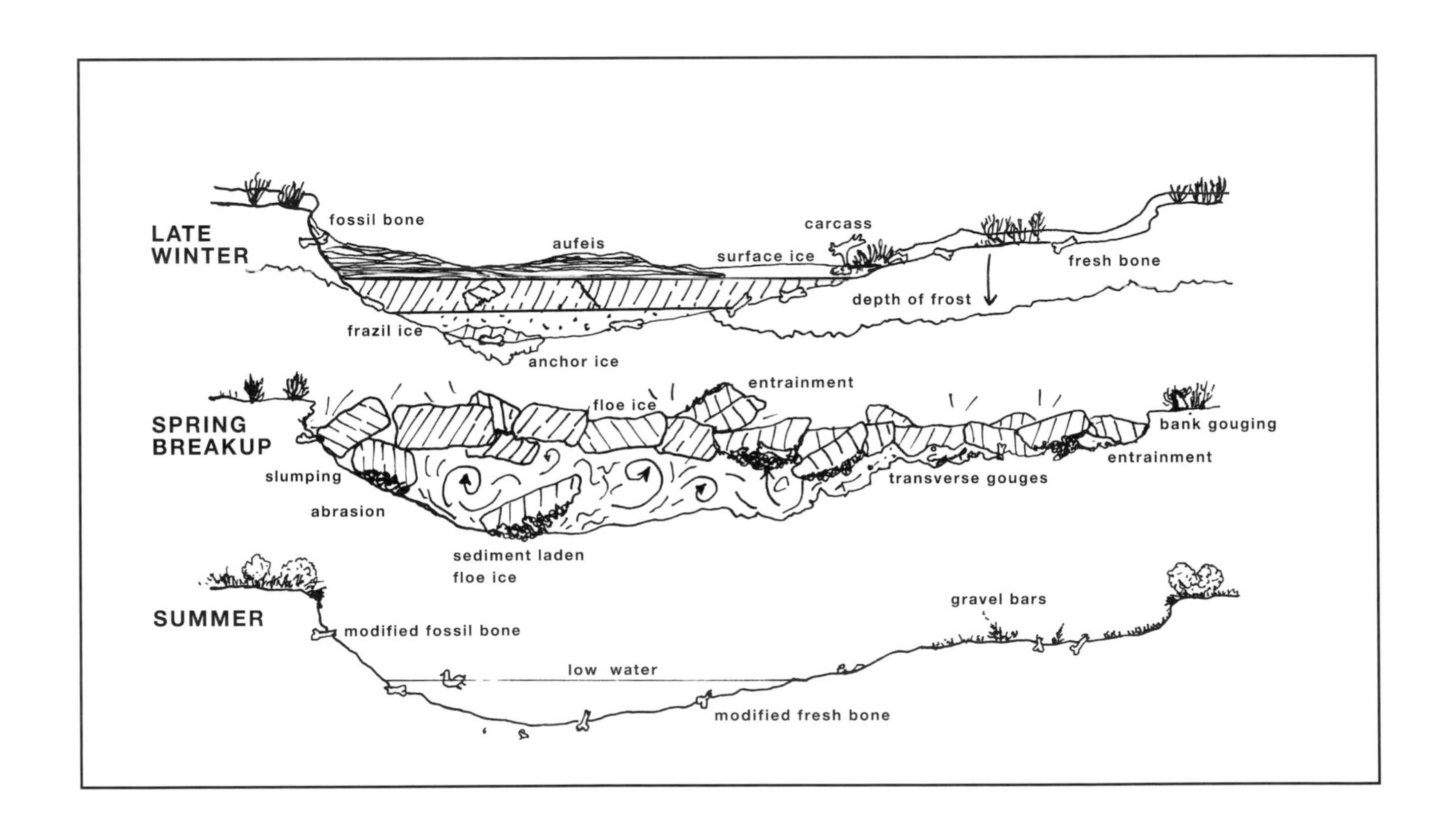
LATE WINTER
fossil bone
aufeis
surface ice
carcass
fresh bone
depth of frost
frazil ice
anchor ice
SPRING BREAKUP
entrainment
floe ice
bank gouging
entrainment
slumping
abrasion
transverse gouges
sediment laden floe ice
SUMMER
gravel bars
modified fossil bone
low water
modified fresh bone

FIGURE 10.7 (left, above) Schematic cross section of sediment exposed in the bluffs along the Old Crow River. Stratigraphic data were obtained by cutting steps and trenches in river embankments, and by jet drilling in the subsurface. Credit: William N. Irving Fonds, Canadian Museum of Nature, IMG2013-0032-0010.

FIGURE 10.8 (left, below) Schematic cross section of a typical northern Yukon river showing the effects of the spring ice breakup from late winter through spring. In 1983 University of Alaska geologists Robert Thorson and Dale Guthrie conducted a series of experiments to determine whether the destructive action of ice floes during the spring thaw in Yukon rivers could produce the kind of bone fractures attributed to human hands seen in Figures 10.6 and 10.7 and similar bones recovered at Old Crow. While they could not rule out the possibility human modification of Old Crow Pleistocene bones, the action of river ice floes produced indistinguishable effects. Courtesy of Robert Thorson.

FIGURE 10.9 (right) Richard Harington (left) and USGS geologist Tom Hamilton excavate the test pit at the site where the flesher seen in Figure 10.10 was found, circa 1968. Courtesy of Richard Harington.

FIGURE 10.10 (below) This caribou bone had been turned into a flesher, used to skin hides. In the early 1970s it was dated to 31,620–30,360 yr BP. It was, however, re-dated in the early 1990s with more advanced methods and turned out to be only about 1,000–1,500 years old. Photograph by William N. Irving Fonds, Canadian Museum of History: J19221-7-Dm.

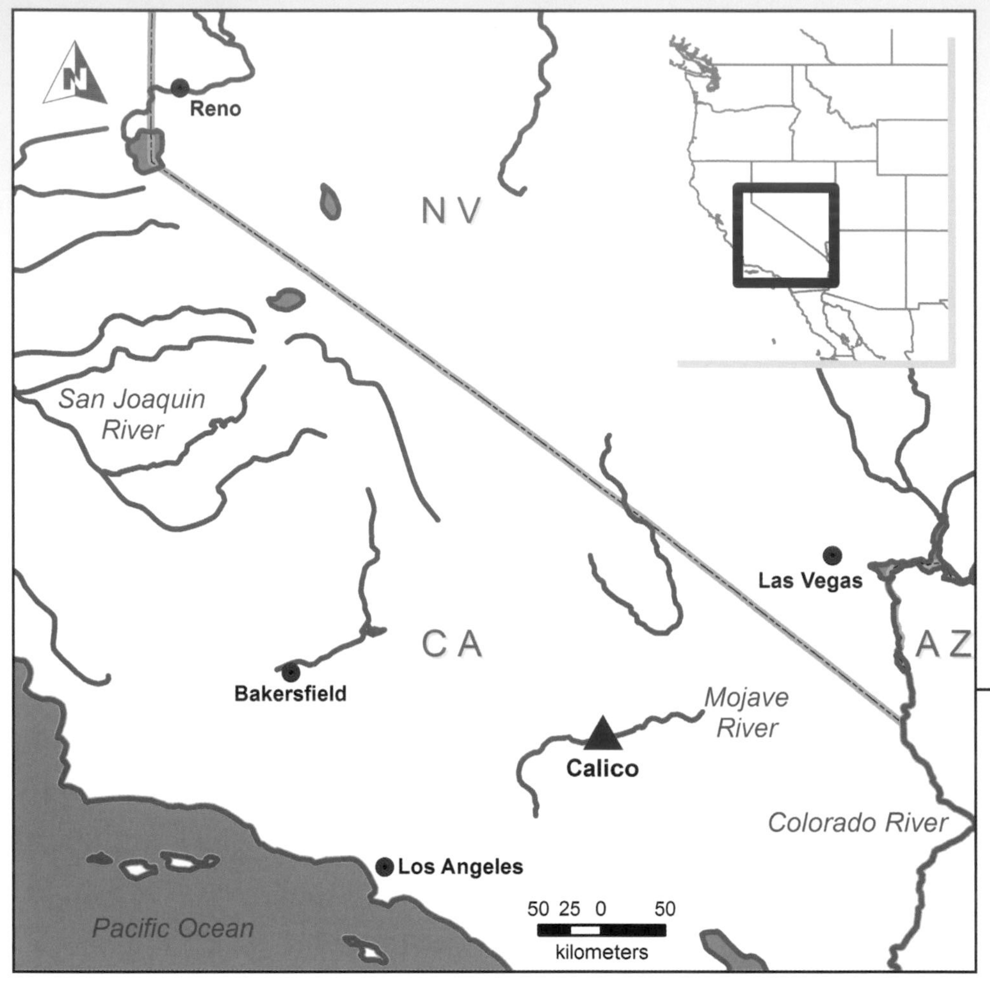

Map by David Pedler

The Calico Mountain site is famous for the decade-long involvement of legendary paleontologist Louis Leakey in the archaeological work there. Leakey concluded that the recovered artifacts were "considerably more than 100,000 years old," and subsequent claims for the site's age have ranged from 200,000 years to as many as 500,000 years.

11 CALICO MOUNTAIN

LOCATION SAN BERNARDINO COUNTY, CALIFORNIA, UNITED STATES

Coordinates 34°56'50.23"N, 116°45'39.57"W.

Elevation 660 meters above mean sea level.

Discovery Ruth Simpson and Louis Leakey in 1963

The Calico Mountain site, also known variously as the Calico Early Man and the Calico Hills Archaeological site, has been claimed by its excavators and proponents to be the oldest archaeological site in the Western Hemisphere, predating all known sites by more than 180,000 years and even the migration of modern humans out of Africa by as many as 80,000 years. Calico is also the only New World archaeological investigation to have actively involved internationally renowned archaeologist and paleontologist Louis Leakey, who lived to see Calico's major excavation phase through to completion, but not its acceptance as a valid archaeological site at the time of his death in 1972.

The Calico site is located in San Bernardino County, southeastern California, about 170 kilometers northeast of Los Angeles, 210 kilometers southeast of Bakersfield, and 200 kilometers southwest of Las Vegas, Nevada. The site lies at the base of the Calico Mountains at an elevation of 660 meters above sea level, 600 meters below the adjacent mountain summits and 100 meters above the channel of the Mojave River. The Mojave River is the largest drainage system in the Mojave Desert, encompassing almost 12,000 square kilometers and extending 135 kilometers east from its headwaters in the San Bernardino Mountains to its terminus at Soda Lake, a dry lake basin about 70 kilometers northeast of

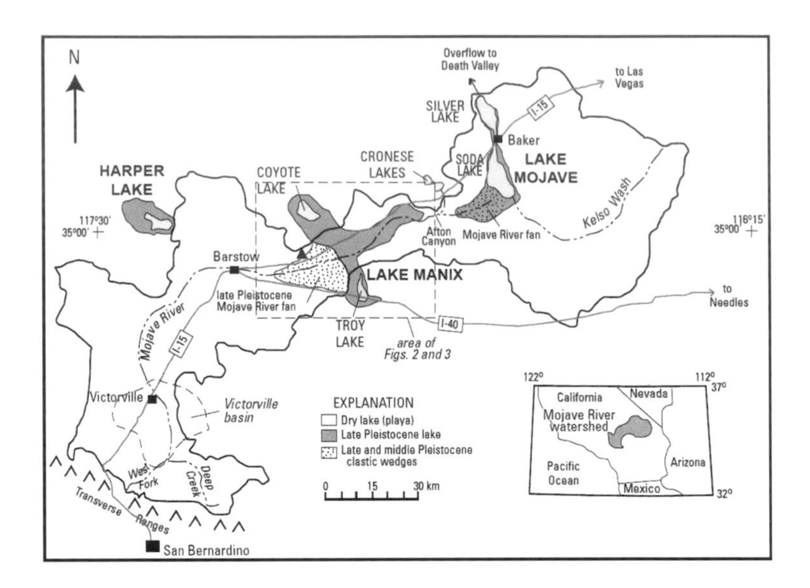

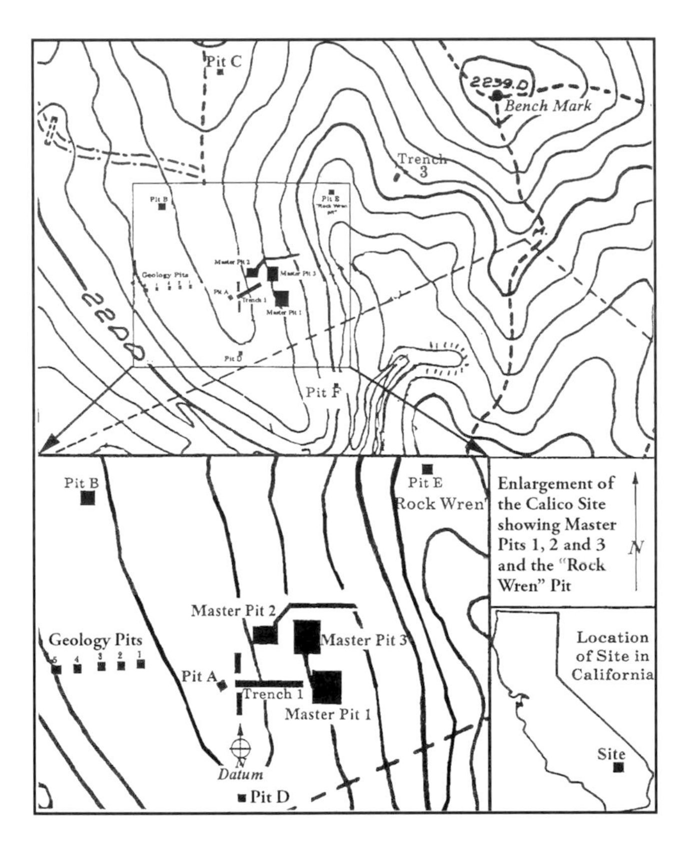

FIGURE 11.1 (above) Mojave River drainage basin in southern California. During especially high-precipitation years, the river flows into Silver Lake playa. In Late Pleistocene times, after breaching of the Lake Manix basin, Lake Mojave episodically discharged northward into Death Valley. The red triangle marks the location of the Calico site in Late Pleistocene Lake Manix alluvial deposits. Courtesy of the United States Geological Survey.

FIGURE 11.2 (left) Topographic maps of the Calico Mountain site's setting (top) and excavations (bottom). Excavation began in November 1964. Twenty-two pits and trenches had been dug by 1980, most of them between 1964 and 1972. The Master Pits penetrate middle-level alluvial fan deposits of mudflows, debris flows, and chalcedony and chert rock fragments. Courtesy of Friends of Calico.

FIGURE 11.3 (right) Louis Leakey and Ruth Simpson at the Calico Mountain site. Simpson began fieldwork at Calico Hills in 1954 and assembled a large assortment of bifacial and unifacial tools. (see Figure 11.6) Advised to bring the material to the attention of Paleolithic specialists in Europe, in 1958 she traveled to England where she was able to arrange a meeting with Leakey. In 1963 Leakey traveled to Calico with Simpson. He climbed hillside above a bulldozer cut and told her "You will dig here." (see Figure 11.4) Courtesy of the Friends of Calico.

the Calico site. The Mojave River does not flow year-round, only reaching Soda Lake for short durations in exceptionally wet periods, typically every ten years or so. [Figure 11.1]

During the much wetter climatic conditions of the Late Pleistocene and prior to the formation of the Holocene epoch Mojave River, this portion of the ancestral Mojave basin was occupied by an internally drained (*i.e.*, with no outlet flowing into another drainage system) body of water known to geologists as Lake Manix. At that time, the Calico location would have been about 3–4 kilometers from the western edge of the lake, whose irregularly shaped shorelines roughly followed the floors of the region's present-day canyons for an east–west length of about 50 kilometers and a north–south width of about 40 kilometers. The site is

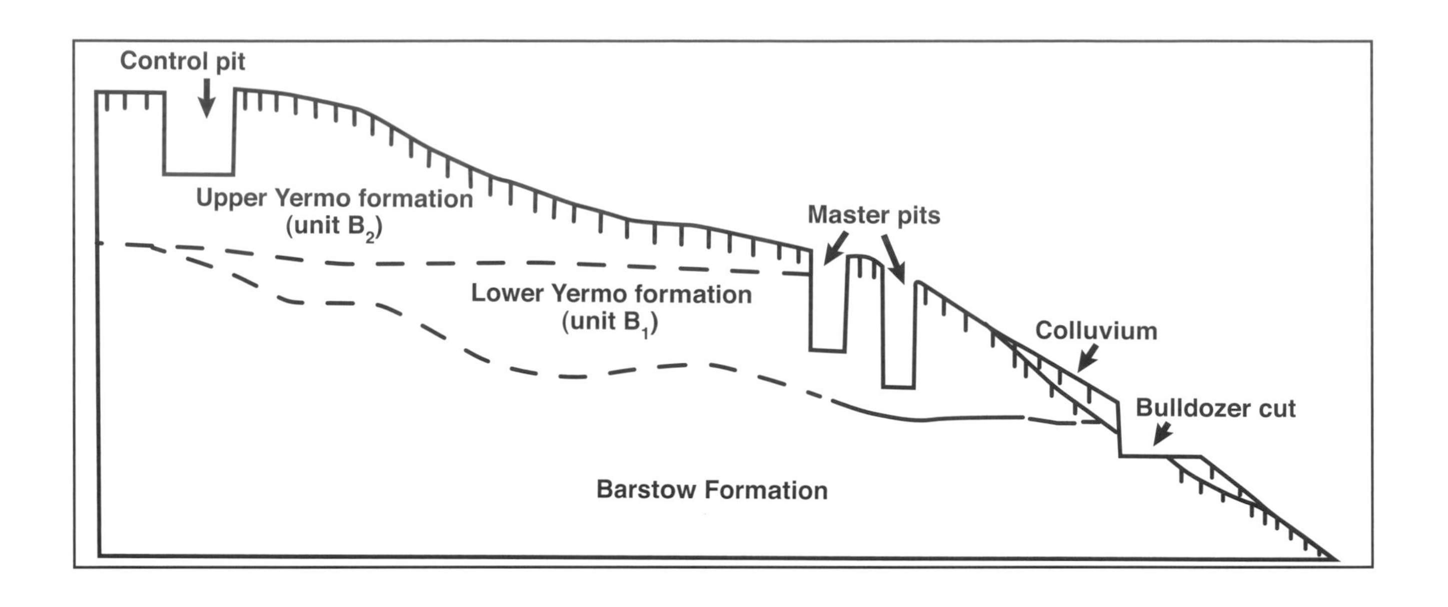
Control pit
Upper Yermo formation
(unit B_2)
Master pits
Lower Yermo formation
(unit B_1)
Colluvium
Bulldozer cut
Barstow Formation

FIGURE 11.4 (left, above) Schematic profile of the Calico site. The bulldozed jeep track made by a prospecting miner exposed a cut bank in the colluvial deposits. It was here in 1963 that Louis Leakey spotted fragments of chalcedony and jasper from which he thought artifacts could be fashioned similar to the 1.85–1.75 million year old tools he had discovered in the Olduvai Gorge, Tanzania. Courtesy of Vance Haynes.

FIGURE 11.5 (left, below) In 1972 the Calico Mountain site was placed on the National Register of Historic Places. Since 1979 it has been administered by the Bureau of Land Management. It is open to the public with a visitor center, gift shop, and guided tours conducted by the Friends of Calico, a volunteer organization which sponsors archaeological work and training at the site. Courtesy of the Friends of Calico.

perched atop an ancient, massive 7 square kilometer, cone-shaped pile of water-transported rock and debris (called an alluvial fan) that flowed out of adjacent Mule Canyon well before the Late Pleistocene. [Figure 11.2] During the Late Pleistocene when Lake Manix was present, local hillslopes probably supported desert shrub vegetation, with juniper woodland and grasses at low elevations and pine at higher elevations.

The Calico site was discovered by Ruth Simpson and Louis Leakey in May 1963 while prospecting for archaeological sites with archaeologist Ruth Simpson, then affiliated with the Southwest Museum in Los Angeles but soon to become Curator of Anthropology at the San Bernardino County Museum in Redlands, California. [Figure 11.3] Simpson had become interested in the archaeology of the Calico Mountains in the early 1940s, and conducted an active investigation there in the mid-1950s while completing her work at the Tule Springs site in Nevada (see pages 58-67). On a trip to England in 1958, she managed to arrange a meeting with Leakey, to whom she showed a collection of artifacts recovered from the ground surface in the Calico area. Upon examining the pieces, Leakey indicated that should she find such items in a buried archaeological deposit (as opposed to material from the ground surface), Simpson would have a very important archaeological project on hand. He recommended that she continue searching, and keep him apprised of her progress. Five years later, Simpson and Leakey found themselves surveying the southeastern Calico Mountains in search of an appropriate location to excavate after Leakey's rejection of Simpson's original site as a secondary deposit. Leakey made the final decision of where the excavation should be conducted.

Funding for the excavation at Calico was provided by the National Geographic Society—against the advice of the society's consulting geologist C. Vance Haynes and archaeologist Emil Haury —and the first field season at the site ran between November 1964 and May 1965. [Figure 11.4] Work resumed in November of that year and continued (with breaks over the extremely hot summer months) until June 1971, when the National Geographic Society ended its sponsorship and Calico was nominated to the National Register of Historic Places. The administration of the site was taken over by San Bernardino County in the mid-1970s and eventually ceded to the Bureau of Land Management in 1979. Sporadic volunteer archaeological investigations and Bureau of Land Management sanctioned archaeological field training programs have been conducted there ever since. [Figure 11.5]

The principal years of investigation at Calico covered a total excavated area of 200 square meters (to depths as great as 9 meters) and recovered tens of thousands of putative artifacts from the site's densely packed, coarse cobble-sized gravel and very coarse sand deposits. The excavators have ascribed these items to a tool stone quarry site containing three discrete temporal horizons: Paleoindian and later, the Lake Manix Lithic Industry, and the Calico Lithic Industry. The possibility that late Paleoindian and later materials should be recovered from Calico is more or less uncontroversial, as Paleoindian materials—though somewhat rare—are not unheard of in the Mohave Desert, and the region has a rich prehistoric archaeological record that begins at the end of the Pleistocene and spans the entire Holocene. The unusually early age of the Manix Lake material (which is thought to date to as early as 24,150–22,400 yr BP) and particularly the profoundly anomalous age

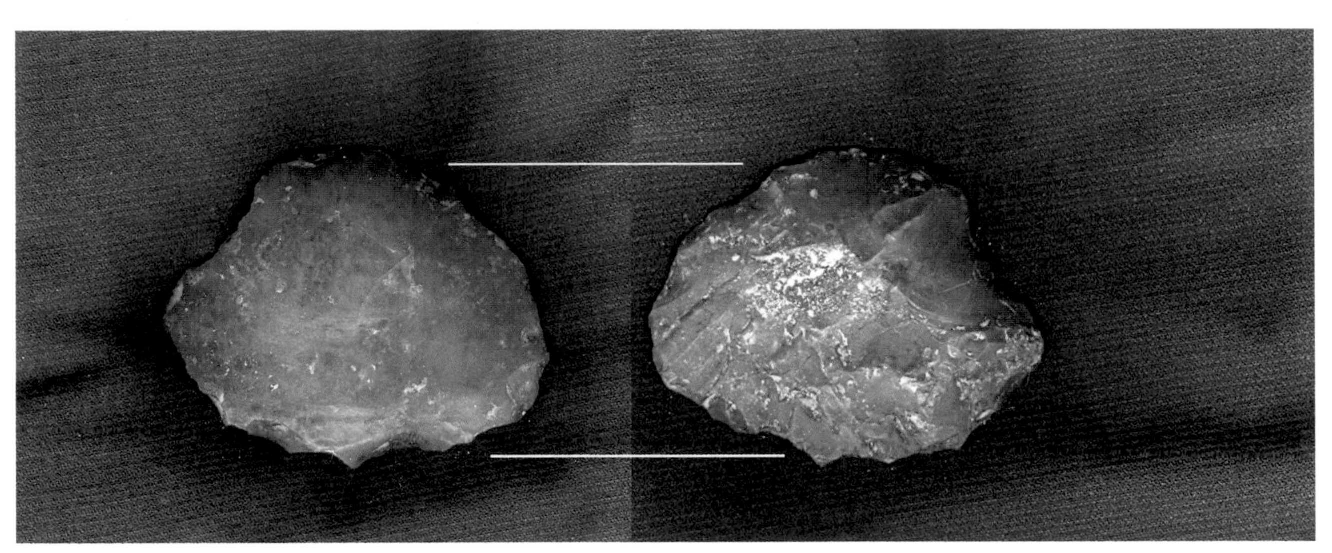

FIGURE 11.6 (top) Bifaces from the Manix Lake Lithic Industry. Although bearing vague resemblance to European Paleolithic tools, no comparable materials have been recovered from any other site in the Western Hemisphere. Courtesy of the Friends of Calico.

FIGURE 11.7 (bottom) A Calico Lithic Industry flake. This one of the more than 60,000 possible tools recovered at the site dated to as early as 200,000 yr BP by the excavators. Anatomically modern humans (*Homo sapiens*) first appear in the fossil record in southern Africa around this time, which would mean that, if these items are in fact tools, they were crafted by archaic humans like *Homo erectus* or Neanderthals. Today the consensus is that they are the result of rock fracturing by tectonic stresses, weather, rock-on-rock percussion in streams and mudflows, pressure retouching of buried cobbles, and cycles of erosion and redeposition. Courtesy of the Friends of Calico.

FIGURE 11.8 The west wall of Master Pit 1. The overall stratigraphy provides evidence that deposition occurred in response to gradual changes from semiarid to arid climatic conditions. Courtesy of the Friends of Calico.

of 200,000 yr BP proposed for the site's Calico horizon specimens, have been widely criticized as being far too old and most likely not associated with the ancient dated geologic deposits. [Figures 11.6 and 11.7]

The Calico Lithic Industry assemblage is composed of "artifacts" whose crude appearance is claimed to be an indication of their deep antiquity. The site's excavators have identified thousands of flakes (various quantities are reported), over 2,700 flake tools of varying configurations, almost 1,600 blades and blade cores, and smaller numbers of chopping tools, axes, hammerstones, and a variety of other miscellaneous forms in the assemblage. The raw materials represented among these items are reported to be 95 percent chalcedony, chert, or agate, all of which are desirable as tool stone. No convincing cultural features (*e.g.*, fire pits, living floors, *etc.*) were identified among any of the three alleged cultural deposits. [Figure 11.8]

The most distinctive artifacts from the Calico site appear to be the so-called Rock Wren Biface, recovered from buried sediments claimed to be about 12,200–16,600 yr BP in age, and an object resembling a hand axe known as the Calico Cutter from the site's lowest temporal horizon. The status of most (if not virtually all) of the lithic items as bona fide artifacts, however, has been at the center of long-standing criticism that began during the second excavation season at the site. The dispute finally came to a head at the *International Conference on the Calico Mountains Excavations* at the San Bernardino County Museum in October 1970, which was sponsored by the Leakey Foundation and the University of Pennsylvania.

At that conference, project geologist Thomas Clements of the University of Southern California "took a viciously critical grilling" from the professional geologists and archaeologists in attendance, as project director Ruth Simpson recalled in a 1980 memoir recounting her work on the Calico project. Several calls for additional artifact and geological analyses were recommended by the attendees, and in the end no consensus could be reached on the validity of the site or even its antiquity. Calico's earliest and most incisive critic, Haynes, somewhat later in 1973 concluded after multiple examinations of the lithic material that no specimen from Calico was obviously an artifact, noting that some specimens found in the massive piles of items rejected as non-cultural by the excavators were essentially indistinguishable from others they had cataloged as artifacts. Haynes also observed the lack of layering in the site's stratigraphy and random distribution of artifacts within a 2–3 meter vertical extent of the deposit, opining that "if these are indeed artifacts, then they have been redeposited." [Figure 11.9] Haynes further speculated that most if not all of the artifacts had been naturally fractured by the alluvial fan at Calico, as that kind of geologic feature is known to tumble and fracture rocks in massive flows of debris that lead to their formation. The view that the Calico artifacts are instead *geofacts* is widely maintained by scholars today.

The Calico site was by no means unanimously discredited at the time—indeed, news of the findings there garnered interest among the growing ranks of archaeological scholars whose criticism of the Clovis-First model had been gathering through the 1970s and 1980s (see pages 23-25). But Calico is now held in widespread disregard among professional archaeologists, largely due to the lack of indisputable artifacts in the Lake Manix and Calico Lithic Industry assemblages, the lack of valid cultural features, a suspicion that the age of any valid archaeological deposit at the location is radically younger than that proposed, and the site's apparent position as a secondary deposit in an alluvial fan.

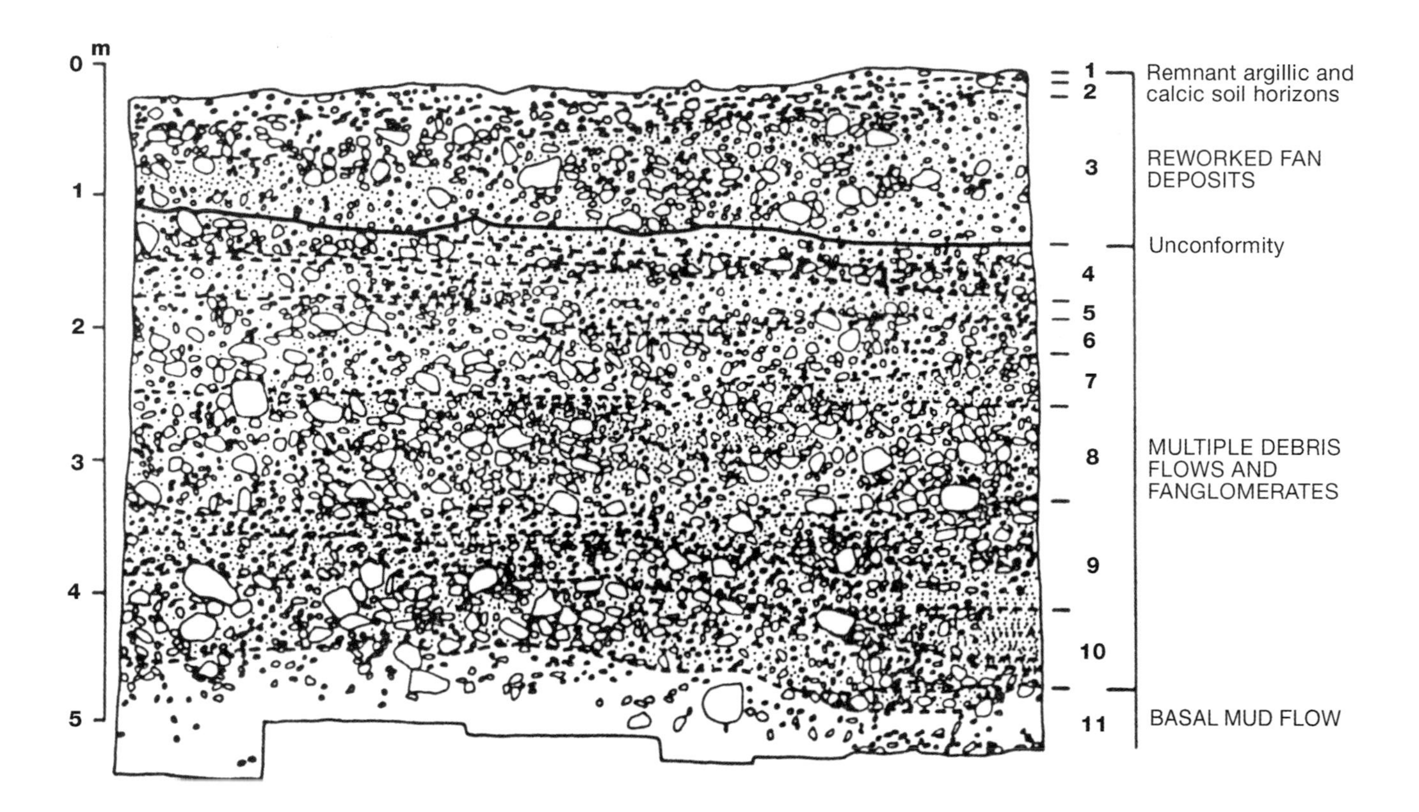

FIGURE 11.9 Stratigraphy of the Master Pit 1 west wall. (see Figure 11.8) The "unconformity" marks the boundary between the upper, more recent deposits which contain Manix Lake Industry artifacts and the lower Pleistocene deposits which contain Calico Industry artifacts. (see Figure 11.8) The "multiple debris flows and fanglomerates" in the Pleistocene deposits are rock fragments washed down from adjacent Calico Mountain slopes. Courtesy of the Friends of Calico.

Map by David Pedler

Pendejo Cave was identified in 1975 as a potentially significant Paleoindian archaeological site. The cave lies within the U.S. Army's Fort Bliss McGregor Firing Range, so it was not until 1990 that access to the cave was permitted under the auspices of the Andover Foundation for Archaeological Research.

12 PENDEJO CAVE

LOCATION OTERO COUNTY, NEW MEXICO, UNITED STATES

Coordinates 32°24'60.00"N, 105°54'60.00"W.

Elevation 1396 meters above mean sea level.

Discovery 1975 by Julio Betancourt.

Pendejo Cave is among a very small class of archaeological sites for which there has been claimed a deep human antiquity that is far in excess of the presently recognized temporal horizon for human settlement in the New World. Though placing third behind the Calico Mountain (see pages 152-161) and Pedra Furada (see pages 184-197) sites, the claim for Pendejo Cave's antiquity nonetheless far outstrips those of all others in the New World, harkening back to the great human antiquity proposed for sites associated with the rise of the notion of an American Paleolithic (see pages 66, 69, and 279). As already noted, the notion of any New World site being at a temporal par with the Upper Paleolithic sites of Europe was roundly discredited over a century ago.

Pendejo Cave is located in Rough Canyon, Otero County, south-central New Mexico, on the Fort Bliss Military Base about 300 kilometers south of Albuquerque, 80 kilometers west of Las Cruces, and 21 kilometers east of the early twentieth century gold rush town of Orogrande. The site lies within the northern reaches of the Chihuahuan Desert, a 520,000 square kilometer semi-arid ecoregion encompassing six Mexican states and portions of Texas and New Mexico, which is the largest desert in North America. In physiographic terms, the site virtually straddles the boundary between the Mexican Highlands and Sacramento sections of the Basin and Range physiographic province, and is less than 100 kilometers from the southeastern boundary

FIGURE 12.1 Originally Pendejo Cave was one of several caves in the area that MacNeish was investigating as part of his long-term research into the origin of maize (*Zea mays*). When on 20 February 1990, the crew excavating test pits at Pendejo Cave uncovered the partial remains of a species of horse (*Equus alaskae*) known to be extinct for more than 11,000 years and, a little later, charcoal dating to 32,500 yr BP, the focus changed from studying early domestication to the highly controversial question of when people first arrived in the Americas. Richard S. MacNeish Collection © Robert S. Peabody Museum of Archaeology, Phillips Academy, Andover, Massachusetts. All Rights Reserved.

of the Great Plains. The southern margin of the Rocky Mountains lies some 350 kilometers to the north.

In the Late Pleistocene, the Pendejo Cave region was covered in pinyon-juniper woodland with an understory of annuals and grasses which were succeeded by Chihuahuan desert scrub as the Holocene progressed, and the climate was generally cooler and wetter than today. As suggested by the floral remains recovered from the site, an abundance of valuable and usable, upland mesa and lowland basin plant resources were available to the site's prehistoric visitors. The present-day vegetation in the region consists principally of desert scrub in the lowlands and desert grassland combined with a variety of shrubs and herbaceous flowering plants in upland locales.

Pendejo Cave is located on the southern wall of a secondary arroyo off the southeastern fork of Rough Canyon, which trends from east to northeast and opens onto the eastern reaches of the Tularosa Basin. This 16,800 square kilometer drainage basin has no external outflow and is bounded on the east by the Sierra Blanca, the Sacramento Mountains, and the Otero Platform, and on the west by the Oscuro and San Andreas Mountains. The canyon is relatively steep and narrow, and the site sits about 65 meters above the canyon floor. [Figure 12.1] The site is a solution cave that formed via the dissolution of bedrock by the movement of ground water.

Pendejo Cave's north-facing entrance is about 6 meters wide, its interior dimensions measure about 11 meters north-south by about 6 meters east-west, and its living space encompasses an area of almost 50 square meters. [Figure 12.2] Before the commencement of excavations, the cave's roof stood 2.5 meters above the cave floor at the dripline (the point at which surface water draining from the overlying rock formation makes contact with the ground surface) and about 1 meter above the floor at the cave's interior southern back wall. The cave's floor sloped about 15 degrees downward from the back wall north to the cave's mouth.

Initially discovered in 1975 by a University of Texas field crew under the direction of geoscientist and ecologist Julio Betancourt, Pendejo Cave was first visited by is principal archaeological investigator, Richard "Scotty" MacNeish, in 1989 upon the easing of access restrictions to Fort Bliss's McGregor Firing Range, within whose boundaries the site lies. [Figure 12.3] The subsequent archaeological excavations took place in the late winter through early spring in three successive field seasons between 1990 and 1992 under the auspices of the Andover Foundation for Archaeological Research, a nonprofit organization based in Andover Massachusetts. The excavation covered more than 40 square meters and removed an estimated sediment volume of about 54 cubic meters. The southernmost portion of the site area adjacent to the cave's back wall has remained unexcavated.

FIGURE 12.2 (right) A 1991 photograph showing Pendejo Cave's stratigraphy outside of the dripline. These zones are interpreted to be extensions of their counterparts inside the dripline (*i.e.*, in the cave's interior), here designated with lower case letters. Uppermost Zone A is the site's most recent stratum, and outside of the drip line it represents the modern ground surface. Zone B appears to have been formed by erosion, and probably dates to around 1250–970 yr BP. This zone contained three hearths and Mesilla phase brownware pottery fragments. Zone C did not contain cultural features, but its correlate zone inside the drip line produced lithic artifacts, plant remains, and a portion of a two-warp *Yucca* sp. sandal dated to 1820–1570 yr BP. Lowermost Zone D also did not contain features, but its correlate zone inside the dripline dates to 21,490–18,390 yr BP. (see Figure 12.4) Photograph by J. M. Adovasio.

FIGURE 12.3 (below) Richard "Scotty" MacNeish (1918–2001), wearing red shirt, directing excavation at Pendejo Cave. He is best known for his research into the development of agriculture in the New World, an interdisciplinary approach to archaeological fieldwork, and claims for evidence of human presence at Pendejo cave over 55,000 years ago, a conclusion disputed by most archaeologists today. Richard S. MacNeish Collection © Robert S. Peabody Museum of Archaeology, Phillips Academy, Andover, Massachusetts. All Rights Reserved.

FB9366
EON1
SOUTH WALL PROFILE
3-1-90
A
b
D
E
E
E
F
G
H
I
J
K
L
M
N

FIGURE 12.4 (left) 1990 view of the south wall at the rear of the cave with markers identifying sixteen of the cave's seventeen zones from uppermost Zone A (loose brown loam 4.19 centimeters thick) to the deepest Zone N (orange brown loam about 8 centimeters thick). A sample of wood recovered in zone N yielded a date older than 55,000 yr BP. Richard S. MacNeish Collection © Robert S. Peabody Museum of Archaeology, Phillips Academy, Andover, Massachusetts. All Rights Reserved.

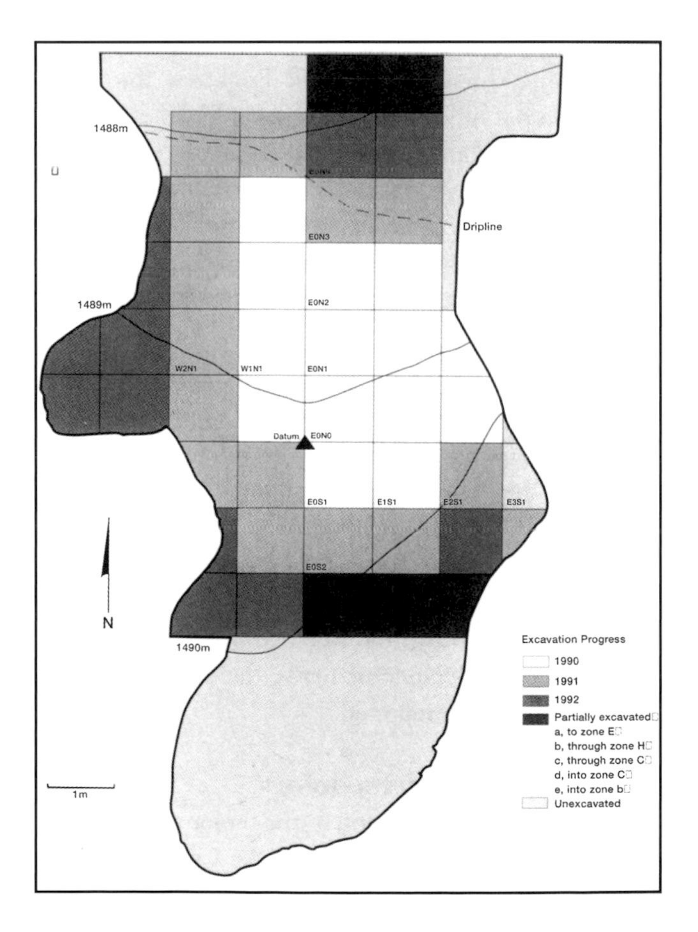

FIGURE 12.5 (below) Map of Pendejo Cave illustrating excavation progress from 1990 to 1992. The east–west axis labelled E0N3 is shown in profile in Figure 12.6. From *Pendejo Cave*, edited by Richard S. and Jane G. Libby.

Pendejo Cave's sediments are composed of wind-deposited, very fine loam and silt loam soils interspersed with human- and animal-introduced organic materials as well as decomposed limestone and roof fall. The excavators identified twenty-six clearly defined, putatively intact and discrete strata, four of which were exposed at the cave's mouth outside the dripline. These strata were lumped into seventeen major stratigraphic "zones" which range about 4–23 centimeters in mean thickness and achieve a mean overall thickness of about 12 centimeters. [Figure 12.4] All but six of these zones produced cultural features. The excavators identified a total of twenty-seven features at the site, ten of which are somewhat tentatively described as "possible features," with the large majority of them appearing to represent hearths. [Figure 12.5] The maximum thickness of the cave's stratigraphic package is about 2.5 meters. The site deposits are dated by a series of seventy-five radiocarbon assays that were run on samples from all of the site's major stratigraphic zones except for the basal Zone O. [Figure 12.6]

The excavation of Pendejo Cave recovered 826 lithic artifacts, fifty worked bone artifacts, thirty-nine ceramic artifacts (attributable to eight or nine ceramic vessels), 976 textile (plant and animal fiber) artifacts, over 41,000 bone specimens identified to the level of species, over 25,000 plant remains (among which fifty-six species have been identified), sixteen pieces of fired clay preserving imprints of human skin, and 200 strands of hair from eight different species—sixty strands of which were identified as human. [Figures 12.7 and 12.8] Most of the artifacts ascribe (based on the projectile point assemblage) to several archaeological complexes that are broadly recognized for the region, including Mesilla (1250–970 yr BP), Hueco (1900–1600 yr BP), Fresnal (4600–2900 yr BP), and Keystone (6300–4600 yr BP). [Figure 12.9] The latter three Archaic levels are underlain by Pendejo Cave's Clovis-age material, dated at this locality to between 14,100–13,440 yr BP and 13,440–13,065 yr BP. If valid, the earlier of these dates would make Pendejo Cave the oldest known Clovis site, though the artifact assemblage's lack of the distinctive Clovis point renders such an ascription rather tenuous.

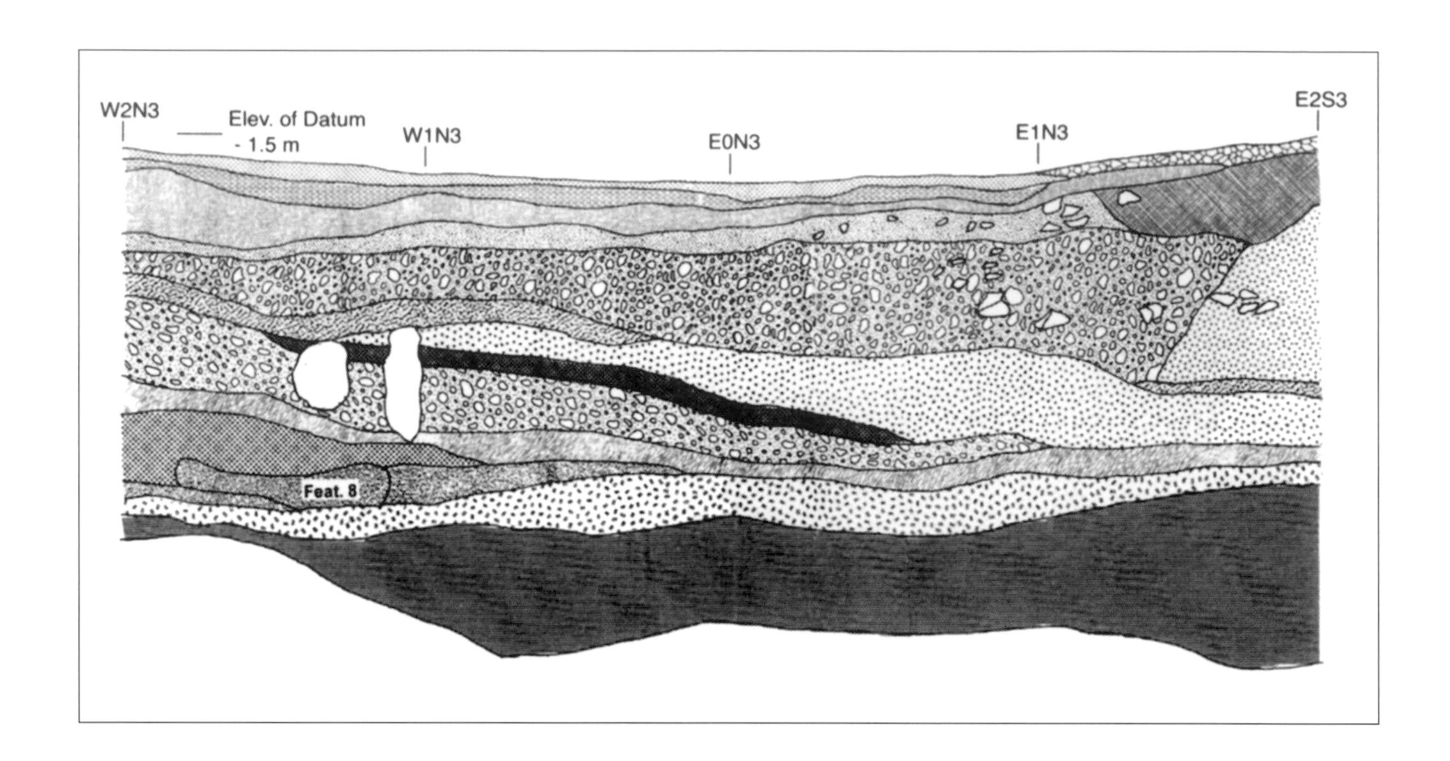

FIGURE 12.6 (above) Stratigraphic profile of the east–west E0N3 axis shown in Figure 12.6. Feature 8 lies in Zone M, which is dates to 52,000–41,000 yr BP, and was described by MacNeish as a pit with fine carbonate ash. Zone O, the lowest stratum illustrated in this profile, has not been dated. From Richard S. MacNeish and Jane G. Libby (eds.) *Pendejo Cave*.

FIGURE 12.7 (left, top) A Pendejo projectile point (above) and a Maiz de Ocho corn cob (below) from Zone C dated to 6650 yr BP. Richard S. MacNeish Collection © Robert S. Peabody Museum of Archaeology, Phillips Academy, Andover, Massachusetts. All Rights Reserved.

FIGURE 12.8 (left, bottom) A fragment of coiled basketry made from willow (*Salix* sp.) rods and woven *Yucca* sp. fiber dated to between 2820 yr BP and 2740 yr BP. Richard S. MacNeish Collection © Robert S. Peabody Museum of Archaeology, Phillips Academy, Andover, Massachusetts. All Rights Reserved.

FIGURE 12.9 (right) Pendejo projectile points found in association with pieces of cordage, cucurbits, cobs of Maiz de Ocho, and squash dated to between 2580 yr BP and 2740 years BP. Richard S. MacNeish Collection © Robert S. Peabody Museum of Archaeology, Phillips Academy, Andover, Massachusetts. All Rights Reserved.

Underlying Clovis-age deposits, the excavators also identified three putative pre-Clovis deposits (collectively termed "Paleoamerican" by MacNeish and his colleagues) from which the vast majority of the site's radiocarbon samples were obtained, including North Mesa (14,875–14,090 yr BP from human hair), McGregor (38,820–33,780 yr BP from charcoal associated with human skin imprints, presumably on fired clay), and Orogrande (older than 55,000 radiocarbon years ago on "large pieces of wood," calibration not possible). The multiple radiocarbon determinations for these Paleoamerican manifestations, moreover, appear to have prompted MacNeish and his colleagues to suggest much broader and earlier date ranges for them based on materials with ambiguous cultural association, expanding the earlier ranges of the North Mesa and McGregor complexes by as much as 25,000 yr BP and 11,000 yr BP, respectively. If valid, these dates would push back the presence of humans into the Western Hemisphere to more than twice the temporal threshold presently accepted by the professional archaeological community. [Figure 12.10]

The validity of the Archaic and later prehistoric archaeological deposits at Pendejo Cave has never been seriously disputed, but since the project's early stages considerable doubt has been cast upon the validity of the lithic artifacts from the site's pre-Clovis levels, with claims that they in fact represent pieces of weathered and fractured dolomite from the cave's roof and walls. Petrographic analyses of the early lithic artifacts by MacNeish and his colleagues concluded that slightly more than half of them were made using raw materials that are not local to the cave, with one source being more than 20 kilometers to the west. Objections to the very small Paleoamerican lithic assemblage, however, have also been raised on morphological grounds. Critics have maintained that the site's Paleoamerican lithic industries, which grade through time from pebble tool (the earliest) through unifacial to bifacial (the most recent) technologies, are not the products of human manufacture and, like the small assemblage of bone artifacts from the site, suggest some form of fracture by natural or at least prehistoric non-human agency.

Perhaps the most controversial—and, quite possibly, the most archaeologically promising—artifacts recovered from Pendejo Cave are the putative "friction-skin imprints" on fired clay and specimens of human hair mentioned earlier. With a few exceptions, the archaeological community has more or less conceded that these objects are indeed attributable to humans, but there nonetheless has been considerable skepticism that they were recovered from deposits which were indeed intact and undisturbed. Former University of Colorado geologist Thomas Stafford, for example, has cited extensive pack rat activity at the site as a probable source of disturbance, and the site's discoverer, geoscientist and ecologist Julio Betancourt, has noted that "the presence of *Prosopis* (mesquite) leaves and pods throughout glacial-age cave sediments suggests temporal mixing through rodent burrows, roof fall events and other disturbances."

Pendejo Cave's excavators concluded that the site's Late Pleistocene "Paleoamerican" populations were composed mainly of micro-bands and/or small task-groups of hunter-gatherers who used the cave with varying degrees of intensity—while never actually residing there—in the course of seasonal visits to obtain plant and animal resources from the environs of Rough Canyon and the adjacent Tularosa Basin. While the site exhibits an analogously ephemeral use by peoples associated with the well-recognized cultural complexes of the succeeding Archaic and Prehistoric periods, few in the professional archaeological community have accepted the archaeological validity of its tenuously defined pre-Clovis levels.

FIGURE 12.10 (right) A sampling of what MacNeish designed McGregor and Orogrande complexes lithics. McGregor complex lithics: (A) projectile points and (B) wedge chopper recovered from zones I to M and dated to between 31,000 and 52,000 yr BP. Orogrande complex lithics: (C) concave uniface, (D) chopper scraper, and (E) pebble chopper recovered from zones N and O dated to between 55,000 and 75,000 yr BP. The dating of these zones has not been seriously questioned. What is in dispute is the nature of the purported lithics recovered from these deposits, which most archaeologist today consider geofacts, naturally formed stones and not man-made artifacts. From Richard S. MacNeish and Jane G. Libby (eds.) *Pendejo Cave*.

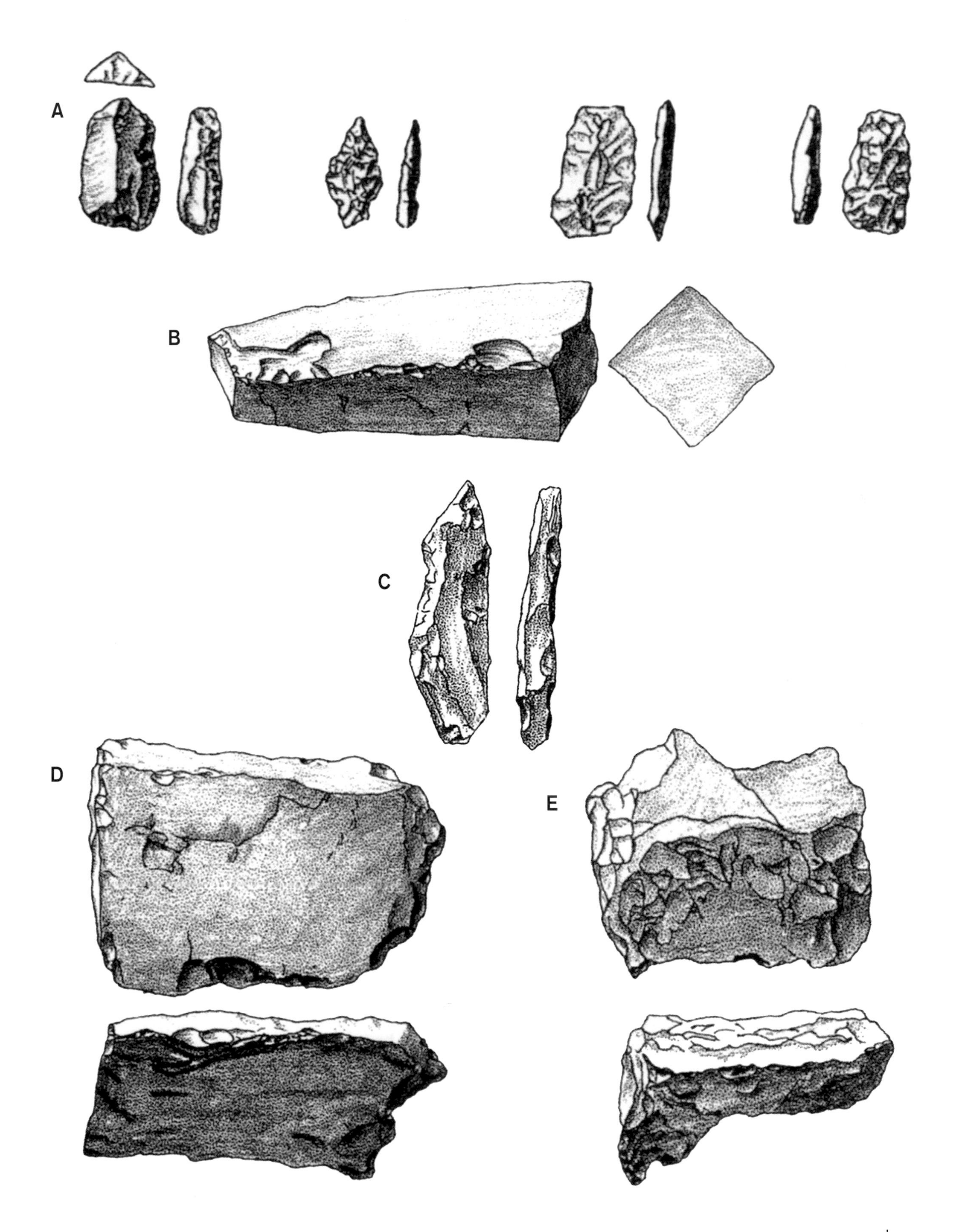
A
B
C
D
E

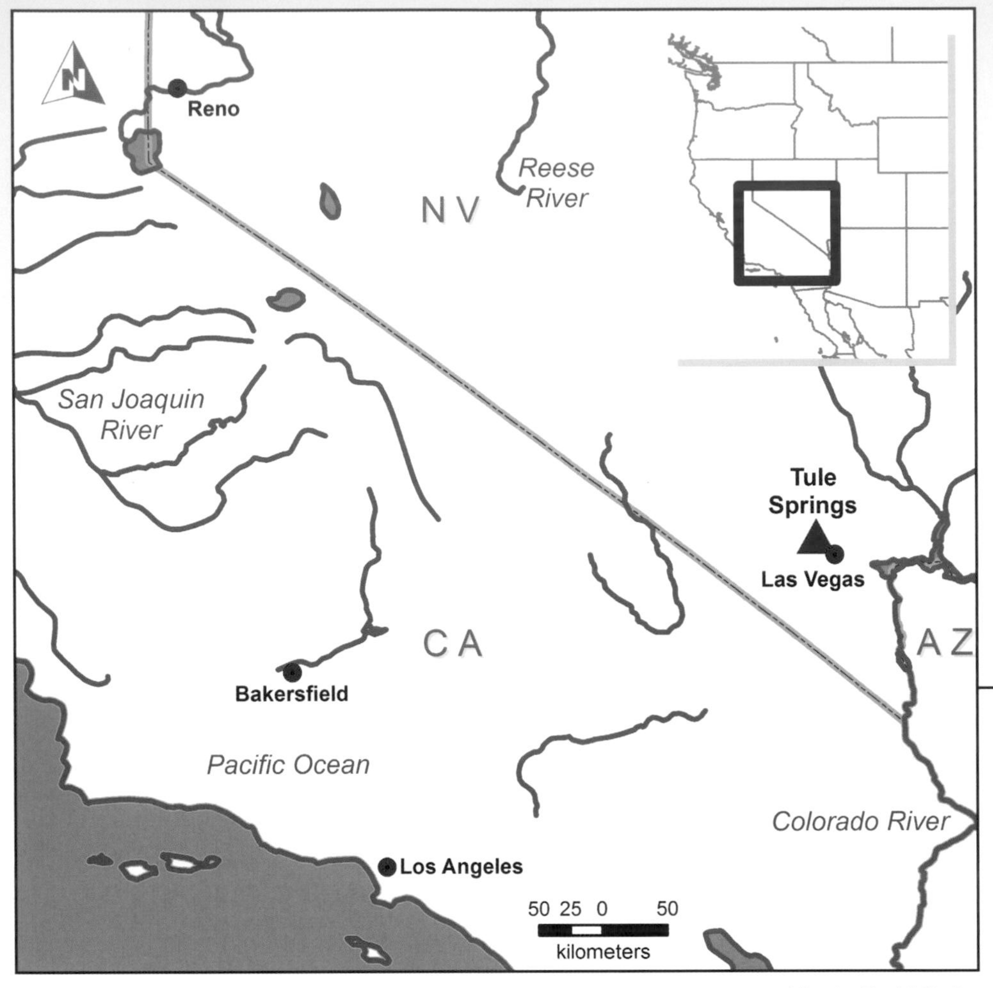

Map by David Pedler

The Tule Springs archaeological site is located in a deeply eroded badland area along he right bank of the Vegas Wash within the Las Vegas Valley 16 kilometers north of Las Vegas. Clovis-age artifacts and fossil bones of Pleistocene mammoth, horse, camel, bison, ground sloth and panther have been found in the site's fine-grained sediments.

13 TULE SPRINGS

LOCATION CLARK COUNTY, NEVADA, UNITED STATES

Coordinates 36°19'20.20"N, 115°16'8.75"W.

Elevation 703 meters above mean sea level.

Discovery Fenley Hunter in 1933.

The Tule Springs site was one of the first archaeological discoveries to produce the remains of extinct Late Pleistocene animal remains in association with apparently solid evidence of a contemporaneous human presence. It was also one of the earliest American archaeological localities whose age was assessed using the radiocarbon dating method, which was developed fifteen years after the site's discovery. And although Tule Springs' ultimate archaeological disposition fell far short of its initial promise, the site nonetheless played a major role in the development and practice of multidisciplinary archaeological investigation through the integrated efforts of the geologists, archaeologists, environmental scientists, and physicists who worked on the project. [Figure 13.1]

The Tule Springs site, named after the community in which it was discovered, is located in Clark County, southeastern Nevada, on the northwestern edge of Las Vegas about 20 kilometers from the city center, 520 kilometers southeast of Reno, and 350 kilometers northeast of Bakersfield, California. The site lies within the 4,000 square kilometer Las Vegas Valley, which is situated just west of Lake Mead, a water impoundment on the Colorado River that houses a large national recreational area, the most voluminous reservoir in the United States, and the Hoover Dam. The Colorado River is the principal drainage system of the southwestern United States and northwestern Mexico.

The valley occupies a basin in the Basin and Range physiographic province, a region that is characterized by steep, narrow mountain chains that tower more than 3,000 meters above the intervening flat, arid valley floors. [Figure 13.2] Tule Springs lies on the northeastern

margin of what is now the Mojave Desert, but at the time of the region's earliest apparent human occupation during the Pleistocene–Holocene transition, the valley abounded in springs surrounded by ash, willow, and sagebrush with pinyon pine, juniper, and sagebrush in higher elevations and sagebrush and saltbush in lower elevations.

The first archaeological discovery at Tule Springs occurred in early 1933, only a few years after the landmark Folsom (see pages 36-45) and Clovis (see pages 46-57) discoveries in New Mexico, when paleontologist Fenley Hunter of the American Museum of Natural History found an obsidian flake in apparent association with a concentration of charcoal and the bones of extinct bison (*Bison occidentalis*), camel (*Camelops hesternus*), horse (*Equus pacificus*), and mammoth (*Mammuthus columbi*) in the arroyo wall of a dry side valley known as Vegas Wash. [Figure 13.3] Hunter subsequently turned his discovery over to the curator of archaeology at the Southwest Museum in Los Angeles, Mark Raymond Harrington, who embarked on a field project at the Hunter locality in October 1933.

FIGURE 13.1 (above) A resin consolidant (hardener) is applied to the jaw of an extinct large species of *Equus* in order to strengthen and maintain its pieces in the correct position. Photograph courtesy of the Nevada State Museum.

FIGURE 13.2 (right, top) "The Tule Springs locality is not spectacular, but it is compelling. Weather worn mountains rise starkly on either hand. Hidden until one stands at its brink, a deep erosion-cut gash slits the wide peaceful Las Vegas Valley. In that arroyo a miniature badlands has developed. Descending into the arroyo, into the shelter and silence of the twisting side canyons, one literally walks into the past—geologically and archeologically." — Ruth Simpson (in *Tule Springs, Nevada, with Other Evidences of Pleistocene Man in North America*, by Mark Harrington and Ruth Simpson). Photograph courtesy Jim Boone.

FIGURE 13.3 (right, bottom) The first artifact found at the Tule Springs in 1933, a 36 mm long obsidian flake. Photograph courtesy of the American Museum of Natural History.

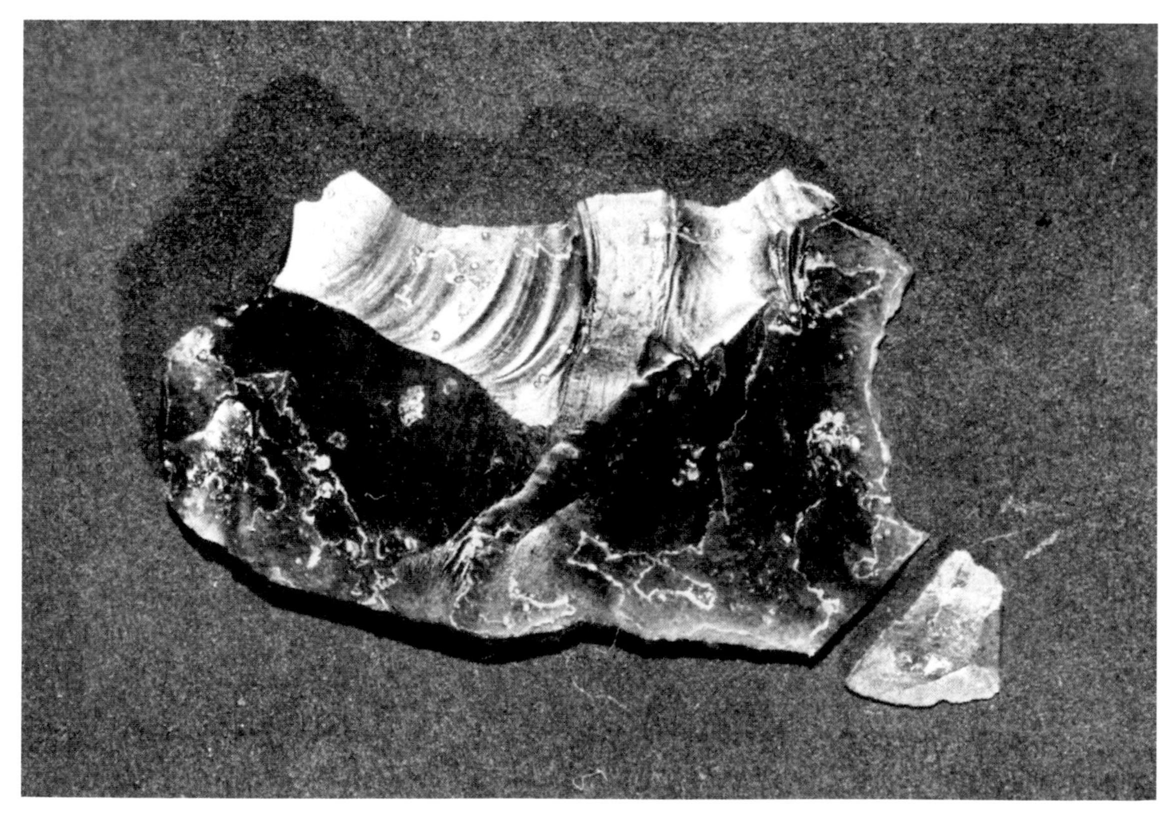

The first expedition at Tule Springs focused on relocating Hunter's site and prospecting for similar nearby localities, which were readily identified in the surrounding arroyo walls by the presence of apparent charcoal deposits (interpreted by the research team as prehistoric campfires) resting on a layer of dense clay and overlain by a 2–3 meter thick layer of looser clay and sand. The Harrington expedition identified two new features, prosaically named Ash Beds 1 and 2, which yielded additional bones—apparently including mammoth (*Mammuthus columbi*), a small number of putatively "cut and polished" bone implements associated with the ash beds, and a small assemblage of crude stone artifacts characterized as bifaces, scrapers, and choppers. These stone artifacts, however, were not recovered in association with the ash deposits, and the obsidian flake recovered by Hunter was the only artifact from the site that was considered by Harrington to be an "unmistakable" artifact.

The investigation of Tule Springs lay dormant for almost twenty years until the advent of radiocarbon dating, which inspired Harrington and his colleague at the Southwest Museum, Ruth Simpson, to submit charcoal samples from Harrington's first expedition to radiocarbon-dating pioneer Willard Libby at the University of Chicago for analysis. Libby's results, which in Harrington's words "thrilled us all," indicated that the charcoal was over 23,800 years old and thus more than twice as old as Clovis. Buoyed by these findings and convinced of the site's potentially tremendous significance, a second expedition was planned for Tule Springs, this time headed by Simpson.

The second expedition at Tule Springs was undertaken in May 1955 with the objectives of obtaining more charcoal samples and artifacts from the Hunter site, documenting the site via comprehensive mapping and photography, and surveying the region for additional archaeological localities. Simpson and her colleagues successfully identified and excavated five new charcoal deposits in addition to those discovered by Harrington in 1933, but only a single unifacial lithic scraper was recovered in putative association with charcoal radiocarbon dated to older than 28,000 yr BP—which was tantalizing, but hardly conclusive. By the end of the project in 1956, the Southwest Museum's archaeological personnel had examined a 5 kilometer long stretch of the Las Vegas Valley.

A third and final Southwest Museum expedition in April 1956 was again largely unsuccessful in archaeological terms. A survey of the greater Las Vegas Valley beyond Tule Springs conducted at that time by the Tule Springs investigators did, however, identify new bone/charcoal deposits (including mammoth remains) and additional archaeological sites. The modest assemblage of lithic artifacts from that survey included diagnostic projectile points, but these materials were clearly attributable to the Archaic period occupation of the valley long after the extinction of the region's megafaunal species. The sole, convincing artifact apparently associated with the extinct megafauna remained the obsidian flake recovered by Hunter in 1933.

Tule Springs came to be widely regarded as a bona fide pre-Clovis locality by its excavators and sympathetic scholars in the years that followed, but lingering disputes about the site's integrity and age—as well as a collective desire on the part of interested scientists to better coordinate and understand the role of radiocarbon dating in archaeology as a discipline—led to the formation of the Tule Springs project in 1962. Conceived at a meeting convened by Willard Libby (who had moved from the University of Chicago to direct the Institute of Geophysics at the University of California at Los Angeles) and attended by a group of ten leading scholars, Tule Springs was advanced as the most suitable location for implementing a multidisciplinary project to resolve complex and interrelated geological, paleontological, and archaeological problems.

The project was conducted under the auspices of the Nevada State Museum in Carson City and the Southwest Museum under the direction of geologist C. Vance Haynes, who would go on to direct the excavation of the Murray Springs Clovis site (pages 58-67) four years later. [Figure 13.4] The project's main goal was to determine, decisively, whether humans and

FIGURE 13.4 C. Vance Haynes brushes dust from what may represent remains of a camp fire or naturally deposited accumulation of carbonized organic matter. Photograph courtesy of the Nevada State Museum.

Pleistocene megafauna were contemporaneous at Tule Springs, and if so, at which specific point in time. The project, in what would be a unique arrangement even today, was planned in such a way that Libby's radiocarbon dating laboratory would serve as a full-time partner and be directly involved in the collection of radiocarbon samples and their processing as the project proceeded. In the end, Libby's facility processed almost eighty radiocarbon samples from Tule Springs.

The scope of this multi-disciplinary investigation, dubbed "The Big Dig," was enormous by any standard, covering a 40 kilometer long stretch of the Las Vegas Valley in an 8 kilometer wide swath. Carried out between October 1962 and January 1963, the project employed bulldozers and wheel tractor-scrapers to remove the 6 meter thick, very hard, compacted sediments [Figure 13.5] that overlaid the bone deposits. [Figure 13.6] The work resulted in the removal of over 180,000 tonnes of material from over 2 kilometers of trenches measuring almost 4 meters wide and up to 9 meters deep. Potential cultural exposures noted in these trenches were further investigated via formal archaeological excavation using hand tools.

As described in the 1967 report of the project's findings, *Pleistocene Studies in Southern Nevada*, Haynes and his colleagues determined that over 40,000 years ago the valley was dotted with springs and occupied by the ancestral Las Vegas River, a Colorado River tributary whose channel was over 200 meters wide. Sometime before 30,000 yr BP, for unknown reasons the river appears to have become blocked, which led to the formation of Late Pleistocene pluvial (rain fed) Lake Las Vegas. Surrounded by pine and sagebrush in the valley's higher areas and wetland plants such as cattail on its

FIGURE 13.5 (above) Allis Chalmers Model TS-360 motor scraper cutting 2 kilometers of trenches, 4 meters wide and up to 9 meters deep in order to expose stratigraphic deposits in the Vegas Wash. Photograph courtesy of the Nevada State Museum.

FIGURE 13.6 (right) In 1962 bulldozer trenches exposed 117 identifiable fossil specimens, including camel (*Camelops hesternus*), Columbian mammoth (*Mammuthus columbi*), horse (*Equus* sp.), pronghorn relative (*Tetrameryx* sp.), rabbit (probably *Lepus californicus*), meadow mouse (*Microtus* sp.), a coyote relative (*Canis* sp.), waterfowl (Anseriformes), and giant condor (*Teratornis merriami*). From *Pleistocene Studies in Southern Nevada*, Nevada State Museum.

shallow margins, the lake attracted mammoths and camels. [Figure 13.7] About 15,000 years later, again for unknown reasons, the lake suddenly discharged and was soon reduced to a small stream surrounded by small springs on the valley floor that were frequented by mammoths, camels, horses, and smaller animals. [Figures 13.8 and 13.9] By around 6000 yr BP, the valley's water table dropped, its springs began do go dry, numerous arroyos were cut, and wind erosion intensified. Around 3,000 years later, the Las Vegas River had become a dry wash, as it remains today. [Figure 13.10]

Though the Tule Springs project recovered a somewhat small (but nonetheless diverse and informative) paleontological assemblage of Pleistocene megafaunal remains, the disappointing results of the project's archaeological investigation conclusively demonstrated that a human presence in the Las Vegas Valley was far more recent than had been presumed. [Figure 13.11] The "hearths" identified at the Hunter site in 1933 and the mid-1950s were ultimately determined to represent organic material associated with a 40,000 year old spring, and evidence for the "burning" of the animal remains was discovered to be the result of staining from iron oxide present in the decayed organic material. (See Figure 13.4) The ancient radiocarbon age of the Hunter site was also cast into doubt, as Haynes and his colleagues determined that the 28,000 yr BP sample was obtained from a mixture of ancient and more modern deposits. Archaeologist Richard Shutler, Jr., an original research team member, noted the remarkably meagre count of only eleven potentially bona fide artifacts from thirty years of excavation in the Tule Springs fossil deposits,

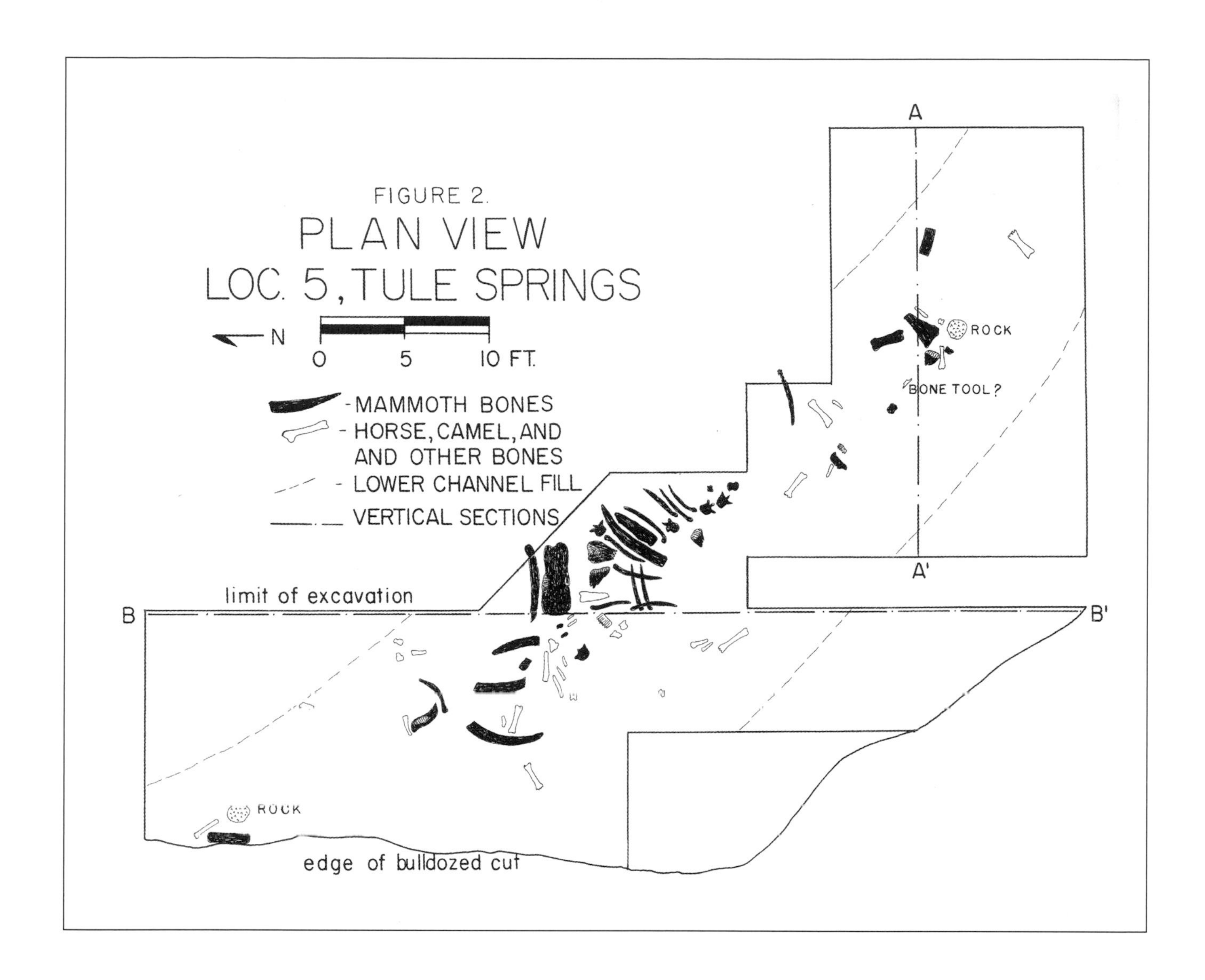

and was convinced that even most of those derived from questionable contexts of uncertain—though undoubtedly much younger—age. [Figure 13.12]

In his final analysis, Haynes concluded that humans *may* have been present in the Las Vegas Valley around 13,300 yr BP (*i.e.,* during early Clovis times) but favored the more conservative interpretation that they had more likely arrived by about 12,800 yr BP, placing Tule Springs firmly in the middle of the age range for Clovis and making it contemporary with sites such as Shawnee-Minisink (see pages 78-89) some 3,500 kilometers to the northeast. In the face of the exceptionally low artifact yield, the presence of many potential geofacts in the assemblage, and total absence of diagnostic artifacts confirming its ascription to the continent-wide Clovis horizon, the prevailing archaeological opinion holds that the artifacts at Tule Springs are actually much later than their presumed contexts, and had been washed onto the ancient layers via erosion. In any case, Tule Springs has remained an important paleontological locality. In December 2014, this 90 square kilometer portion of the Las Vegas Valley was formally designated the Tule Springs Fossil Beds National Monument and is administered by the United States National Park Service.

FIGURE 13.7 Romanticized depiction of the now-extinct North American camel and her calf being hunted by Paleo-indians in the Las Vegas Valley 10,000 yr BP. Painting by Jay Matternes.

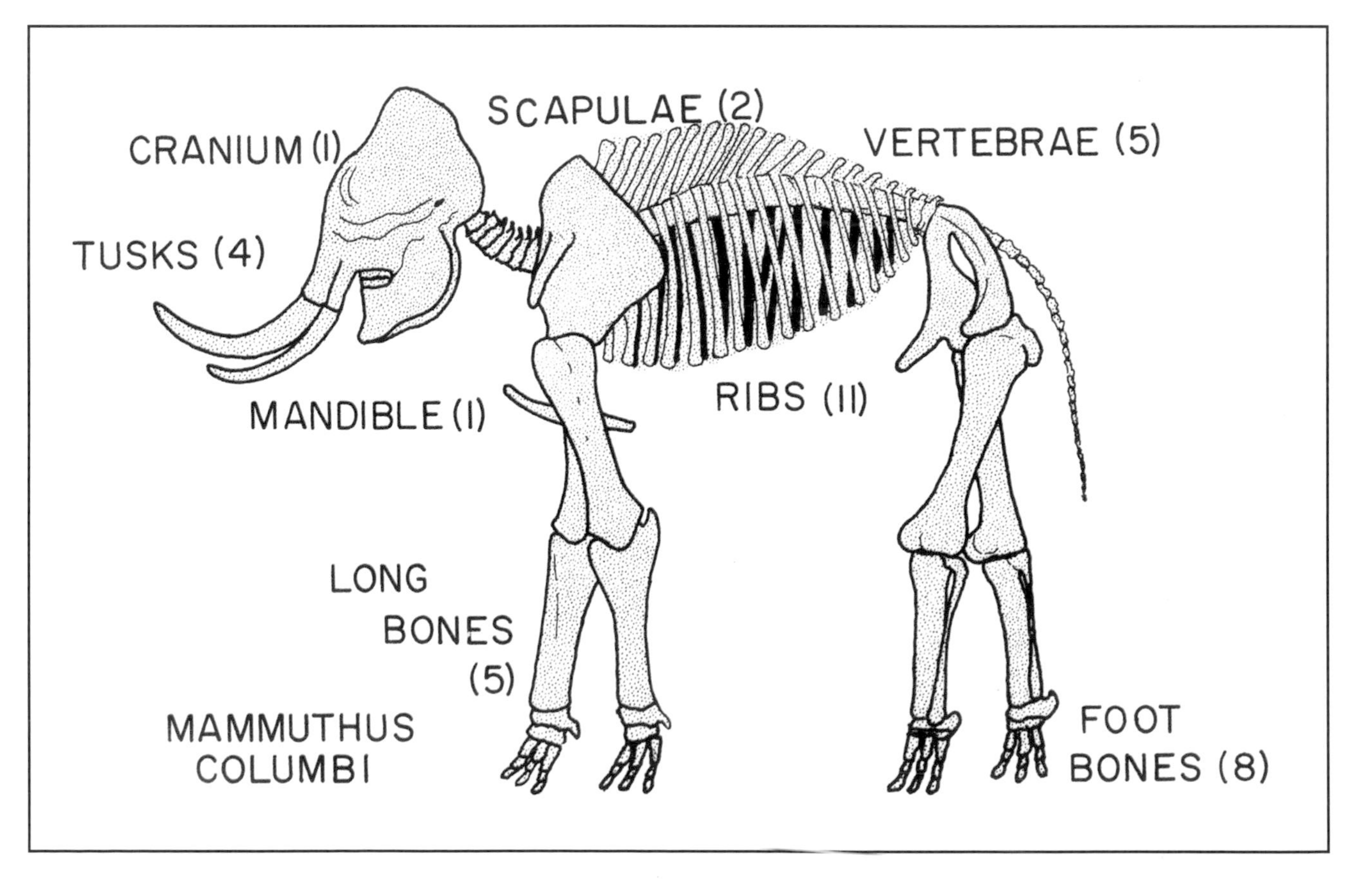
SCAPULAE (2)
CRANIUM (1)
VERTEBRAE (5)
TUSKS (4)
MANDIBLE (1)
RIBS (11)
LONG
BONES
(5)
MAMMUTHUS
COLUMBI
FOOT
BONES (8)

FIGURE 13.8 (left, top) Mammoth long bone being encased in a jacket of plaster of Paris for safe transport to the Southwest Museum in Los Angles. Photograph courtesy of the Autry National Center.

FIGURE 13.9 (left, bottom) Identified mammoth bones from the lower channel deposits diagramed in Figure 13.6 include a maxilla and part of the cranium, mandible, four tusks, three molars, numerous tooth plates, two scapulae, five limb bones, five vertebrae, eleven ribs, an astragalus (anklebone), a calcaneum (heel), metacarpal (front foot), metapodial (rear foot), sesamoids (kneecap), phalanxes, and several tarsals and carpals. Several of the bones, possibly from one individual, were very nearly articulated. The total number of mammoths present in these deposits is estimated at three. From *Pleistocene Studies in Southern Nevada*, Nevada State Museum.

FIGURE 13.11 Shortly before the completion of field work in 1963 at Tule Springs, this apparent quartzite scraper was discovered *in situ* in a context that is reported to be roughly contemporary with Clovis. Photograph courtesy of the Nevada State Museum.

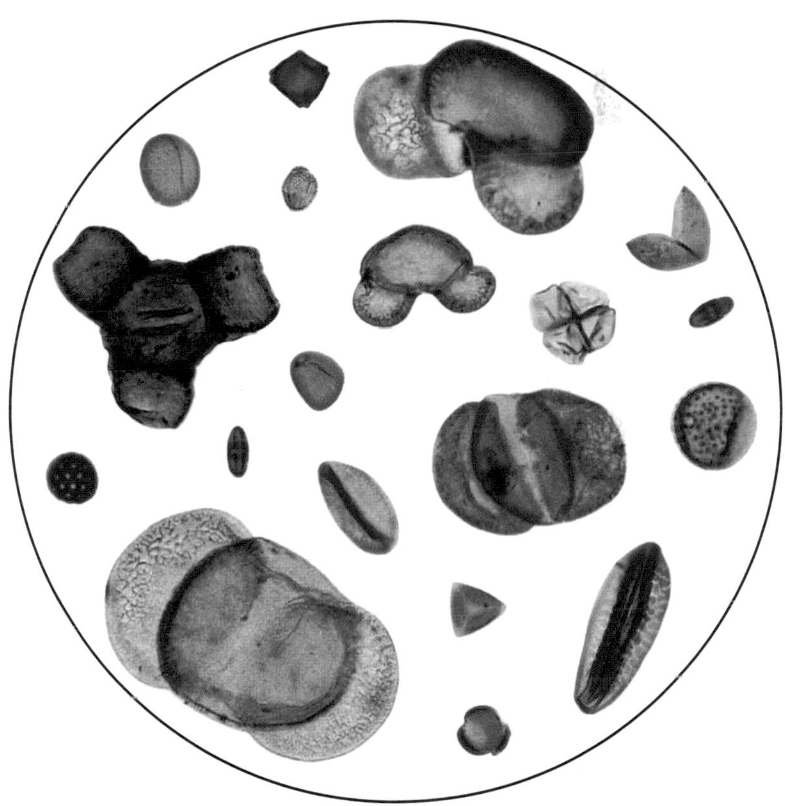

FIGURE 13.10 Pleistocene pollen from the Las Vegas Valley. Paleobotanists can use pollen from successive strata to reconstruct ancient environments through time. Based on the vegetation represented by pollen, it appears Tule Springs and surrounds experience a trend toward warmer and dryer conditions starting about 12,000 yr BP. By 6,000 yr BP the vegetation was probably much as it is today. From *Pleistocene Studies in Southern Nevada*, Nevada State Museum.

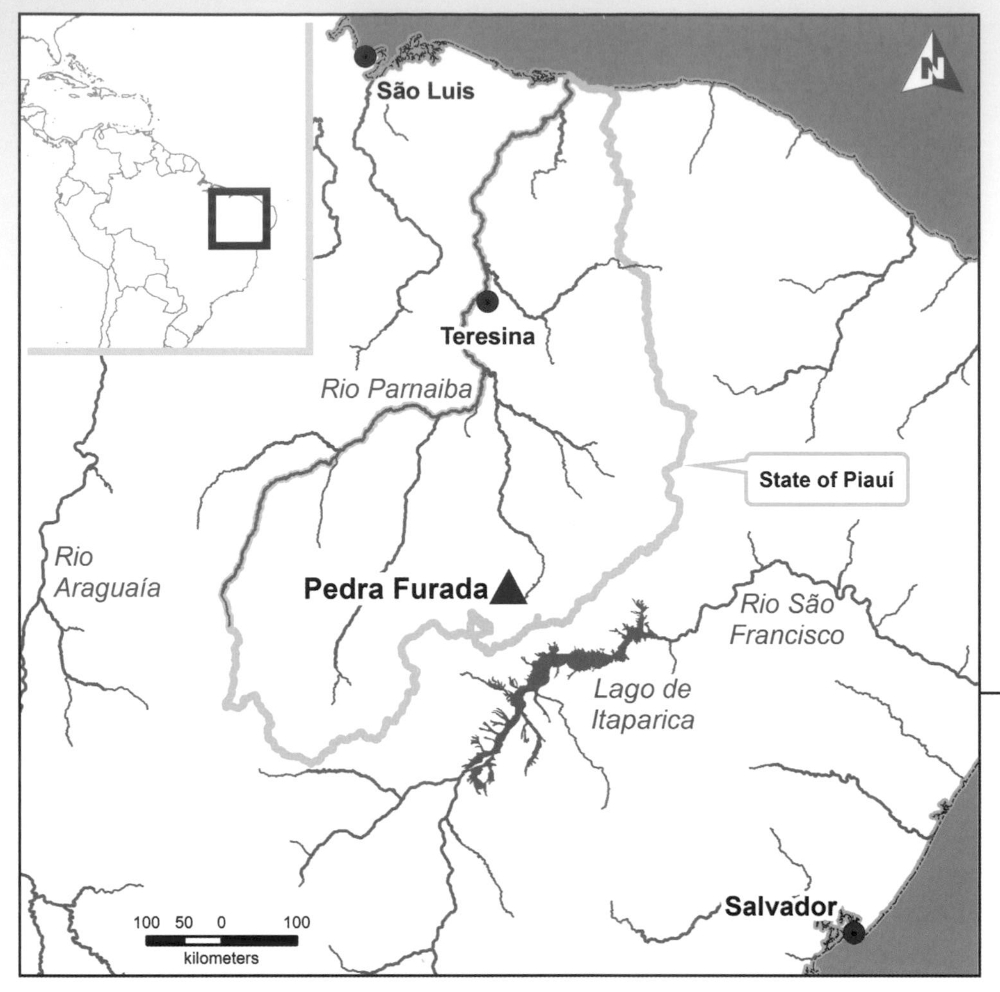

Map by David Pedler

Pedra Furada (meaning "pierced rock") is located within the 1291 square kilometer Serra da Capivara National Park in northeastern Brazil. It is a rock art shelter with over 1,150 images and from which thousands of artifacts have been recovered, some of which are claimed to be over 50,000 years old.

14 PEDRA FURADA

LOCATION PARQUE NACIONAL SERRA DA CAPIVARA (SERRA DA CAPIVARA NATIONAL PARK), PIAUÍ, BRAZIL

Coordinates 8°50'10.11"S, 42°33'21.68"W.

Elevation 420 meters above mean level.

Discovery Niède Guidon in 1973.

With the relatively recent acceptance of the pre-Clovis MV-II component at the Monte Verde site in southern Chile (see pages 216-227), Pedra Furada may have reclaimed its status as the most controversial archaeological site in South America. As with the distant sites of Calico Mountain in California (see pages 152-161) and Pendejo Cave in south-central New Mexico (see pages 162-171), claims have been made for a human presence at Pedra Furada that extend well into the Pleistocene at a time when anatomically modern humans are thought to have just recently migrated out of Africa and would not be present in Arctic Siberia for another 30,000 years.

Pedra Furada, formally known as *Toca do Boqueirão do Sítio da Pedra Furada*, is located in southern Piauí state within Brazil's Northeast Region, about 640 kilometers northwest of Salavador, 725 kilometers south of São Luis, and 415 kilometers south of Teresina, the capital city of Piauí. The site lies on the southern boundary of the Serra da Capivara National Park, a 1,290 square kilometer heritage preserve that contains at least 300 recorded archaeological sites, over 200 of which are distinguished by spectacular prehistoric paintings in ochre on rock walls. [Figure 14.1] This portion of the park marks the division between an upland zone to the north and a lowland plain to the south, while also effectively

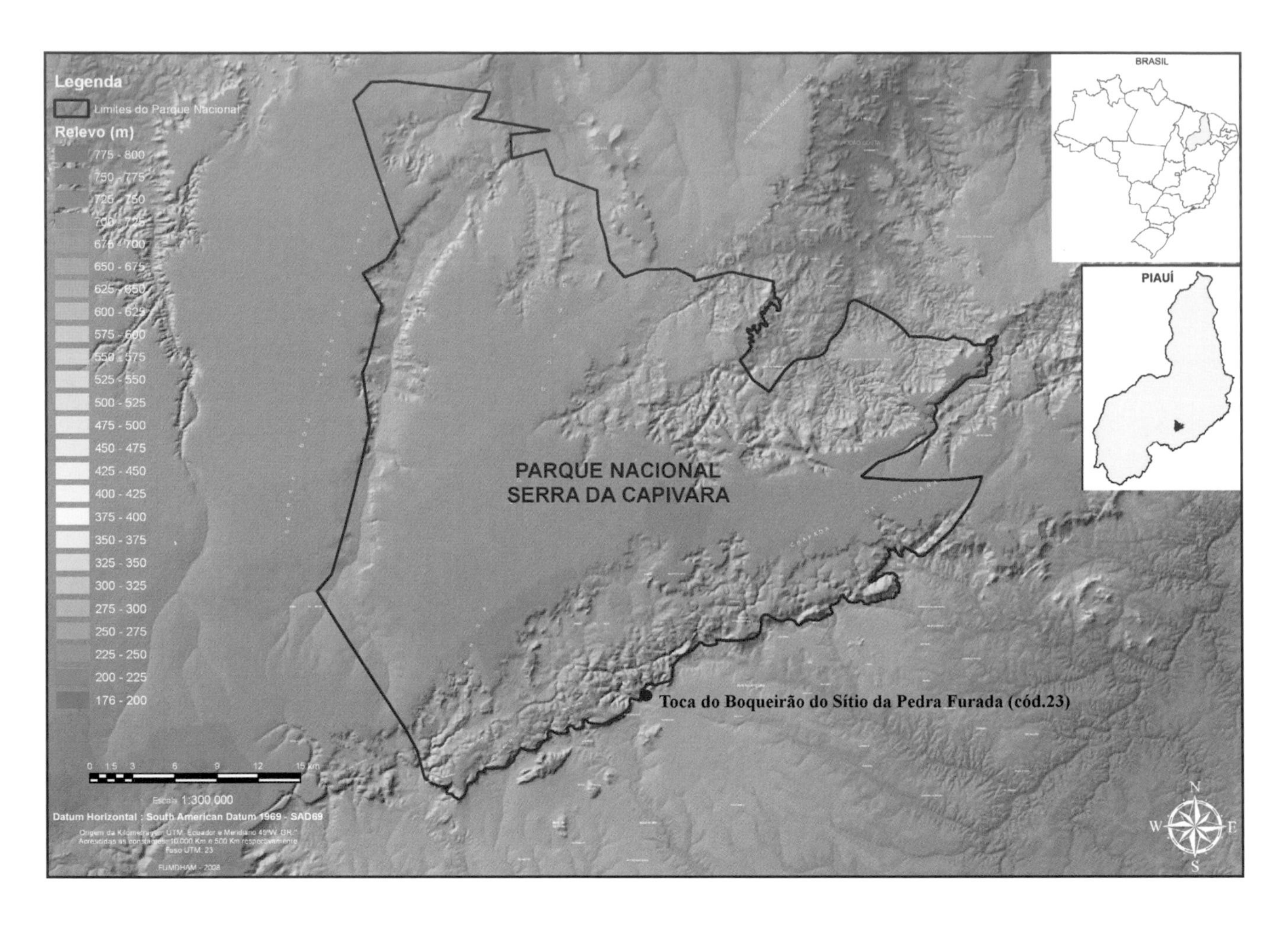
Legenda
Limites do Parque Nacional
Relevo (m)
775 - 800
750 - 775
725 - 750
700 - 725
675 - 700
650 - 675
625 - 650
600 - 625
575 - 600
550 - 575
525 - 550
500 - 525
475 - 500
450 - 475
425 - 450
400 - 425
375 - 400
350 - 375
325 - 350
300 - 325
275 - 300
250 - 275
225 - 250
200 - 225
176 - 200
0 1.5 3 6 9 12 15 km
Escala 1:300.000
Datum Horizontal : South American Datum 1969 - SAD69
BRASIL
PIAUÍ
PARQUE NACIONAL
SERRA DA CAPIVARA
Toca do Boqueirão do Sítio da Pedra Furada (cód.23)
N
W
E
S

FIGURE 14.1 (above) Relief map of the Serra da Capivara National Park, a 1,290 square kilometer heritage preserve that contains hundreds of recorded prehistoric archaeological sites. Pedra Furada's location is shown on the southern boundary of the park in the center of the map. Map courtesy of Niède Guidon.

FIGURE 14.2 (below) View of the lowland and upland boundary in the vicinity of Pedra Furada. The site lies at the base of the escarpment shown here amid a shrub and scrub to forest community known to environmental biologists as *caatinga*. The area is hot and dry today, but in the Late Pleistocene it was hot and humid, its vegetation was dominated by grasslands, and its landscape was dotted with small ponds and lakes. Photograph courtesy of Niède Guidon.

serving as the boundary between the Rio Parnaiba and Rio São Francisco drainage basins. The upper tributaries of the Rio Parnaíba drain the uplands to the north as part of the 330,000 square kilometer Rio Parnaíba basin, which enters the Atlantic Ocean on Brazil's northern coast near Parnaíba. The lowland plain is drained by the Rio São Francisco, whose 640,000 square kilometer basin enters the Atlantic on Brazil's east coast.

The regional vegetation is dominated by a diverse, complex, and highly variable semi-arid forest regime known as *caatinga*. Botanists have recognized as many as ten distinct types of *caatinga* that range from low (less than 1 meter tall) shrub and scrub to forest composed of thorny trees that attain heights of 25–30 meters. [Figure 14.2] *Caatinga* is also known to form very complex boundaries with neighboring vegetation communities, and the Pedra Furada region is no exception. The site is less than 100 kilometers away from tropical–subtropical grassland and dry broadleaf forest ecological zones that lie to the west and south. Presently hot and dry, the region's Late Pleistocene climate instead appears to have been hot and humid, with rain forest occupying the uplands and grasslands dotted with ponds and lakes in the lowlands. With the onset of more-arid conditions, this regime began its shift to *caatinga* around 9000 yr BP, and by 6,000 yr BP the region's formerly rich and varied animal species had disappeared.

Pedra Furada is a southeast-facing rockshelter at the base of an escarpment that rises about 160 meters above the valley floor's elevation of 420 meters above sea level. [Figure 14.3] The mouth of the rockshelter is about 70 meters wide, its interior floor is about 12–18 meters deep, and the site's now fully excavated deposits extend to a depth of at least 5 meters. [Figure 14.4] The rockshelter opening is flanked on both sides by naturally formed clefts or "chutes" that apparently have transported material from the rock face above the site down to its floor for thousands of years, leading to the accumulation of large piles of broken quartzite cobbles.

An intensive ten-year long investigation of the site began in 1978, led by Serra da Capivara National Park archaeological director Niède Guidon, who had first visited the region in the 1960s to study its compelling rock art. [Figure 14.5] The site's stratigraphy has been variously described as analysis of the site has proceeded. [Figure 14.6] Guidon first described five strata, labeled A through E, mostly composed of sand with some silt and containing roughly similar proportions of gravel and pebbles (4–8 percent) throughout the entire deposit. Formally refining this scheme fifteen years later in 2001, project archaeologist Fabio Parenti discerned eleven layers which he assigned to three discrete groups or units labelled upper, middle, and lower. [Figure 14.7]

In the upper unit, the excavators identified a cultural zone they designated the Serra Talhada phase, which yielded bifacial and unifacial lithic artifacts of quartzite and chert, well-defined hearths and living surfaces, organic remains, and the associated rock art. [Figure 14.8] This deposit is dated by fourteen radiocarbon assays to as early as 12,700–11,610 yr BP and appears to have remained in place until the full encroachment of *caatinga* vegetation some 6,000 years later. Both this age range and the accompanying cultural assemblage are compatible with those of other rock art sites identified to date in northeastern Brazil, and the Serra Talhada phase not been disputed.

The middle and lower units at the site contain what Guidon and her colleagues have designated the Pedra Furada phase. The assemblage from this phase is composed exclusively of about 600 lithic artifacts,

FIGURE 14.3 (left) The escarpment above the site of Pedra Furada, with the rockshelter shown at its base. One of two naturally formed clefts or "chutes" is visible in the left side of the photograph. These clefts in the escarpment have apparently guided rock eroding from the rock face above the site to be smashed on the floor, forming piles of fractured cobbles. Photograph by J. M. Adovasio.

FIGURE 14.4 (above) Map of the excavation at Pedra Furada. The mouth of the rockshelter is about 70 meters wide and its interior floor is about 12–18 meters deep under the rock overhang. The excavation extends to a depth of at least 5 meters. Courtesy of Niède Guidon.

FIGURE 14.5 (above) Example of the prehistoric rock art that first drew Niède Guidon and her research team to the Pedra Furada region in the 1960s. Painted in ochre on the rockshelter's walls and depicting various scenes involving animals and humans in leisure, hunting, and various ceremonial activities. These paintings date to as early as 12,700–11,610 yr BP and continued to be created over the following 6000 years. They have been attributed to a cultural phenomenon known as the "*Nordeste* (Northeast) Tradition" of northeastern Brazil. Photograph courtesy of Niède Guidon.

FIGURE 14.6 (right) The eleven archaeological strata at Pedra Furada are mostly composed of sand with varying amounts of gravel, pebbles, and cobbles. The cultural levels among them range in age from 12,700–11,610 yr BP, which is compatible with the known archaeological record of northeastern Brazil, to an apparently anomalous age of 50,000–42,340 yr BP to older than 59,000 yr BP. No criteria have been formalized to differentiate cultural artifacts from amidst the coarse matrix of broken quartzite seen here, which has led archaeologists to question the validity of the site's older artifacts, dismissing them as naturally occurring geofacts. Photograph courtesy of Niède Guidon.

FIGURE 14.7 (above) Pedra Furada field crew excavating the site in 1985. The excavators have distinguished a total of eleven discrete strata whose purported cultural materials range in age from older than 59,000 yr BP to as late as 6000 yr BP. Photograph courtesy of Niède Guidon.

FIGURE 14.8 (right) An unequivocal bifacial stone tool recovered from Pedra Furada's more-recent Serra Talhada phase deposits, which date from as early as 12,700–11,610 yr BP to about 6000 yr BP. Photograph courtesy of Niède Guidon.

FIGURE 14.9 (above) Putative quartzite pebble-tool chopper from the oldest of the Pedra Furada phase (PF1) levels at the site, which range in age from about 50,000–42,340 yr BP to older than 59,000 yr BP. Photograph courtesy of Niède Guidon.

FIGURE 14.10 (below) The area in the blue box in Figure 14.9 at 100x showing linear use ware the edge in perpendicular groups (yellow arrows). Note the areas of abrasive polish (red arrows). Photograph courtesy of Niède Guidon.

which include so-called "blunt points," pebble-tool choppers, saw-like denticulate tools, burins, notched pieces, retouched flakes, and bifacial flake tools. [Figures 14.9 and 14.10] The Pedra Furada I assemblage also includes smaller numbers of items described by the excavators as pebble-hammers, flaked pebbles, debitage flakes, and fragments of human-painted rock. [Figure 14.11] A large number of large, circular "well-made" hearths are also reported for the Pedra Furada, which were defined by the presence of thermally altered rock, charcoal, and ash. There appear to have been no living surfaces defined in this phase, however, nor were any organic remains other than charcoal recovered.

Initially thought to be composed of two stages, in 2003, the Pedra Furada research team has revised its interpretation to designate three discrete stages in this earlier phase. The earliest of these, PF3, is bracketed above and below by radiocarbon ages of 17,940–16,800 yr BP and 26,490–24,630 yr BP, respectively. The youngest age for PF2 is 30,170–28,590 yr BP and the oldest is 37,840–33,420 yr BP. Finally, lowermost PF1 ranges in age from 50,000–42,340 yr BP to older than 59,000 yr BP (the latter age exceeds the range of radiocarbon dating). While some of the older ages in Pedra Furada phase are exceptionally broad due to very high degree of error calculated for the sample measurements, Guidon and her colleagues point out that all are in perfect depositional and chronological order, and that no evidence of stratigraphic disturbance can be detected.

Perhaps the most distinctive artifact forms in the Pedra Furada phase assemblage are the so-called blunt points, which are putatively fashioned by striking between two and four small flakes from larger flakes or pebbles. [Figure 14.12] These items are thought by Guidon to have been used as awls for piercing leather or wood and, because their sides have been retouched, they may have been used as scrapers. Additional distinctive forms are variously described as centripetal cores (a term typically associated with Middle Paleolithic technology dating 300,000–40,000 yr BP) and bifacial trimmed pebbles. Unlike the mixture of lithic raw materials from the Talhada phase of the site's upper layers, these objects are exclusively made from the same variety of quartzite that appears in the cliff face some 100 meters above the site. The materials were judged to be artifacts (rather than naturally fractured rocks) on the basis of their flaking patterns and their position in the deposit.

During an archaeological conference and site visit in 1993, David Meltzer, coauthor Adovasio, and Tom Dillehay raised a series of highly detailed questions and concerns about the site's stratigraphy and formation, the defined cultural horizons, the archaeological integrity of the identified cultural features (especially the hearths), and the validity of the artifacts. Of these, perhaps the most enduring and widely acknowledged concern among archaeologists is whether the artifacts from the Pedra Furada phase—or, for that matter, even the quartzite artifacts from the accepted Serra Talhada phase—are indeed the products of human manufacture. The three visiting archaeologists made the case that the quartzite "artifacts" had most likely eroded from the cliff face above and been transported through the chutes on either side of the rockshelter (see above) to be fractured upon landing. They further reasoned that this natural process would account for the presence of such items throughout the entire deposit, from Late Pleistocene through Holocene times. It is indeed curious that a lithic technology would remain unchanged for 60,000 years and co-occur with an otherwise radically different adaptation such as the Serra Talhada phase.

For many South American archaeologists, the issue of the Pedra Furada phase artifacts—insofar as there ever was one—has long been settled in the affirmative. It remains uncertain whether the validity of the Pedra Furada phase will ever be resolved to the satisfaction of the North American archaeological community. After all, the Pedra Furada phase artifacts may well indeed be artifacts. Compelling cases have been made for the archaeological validity of pebble tool artifacts in South America at Huaca Prieta on the north coast of Peru and Monte Verde on the south coast of Chile (see pages 216-227), and with a far lesser degree of certainty even elsewhere in Piauí. But those localities have not borne the profound site formational and stratigraphic ambiguities that have dogged Pedra Furada, whose geological setting has been described by Meltzer as a "geofact factory."

16873

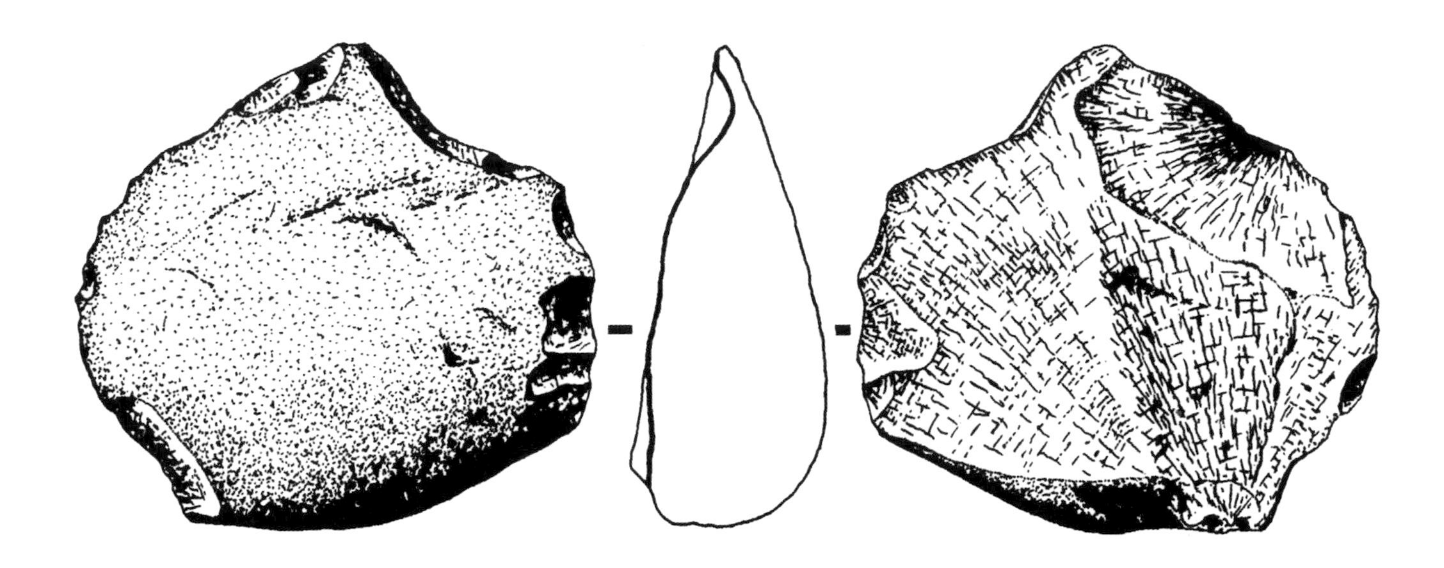

FIGURE 14.11 (left) Putative pebble-tool chopper from the youngest of the Pedra Furada phase (PF3) levels at the site, whose ages range between 17,940–16,800 yr BP and 26,490–24,630 yr BP. Critics of Pedra Furada's claim of deep antiquity have noted quartzite specimens like this one found in the youngest strata are virtually indistinguishable from to specimens from the oldest strata. Photograph courtesy of Niède Guidon.

FIGURE 14.12 (above) A so-called "blunt" point manufactured on a primary quartz flake, recovered from the oldest of the Pedra Furada phase (PF1) levels at the site. Such artifacts, which are also referred to as "beaks" (or "*becs*" in French) by some archaeologists, are broadly similar to what others call burins or gravers. These forms display a pointed, chisel-like edge for engraving/carving wood, bone, or other organic materials. Illustration courtesy of Fabio Parenti.

Fieldwork at Gault, Buttermilk Creek, Texas.

LEGITIMATE PRE-CLOVIS SITES

The continent-wide appearance of Clovis archaeological sites and the conspicuous absence of an earlier bona fide cultural entity of similarly wide distribution and high visibility eventually led to agreement among archaeologists that Clovis was the earliest sole progenitor of all other lithic tool kits—and, by extension, all cultural complexes—that occurred throughout the later archaeological record of the entire western hemisphere. Known by late twentieth century archaeologists as "Clovis First," this consensus had become entrenched as a theory which held that small bands of Paleoindians crossed unglaciated Beringia (see pages 24-25) several hundred years before 13,300 yr BP and proceeded south through a narrow corridor between the continental glaciers and reached the hemisphere's southern extremes in a matter of about 500 years, ultimately leading to the establishment of North America's Clovis cultural horizon and the mass extinction of numerous forms of Late Pleistocene wildlife. But the same rigorous archaeological site evaluation criteria that consigned the voluminous disputed pre-Clovis age sites to scientific obscurity also proved to be instrumental in the verification of once-inconceivable pre-13,300 yr BP archaeological localities. The following entries describe the most-prominent archaeological sites to definitively break through the Clovis First barrier and thereby demonstrate that Paleoindian peoples had long populated the entire hemisphere thousands of years before the Clovis cultural efflorescence.

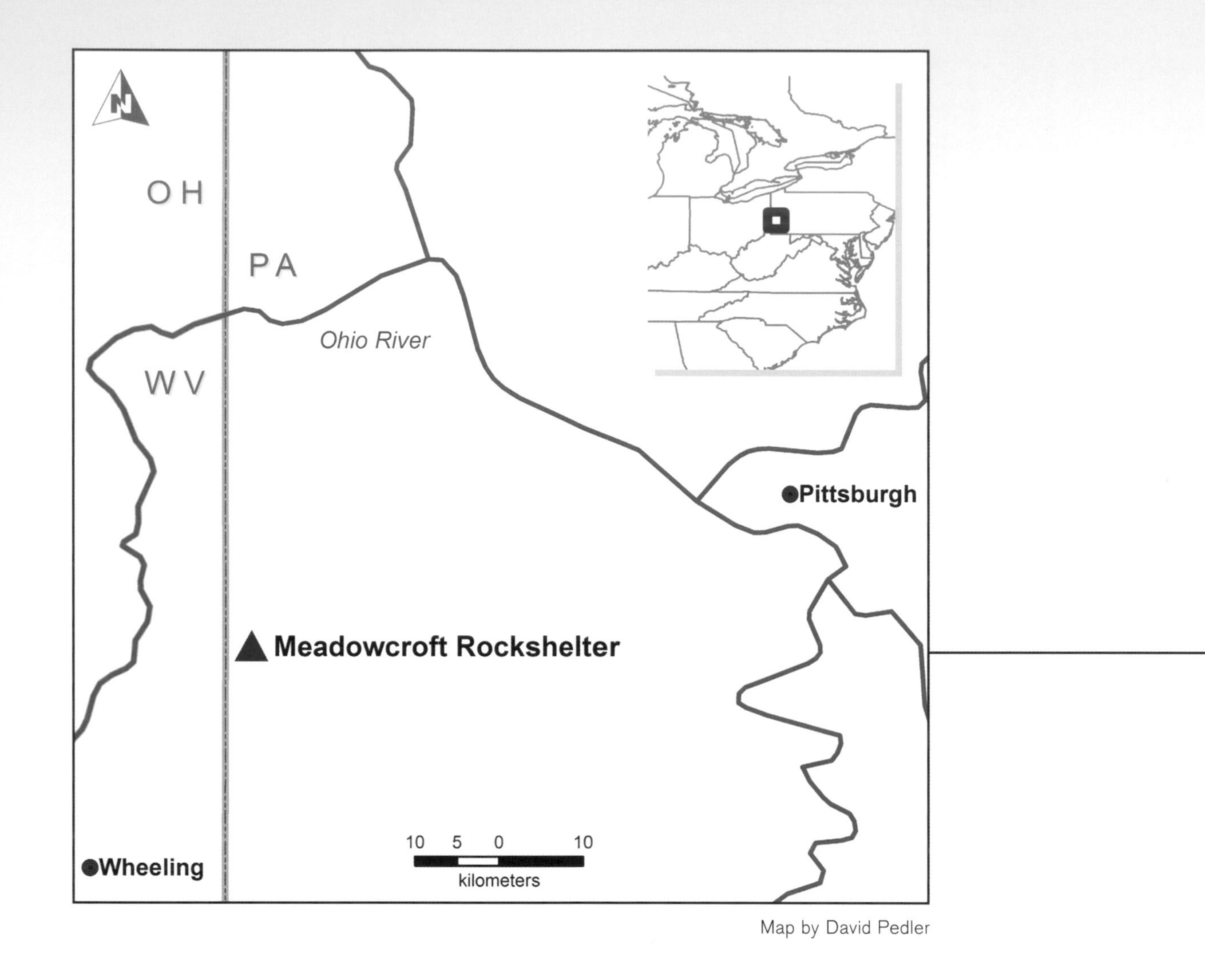

Map by David Pedler

Meadowcroft Rockshelter holds the distinction of evidencing the longest occupational sequence of humans in the Americas. By a very conservative estimate, the site was occupied by 15,000 yr BP, which is 2,000 years before Clovis points first appeared in North America.

15 MEADOWCROFT ROCKSHELTER

LOCATION WASHINGTON COUNTY, PENNSYLVANIA, UNITED STATES

Coordinates 40°17'10.81"N, 80°29'27.56"W.

Elevation 260 meters above mean sea level.

Discovery Albert Miller in 1955.

Excavations at Meadowcroft Rockshelter, nested in the rolling hills of southwestern Pennsylvania just a few kilometers from the border of West Virginia's western panhandle, have documented the longest continuous archaeological sequence in eastern North America and one of the longest in the New World. Dating from the waning millennia of the last glacial period to the American Revolutionary War, Meadowcroft's culture-bearing strata contain materials from all of the region's known prehistoric periods and sub-periods, from Paleoindian through Late Woodland. The site's oldest cultural levels produced about 700 stone artifacts and thirteen radiocarbon dates that posed the first credible challenge to the then widely accepted and very well entrenched Clovis-First hypothesis.

Meadowcroft is located about 50 kilometers southwest of Pittsburgh in western Washington County on the north bank of Cross Creek, a small tributary of the Ohio River, about 11 kilometers east of their confluence. [Figure 15.1] The Cross Creek valley, like many of those in this very hilly region, is steeply sloped and narrow, and the surrounding hilltops are relatively narrow and sinuous. The valley, on the other hand,

FIGURE 15.1 A view of Cross Creek and the rockshelter. Cross Creek was an important route for Paleoindians traveling west to the Ohio River and east into the interior. Flint from Ohio and West Virginia, as well as jasper from Pennsylvania and marine shells from the Atlantic recovered at Meadowcroft Rockshelter suggest the people inhabiting the area were mobile and involved in long distance trade. Situated 11 kilometers upstream from the Ohio River, the rockshelter would have been a convenient temporary campsite. Courtesy of J.M. Adovasio.

maintains a relatively consistent elevation throughout its course and, unlike many other streams in this portion of the Ohio River watershed, its low relief doubtlessly provided one of the region's most efficient routes for east-west foot travel in prehistoric times. If not for the presence of modern roads, this would still be the case today.

The site is considered a rockshelter (or, in geologic terms, a sandstone re-entrant) owing to the presence of a characteristic overhang of erosion-resistant rock that, long before its human occupation, had been underlain by less-resistant, presently eroded rock. Erosion of this lower, less-resistant rock through time created a void space or "shelter" from the elements in its place which proved ideal for human use. Meadowcroft's position on the north side of the Cross Creek Valley about 15 meters above the stream ensured that the site would be high and dry, and its southern exposure ensured that it would have shade in the summer and a high potential for absorbing solar radiation, and hence warmth, in the colder parts of the spring and fall. The prevailing wind, which blows from west to east across the mouth of the rockshelter, would have efficiently ventilated smoke from any fires in the interior. The area of the site that is protected by the rock overhang covers about 65 square meters. [Figure 15.2]

Meadowcroft Rockshelter was originally discovered in 1955 by the landowner, the late Albert Miller, whose family had owned the property continuously since 1795. Miller resolved to protect the site from amateur collectors and looters by keeping quiet about his discovery until he could make arrangements for its excavation by professional archaeologists. In the early 1970s, Miller and the late Phil Jack, an historian at California State College in California, Pennsylvania, brought the site to the attention of co-author Adovasio, who at the time was seeking a site for archaeological field training in the early years of his teaching career at the University of Pittsburgh. Meadowcroft ultimately became the focus of an intensive long-term multidisciplinary project that was complemented by an archaeological survey of the entire 14,000 hectare Cross Creek watershed. That survey identified an additional 236 prehistoric sites, of which twenty-two were tested and two were intensively excavated. [Figure 15.3]

The project's most intensive phase began in the summer of 1973 and concluded at the end of the 1978 field season. Sporadic but often intensive field work was conducted throughout the 1980s, and a major re-excavation of the site's post-Paleoindian Holocene deposits occurred from 1994 to 1995. [Figure 15.4] In early 2008, the site became accessible to the public with the construction of an enclosure and observation deck. The site is now owned by the Senator John Heinz History Center based in Pittsburgh, which also maintains a living museum of nineteenth century rural life and a replicated sixteenth century Indian village on the hilltop above the site.

The ten culture-bearing strata identified at Meadowcroft, which were excavated to a maximum depth of almost 5 meters, yielded over 20,000 artifacts (predominantly flaked stone), well over 2 million animal and plant remains, more than 150 fire pits, sixteen specialized activity areas, almost thirty refuse or storage pits, and about thirty-three extensive, burned areas known as fire floors. [Figure 15.5] These burned surfaces resulted

FIGURE 15.2 (on following page) Not only were there humans in the New World before Clovis, but evidence from sites like Meadowcroft Rockshelter and Monte Verde clearly demonstrate their lifeways were very different from those postulated for Clovis. Instead of solely being highly focused, mobile big game hunters, the earliest Americans were also generalized foragers subsisting on small game and edible wild plants. Tableau by Anatomical Origins.

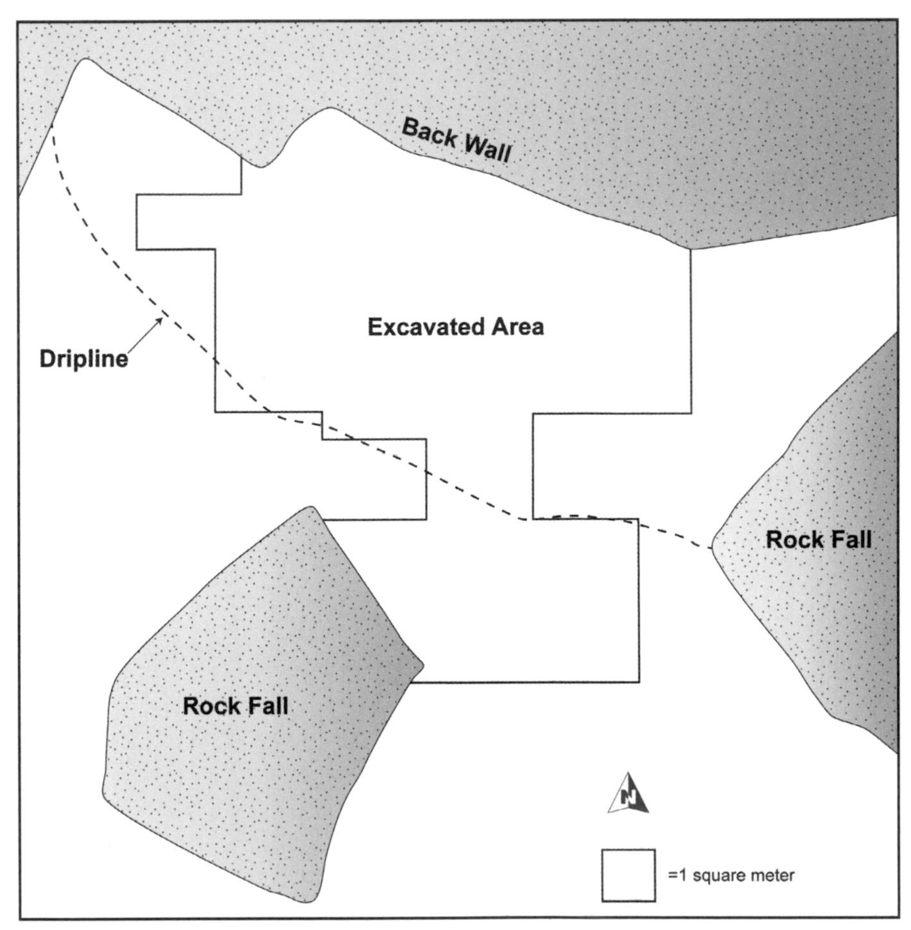
Back Wall
Excavated Area
Dripline
Rock Fall
Rock Fall
N
=1 square meter

FIGURE 15.3 (left, top) The excavation protocols employed at Meadowcroft Rockshelter are considered state-of-the-art and the site is widely regarded as one of the most carefully excavated localities in North America. The recovery of artifacts used for food acquisition, transport, and/or processing suggest that throughout its history Meadowcroft served as a temporary camp for hunting, collecting, and processing food. Meadowcroft Rockshelter holds the distinction of demonstrating the longest continuous human occupational sequence in the Americas. Humans were present at the site by at least as early as 15,000 yr BP and as late as AD 1775. Courtesy of J.M. Adovasio.

FIGURE 15.4 (left, bottom) Plan map of the excavation at Meadowcroft Rockshelter. During the 600+ working days of the 1973–1978, 1983, 1985, 1987, 1993–1994, and 2007 field seasons, approximately 60 square meters of surface area inside the drip-line and 46 square meters outside the dripline (the point at which surface water draining from the overlying rock formation makes contact with the ground surface) were excavated. Map by David Pedler.

from the accidental or intentional incineration of flammable trash by the site's aboriginal visitors. [Figure 15.6]

The oldest recovered material ascribes to the Miller complex, which appears to represent a—and perhaps *the*—pioneering human population that entered the upper Ohio Valley at the close of the Pleistocene. [Figure 15.7] Almost exclusively composed of stone artifacts, the raw materials identified in Miller complex assemblage appear to have been obtained from local sources such as the Monongahela chert outcrops in the Cross Creek watershed and, either through exchange or direct exploitation, from sources as far away as Flint Ridge in central Ohio (160 kilometers to the west), the Kanawha Valley of south-central West Virginia (240 kilometers to the southwest), and Pennsylvania jasper quarries well (240–320 kilometers) to the east.

Although far less numerous than those from succeeding levels, the Miller artifacts nonetheless provide a distinct picture of a lithic technology whose blades and cores closely resemble artifact assemblages from Upper Paleolithic Eurasian sites while also being quite dissimilar from Clovis artifacts. The debitage flakes that dominate the assemblage are the byproduct of middle- to late-stage core reduction (for the production of expedient tools) as well as the late-stage manufacture and maintenance of finished bifacial implements. As would be expected from the debitage produced from the first of these byproduct classes, the stone tool assemblage includes blade flakes that could only have been struck from small, prepared cores. Unfortunately, no complete cores were recovered from Meadowcroft, but artifacts recovered from the Krajacic site some 16 kilometers to the southeast—also a Miller complex locality—include a number of the same distinctive blade flakes along with small, cylindrical polyhedral cores that, considered together, precisely follow the technological strategy that has been posited for the blades from Meadowcroft. [Figure 15.8]

But by far the most distinctive artifact in the site's Miller complex assemblage is the Miller Lanceolate projectile point, an unfluted bifacial implement which was found *in situ* on the uppermost living floor of lower Stratum IIa, the site's lowest artifact-bearing stratum. [Figures 15.9 and 15.10] This living surface is bracketed above and below by radiocarbon assays of 15,160–11,250 yr BP and 17,580–13,060 yr BP, respectively, which obviously makes it older than Clovis—or, at the very least, contemporaneous with the earliest Clovis sites. While its date is conclusive, however, this artifact's original form is unfortunately somewhat more difficult to determine, as repeated re-sharpening has considerably changed the shape of point from its original configuration. The project's lithic analysts nonetheless agree that in its original form the Miller Lanceolate was almost certainly longer and that it may have attained its maximum width toward its distal (or pointed) end.

One fragmentary Miller specimen, again from the nearby Miller complex contemporary Krajacic site, closely corresponds to what analysts believe the Miller Lanceolate might have looked like in its original form. But recently, a complete Miller point that had not been re-sharpened in antiquity was recovered from the Mungai Farm site on the upper reaches of Cross Creek. This artifact conforms almost exactly to the form proposed for unmodified Miller points. The Krajacic

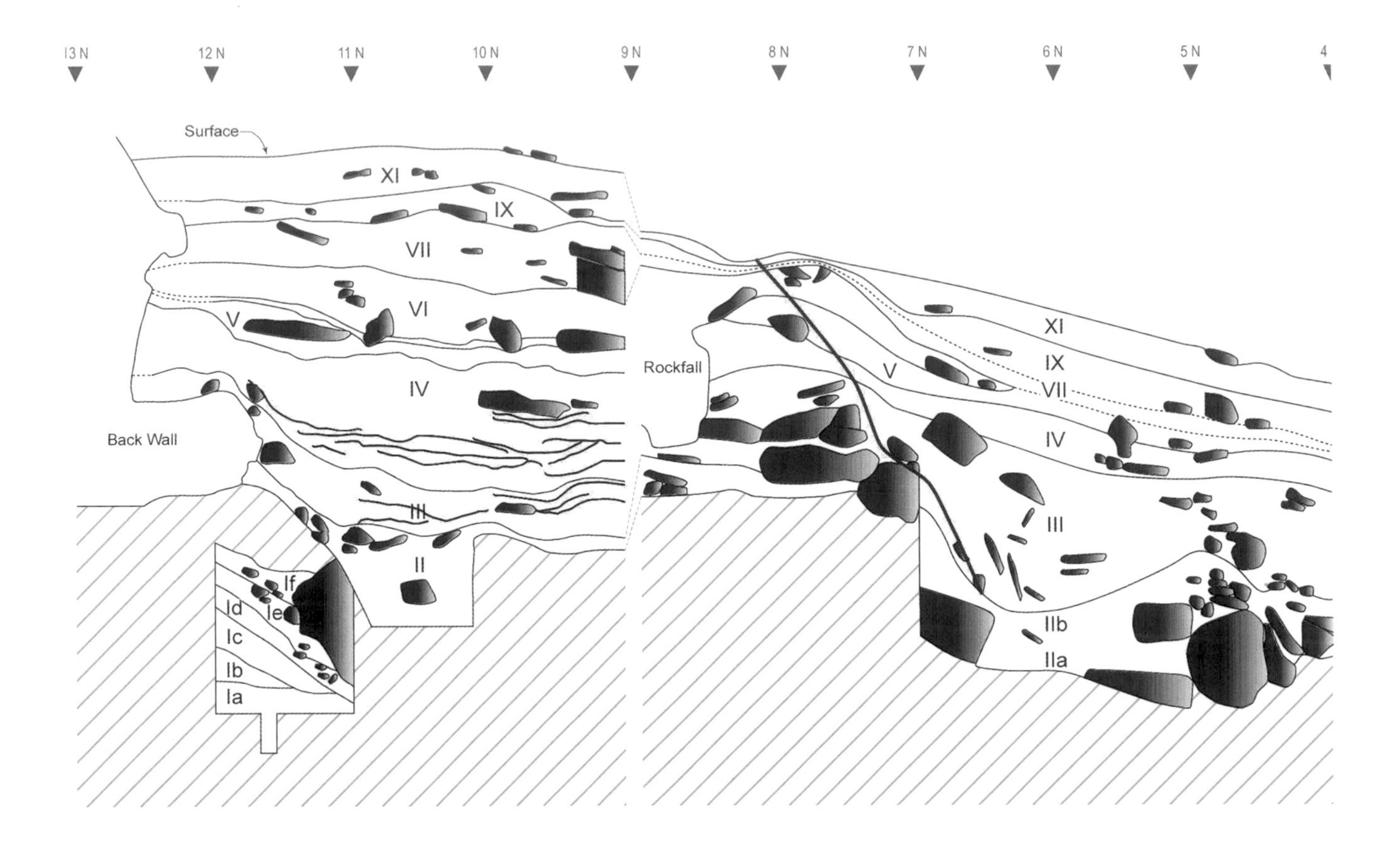

and Mungai Farm site specimens, however, like all other examples of the Miller point recovered from elsewhere in the Cross Creek watershed and unlike the Meadowcroft specimen, were not recovered from a directly dated stratigraphic context. Without such dates, and despite the striking and obvious similarities between all of the recovered Miller points, their precise relationships in time can only be inferred and not strictly proven.

A very large and diverse array of plant and animal remains were recovered from all levels at Meadowcroft Rockshelter, including 115,166 identifiable bones and bone fragments that represent more than 140 species. A similarly diverse array of bone and plant-based artifacts were also recovered, including items fashioned from wood, mammal and bird bone, antler, and shell. [Figure 15.11] Unlike the overlying Archaic and Woodland levels, however, the plant and animal remains from the Miller complex level are very few in number, consisting of only 278 bone fragments and about 12 grams of plant matter.

The eleven identifiable bones indicate the presence of white-tailed deer (*Odocoileus virginianus*), southern flying squirrel (*Glaucomys volans*) and passenger pigeon (*Ectopistes migratorius*), while the plant remains indicate a mixed conifer hardwood forest environment dominated by oak, hickory, pine, and perhaps walnut, with hackberry occurring as an understory species. [Figure 15.12] This plant and animal species composition, which was more or less maintained throughout Meadowcroft's entire occupational sequence, indicates that the Cross Creek Valley underwent only moderate, low-order environmental change in its transition from the Late Pleistocene to the more stable Holocene environmental conditions that are in place today. This pattern

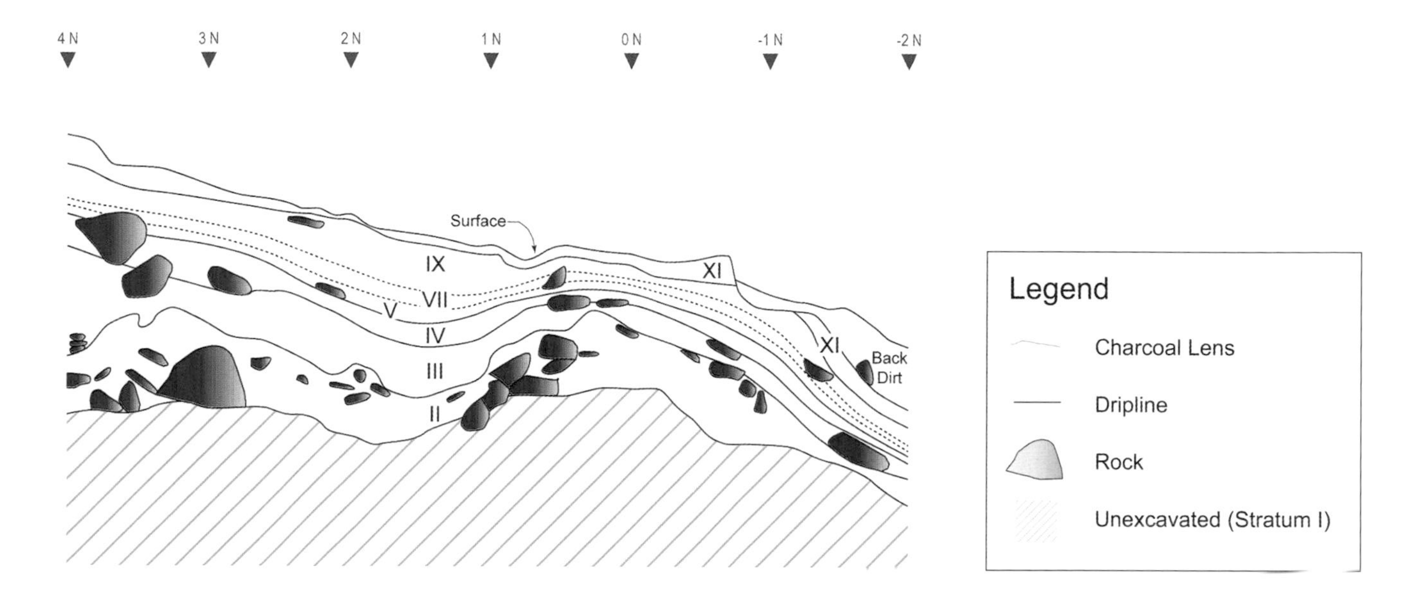

FIGURE 15.5 North-south composite stratigraphic profile of Meadowcroft Rockshelter along the 17.5 meter, 20 meter, and 23 meter west grid lines, facing east. The red line shows the position of the drip line through time. Profile by David Pedler.

is consistent with geologists' view that Late Pleistocene environmental change was rapid, uneven, and patchy south of the glacial front, and that a great variety of environments and food sources would have presented themselves to the continent's earliest colonizers.

The later Miller complex materials at Meadowcroft are underlain by even older deposits, which notably include hearths and burning episodes dated by five radiocarbon assays ranging in age from 21,980–17,250 yr BP to 18,360–13,220 yr BP. [Figure 15.13] These features are in turn underlain by materials of uncertain, but possible, cultural origin that range in age from 26,210–24,140 yr BP to 25,040–21,090 yr BP, and include a piece of cut bark dated (with a very large degree of error of ± 2,400 years in the uncalibrated radiocarbon date) to 28,510–17,960 yr BP that appears to represent a basket fragment. [Figure 15.14] However, in the absence of indisputable artifacts and the rather broad time range exhibited by the individual calibrated ages, the definition of a "pre-Miller complex" cultural horizon remains very tentative.

As many as 15,000 years of human history (and perhaps many more) are preserved in Medowcroft's deposits, beginning with visits by exceptionally adaptive colonizers spreading across a varied landscape and ending in historic times with visits by yet another wave of colonizers, with millennia of visits by successive peoples in between. With the establishment of the Meadowcroft Rockshelter Museum, this National Historic Landmark continues to be visited to the present day. [Figure 15.15]

FIGURE 15.6 One of the fire pits identified at Meadowcroft Rockshelter. Courtesy of J.M. Adovasio.

FIGURE 15.7 Stratigraphic profile of the east face at Meadowcroft Rockshelter showing the stacked cultural layers at the site. The ten culture-bearing strata identified at Meadowcroft yielded over 20,000 artifacts (predominantly flaked stone), well over 2 million animal and plant remains, more than 150 fire pits, about thirty-three extensive burned areas known as "fire floors," sixteen specialized activity areas, and almost thirty refuse or storage pits. The white circular tags indicate the archaeological strata and arbitrary excavation levels or microstrata within them. Courtesy of J.M. Adovasio.

FIGURE 15.8 (above) The Miller Lanceolate projectile point shown as found *in situ* at Meadowcroft Rockshelter, on the uppermost living floor of lower Stratum IIa. This living surface is bracketed above and below by radiocarbon assays of 15,160–11,250 yr BP and 17,580–13,060 yr BP, respectively. This unfluted biface is the only Miller Lanceolate point thus far recovered from a directly dated stratigraphic context. Courtesy of J.M. Adovasio.

FIGURE 15.9 (below) Three Miller complex blades struck from multidirectional polyhedral cores recovered from the Krajacic site, located on a hilltop 16 kilometers southeast of Meadowcroft Rockshelter. Miller complex-like artifacts have also been reported at the nearby Mungai Farm site. Miller complex lithic technology bears varying degrees of similarity to artifacts from sites as far ranging as Cactus Hill in Virginia, on Buttermilk Creek in Texas, and the Delmarva Peninsula in Maryland. Courtesy of J.M. Adovasio.

FIGURE 15.10 (above) The Miller Lanceolate projectile point was produced from fine-grained local chert. While the exact geologic provenance for the chert is uncertain, it is found locally and is opaque and light gray, with yellow and purple striations on both faces. The point as has a distinctive base with straight margins that articulate with the straight basal attribute margin at angles of 97 degrees. The point was produced via the removal of parallel, overlapping biface thinning flakes, most of which traverse the centerline to create a lenticular cross section. Courtesy of J.M. Adovasio.

FIGURE 15.11 (right, top) Bone artifacts from the site's younger Archaic and Woodland period levels. Left to right: bone button fragment, bird bone bead stock, bone fish hook fragment, bone trigger from a snare trap. These bone artifacts from Meadowcroft were those of animals used for food. Courtesy of J.M. Adovasio.

FIGURE 15.12 (right, bottom) *In situ* broken deer antler punch and to its right a snapped blade in the Miller complex deposits at Meadowcroft. They were recovered from a buried surface directly adjacent to a fire pit. Courtesy of J.M. Adovasio.

FIGURE 15.13 (below) A wooden bipoint artifact from the very base of the Miller complex deposits in lower Stratum IIa. This level produced the Miller point as is dated between 15,160–11,250 yr BP and 17,580–13,060 yr BP. The function of this item is not clear. Courtesy of J.M. Adovasio.

F346
F46
F343

FIGURE 15.14 (above) Carbonized simple-plated wall and rim basket fragments recovered from the Late Archaic levels at Meadowcroft. These items are structurally identical to the Miller complex basket fragments from middle and lower Stratum IIa. Courtesy of J.M. Adovasio.

FIGURE 15.15 (right) Present-day exterior view of Meadowcroft Rockshelter, facing northeast from the north bank of Cross Creek. In 2008, an observation deck and attachment were built in the rockshelter. Prior to this $4 million renovation, there was only limited public access to the site. The new enclosure protects the site and permits visitors to view the entire excavation without interfering with the ongoing scientific studies. The rockshelter was named a National Historic Landmark in 2005. Courtesy of J.M. Adovasio.

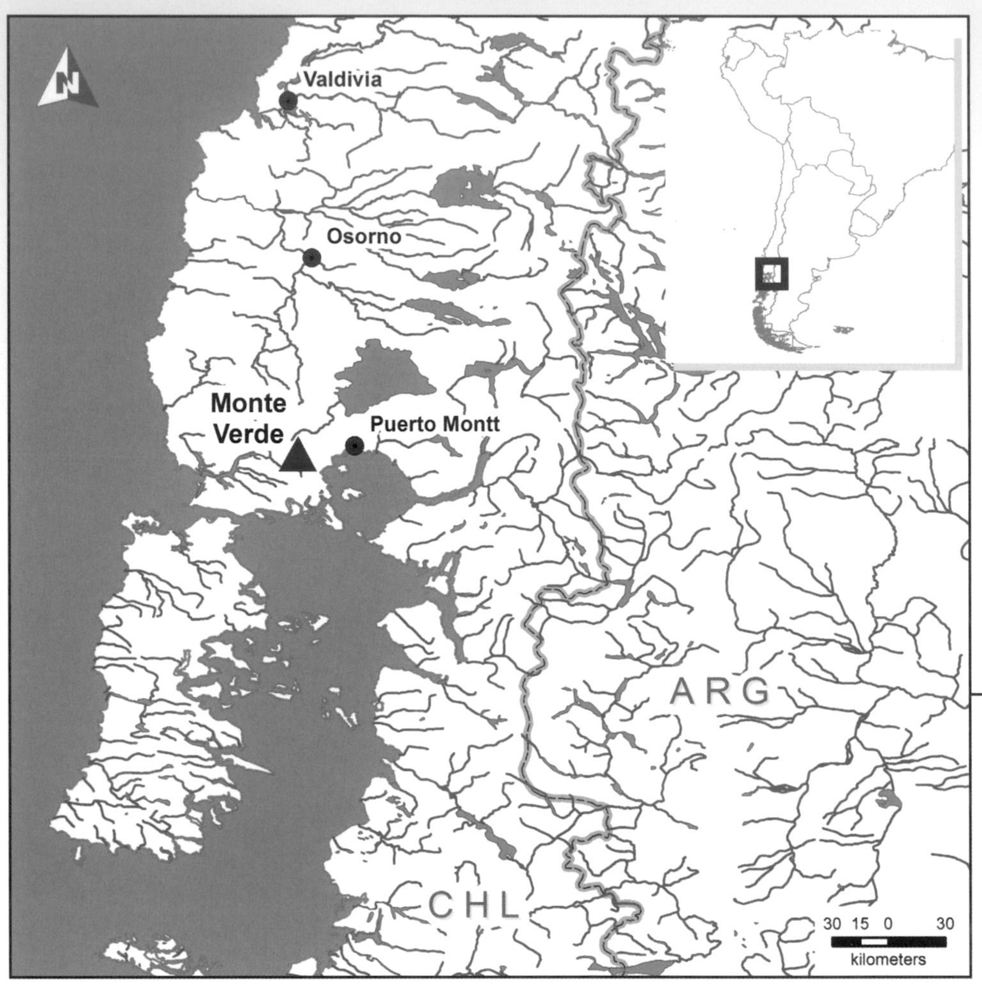

Map by David Pedler

Archaeologists at Monte Verde, located 800 kilometers south of Santiago, Chile, and over 14,000 kilometers south of the Bering Strait, have recovered evidence of human occupation dating to at least 14,500 yr BP and perhaps as far back as 38,800 yr BP. Cultural materials, some of which have never before been seen in early American sites, include remnants of hide-covered huts; a scrap of animal meat that DNA analyses identified as gomphothere; digging sticks, plant material, and more than 700 bone tools; and child's footprint.

16 MONTE VERDE

LOCATION LLANQUIHUE PROVINCE, LOS LAGOS REGION, CHILE

Coordinates 41°31'17.00"S, 73°12'16.00"W.

Elevation 50 meters above mean sea level.

Discovery Gerardo Barria family in 1976.

Monte Verde is one of the most ancient archaeological sites in the Western Hemisphere and the most distant pre-Clovis locality from the presumed point of entry for humans in the New World. The site's deepest cultural level has been very tentatively dated to as early as 38,800 yr BP, and the overlying, far more extensive site occupation has been confidently dated to around 14,500 yr BP. The earlier end of this date range is extraordinarily ancient for the Americas and more in line with the earliest sites that have been discovered in northwestern Beringia, some 17,000 kilometers to the north. The subject of a decades-long multidisciplinary research investigation, Monte Verde is currently recognized by archaeologists as the most definitive, unequivocal evidence of a human presence in the New World at least 1,000 years before the appearance of Clovis.

Monte Verde is located on Chinchihuapi Creek, a tributary of the Maullín River, about 20 kilometers from Puerto Montt, a port city on the Gulf of Ancud's Reloncaví Sound, in the Los Lagos Region of southern Chile. Situated in the low, rolling hills of Chile's Central Valley between the Coastal Mountain Range and the Andes Mountains, Monte Verde lies at the southern limit of continental Chile. [Figure 16.1] In prehistory, this

FIGURE 16.1 An aerial view of Monte Verde with the Chinchiuapi Creek bisecting the site. Analysis of organic items found at Monte Verde suggests the environment of 14,500 yr BP was broadly similar to that of the Holocene before modern development. Courtesy of Tom Dillehay.

area would have been the southern limit of any possible human terrestrial migration along Chile's inner coast, as to the south lies a complex system of channels, islands, and fjords that could only have been negotiated by seafaring populations. Four centuries of human resource exploitation have radically altered the region's native habitat, but at the time of its ancient human occupation Monte Verde would have been surrounded by a varied landscape dotted with a mosaic of patchy deciduous and coniferous forest, vernal pools, wetlands, and rainforest interspersed with openings of rushes, reeds, and grasses. Abundant and unusually diverse natural resources such as plants for food, medicine, residential hut construction, and fuel for fires, as well as a broad variety of both large and small animal resources appear to have been readily available to the site's inhabitants.

Described by principal investigator Tom Dillehay as an "unusual open-air, wetland residential site," Monte Verde was found to contain well-preserved architectural remains, plant resources, the bones of extinct animal species, and artifacts composed of wood and stone, all overlain by a layer of peat and a meter-thick zone of sand, gravel, clay, and soil. The site is divided into four distinct analytical zones labeled Areas A through D that stretch for a total distance of about 360 meters along the ancient banks and terraces of Chinchihuapi Creek. The largest and densest archaeological deposits occur in Areas A and D on the creek's north side. [Figure 16.2] Over the course of several field seasons, Dillehay and his research team excavated about 450 square meters of the site.

The site was first brought to the attention of anthropologists at the Universidad Austral de Chile, Valdivia, in 1975, when presumed cow bones recovered by the family of landowner Gerardo Barria during a lumbering operation along the creek were presented for scientific examination. A field inspection of the site the following year recovered over forty large bone specimens that were ultimately identified as the partial remains of at least four or five individual proboscideans originally identified

FIGURE 16.2 Map of the Monte Verde and Chinchihuapi sites showing the different site sectors, block excavation, test pits and cores carried out during the 2013 excavation season.

as mastodons but now known to be gomphotheres. Initially, Dillehay was particularly interested in the very high proportion of rib fragments among the proboscidean assemblage, which suggested to him that Monte Verde might have served as a kill site. After a field trip to the site in late 1976 and relatively brief test excavations in 1977, biannual field work began in earnest in 1979 and concluded in 1987. The principal phase of work for the project culminated in the watershed publication of Dillehay's *Monte Verde: A Late Pleistocene Settlement in Chile*, a report of investigations whose two volumes appeared in 1989 and 1997. More recently, Dillehay and his colleagues returned to the site in 2013 to resume excavation of the Monte Verde I site component and to test for the presence of the hitherto only preliminarily documented archaeological remains upstream from Monte Verde at the Chinchihuapi I and II sites.

Two cultural episodes were identified in the Monte Verde site deposits. [Figure 16.3] The older and deeper of the two, identified in Area C and designated MV-I, produced a radiocarbon age of 38,800–36,260 yr BP and is considered by the excavators to be a "possible" habitation surface based on the presence of three clay-lined pits within a 50 square meter area (out of a maximum estimated area of 1,200 square meters for the deposit), twenty-six associated lithic artifacts, and eight scatters of charcoal. [Figure 16.4] The lithic artifacts are clearly products of human manufacture and include one core, three percussion flakes, two hammerstones, and twenty sharp-edged pebble flakes, all of which appear to

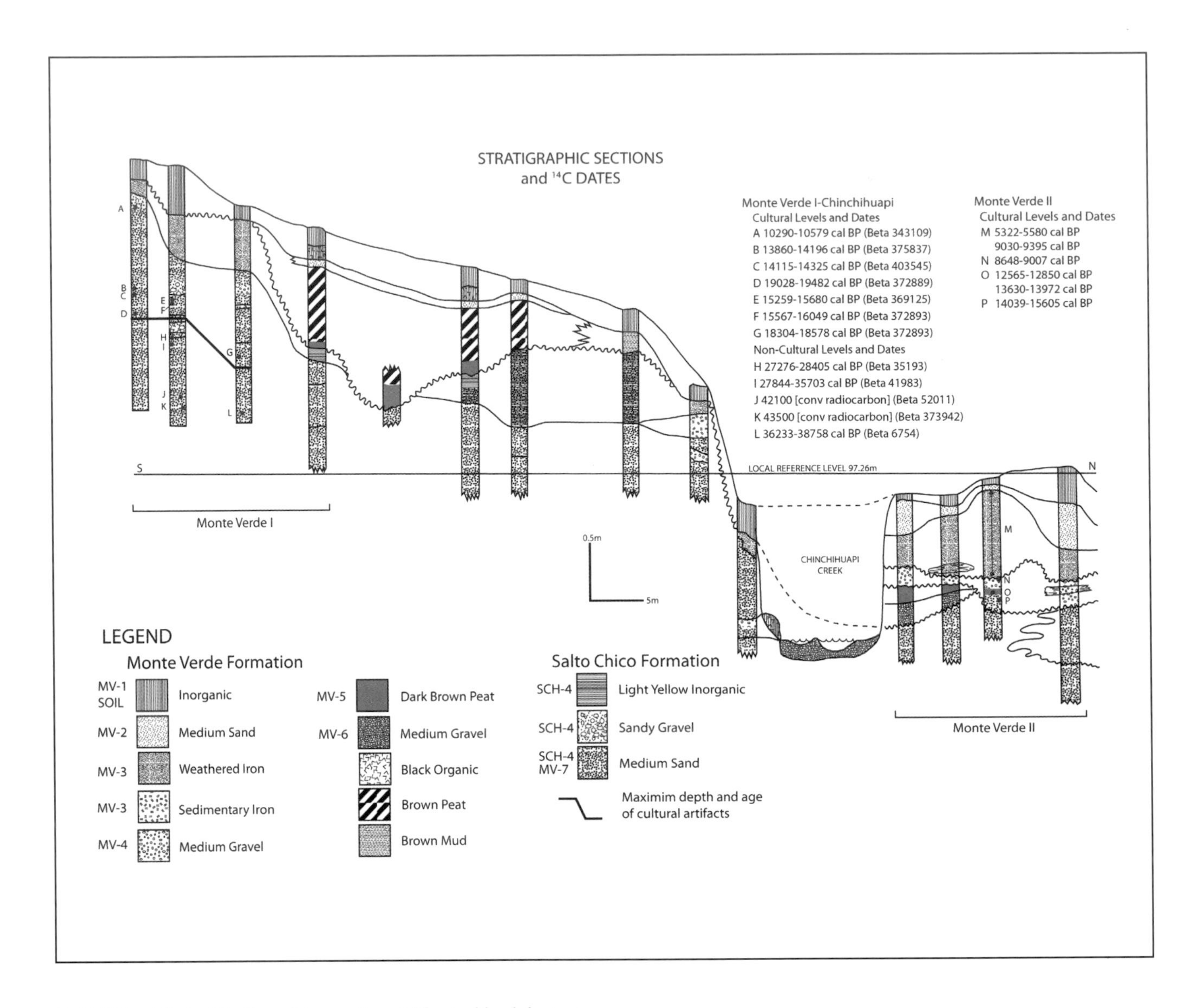

FIGURE 16.3 Stratigraphic profile of Monte Verde's two cultural levels: MV-II, dated to about 14,500 yr BP and MV-I, dated to 38,800 yr BP. Monte Verde is an open site with remarkable organic preservation of bone, wood, twigs, seeds, leaves, and animal tissue in direct association with evidence of human occupation including stone and bone tools and the remains of domestic (house) structures. Selected calibrated radiocarbon ages shown are presented alongside their conventional radiocarbon ages in Radiocarbon Dating (page 321-329). For ease of reading and convenience, these ages have been rounded to the nearest decade in the text. Courtesy of Tom Dillehay.

FIGURE 16.4 Charcoal scatters in a hearth-like basin with a flake core made of fine grained basalt from the MV-I zone, dated to 38,800–36,260 yr BP. Courtesy of Tom Dillehay.

have been made from locally available glacial outwash deposits. The relationship between MV-I and the much younger MV-II, which spans a very long period about 20,000 years, is likely to remain unclear. Dillehay has essentially dismissed the possibility that a culture-historical connection exists between the two. He also believes that while there is no compelling evidence to reject outright the validity of the earlier and admittedly ephemeral MV-I deposit, its status will remain ambiguous in the absence of additional evidence from Monte Verde and other as yet undiscovered sites in the region. [Figure 16.5]

The younger and massively more substantial MV-II cultural episode, which ranges in age from 15,120–14,030 yr BP to about 14,560–13,400 yr BP and occupies about 3,000 square meters throughout site Areas A and D, lies about 80 meters northwest of MV-I and sits about 1.5 meters higher in the site's stratigraphic sequence. The site's investigators discovered and excavated a total of 167 cultural features in undisturbed contexts within MV-II, including forty-one pits (mostly braziers and hearths) and 112 posts from architectural structures. The most intriguing discovery at the site, however, are the wooden architectural remains of thirteen structures that were directly associated with the features (braziers, hearths, and posts), the bulk of the site's artifacts (lithic, bone, wood, and plant fiber), plant remains of more than seventy species, four "workshop" areas (indicated by the presence of well-defined, concentrated artifacts and cultural features), and a refuse disposal area. Twelve of these structures (termed "huts") were roughly rectangular in floor plan, each measuring between about 4–16 square meters, and laid out in adjacent rows on the site's eastern reaches to collectively form either one or two tent-like structures that were probably covered with animal hides at the time of their occupation. [Figure 16.6]

Some 40 meters to the east of this residential area lay the thirteenth architectural feature, dubbed the "wishbone structure" due to its distinctive shape,

FIGURE 16.5 Assortment of stone artifacts recovered from the significantly older MV-I deposits. Courtesy of Tom Dillehay.

which appears to have been constructed and used for some special purpose in what Dillehay describes as the site's "most unusual and complex occupational zone." [Figures 16.7 and 16.8] Unlike the essentially residential character of the site's eastern portion, this isolated structure—along with its associated hearths and clay-lined braziers—appears to have been used for specialized activities, especially the preparation and use of medicinal plants, as well as for tool maintenance and the processing of hides and meat. Dillehay interprets these two portions of the site to be associated, contemporaneous, and occupied by the same group of people.

FIGURE 16.6 The remnants of 112 posts used to frame twelve animal hide-covered, tent-like huts were uncovered at Monte Verde. The huts were joined on their sides, arranged in two parallel rows. Knotted plant-fibor cordage has been found tied to some of the stakes. Courtesy of J. M. Adovasio.

The MV-II lithic artifact assemblage, numbering around 1,500 pieces in all with fewer than 700 being classified as "culturally modified," is radically different from any assemblage collected from the other sites discussed in this volume. [Figure 16.9] At first glance, Texas State University lithic specialist Michael Collins suspected that the vast majority of the lithic artifacts from Monte Verde were not demonstrably cultural. Over the course of his examination, however, he eventually determined that the artifacts were indeed cultural and exhibited an effective and sophisticated use of available stone resources, adding that they "simply *look* clumsy and ineffective." The assemblage contains very few bifacial implements and only three fragmentary projectile points which appear to display affinity with the El Jobo type encountered at Taima-taima, a Late Pleistocene mastodon kill site in Venezuela (see pages 296-305). The remaining lithic items in the assemblage are mostly expediently prepared pebble tools, along with a relatively small collection of ground/polished stone implements such as bolas and perforators. [Figure 16.10]

The natural preservation of the MV-II cultural deposits through time was sufficient to have permitted the recovery of a quantity and range of organic and inorganic materials that is unparalleled for a Late Pleistocene archaeological site in South America. In addition to the architectural remains, hearths, braziers, posts, and lithic artifacts, the site's excavators also recovered the kind of evidence that is usually destroyed in many ancient sites, including items such as animal bones, animal hide and tissue, gomphothere bone and ivory, bone artifacts, wooden artifacts, plant-fiber artifacts, plant leaves

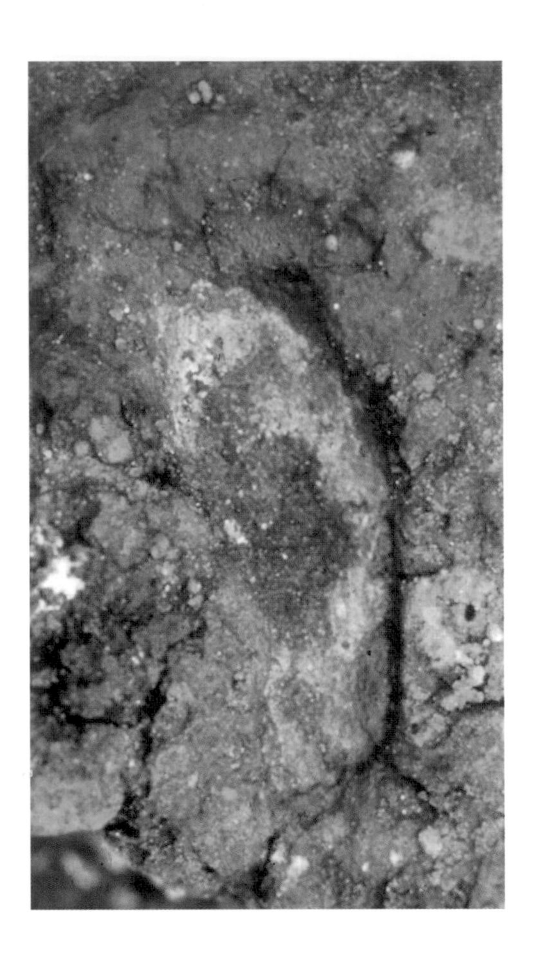

FIGURE 16.7 (left) Close-up of a preserved child's footprint impressed in hardened muddy sand. It was discovered in the so-called "wishbone structure" seen in Figure 16.9. Courtesy of Tom Dillehay.

FIGURE 16.8 (below) Remnants of wishbone structure, which was isolated from the huts at Monte Verde and is interpreted as a nonresidential public area that may have been a place of healing. The footprint impression seen in Figure 16.7 was found here. Courtesy of Tom Dillehay.

FIGURE 16.9 (below) A pebble tool, described by Dillehay as a "basalt wedge," showing modification in the form seven facets removed from its obverse face (above) and three from its reverse face (below). Recovered during the recent 2013 investigation of Monte Verde, this item is similar to others identified within the MV-II portion of the site but dates slightly older, to about 16,000–15,000 yr BP. Courtesy of Tom Dillehay.

FIGURE 16.10 (right) Lithic artifacts recovered from the MV-II deposits at Monte Verde, including two chert El Jobo projectile points (left) and a ground stone implement made of slate (right), all dated to around 14,500 yr BP. The slate bayonet-like artifact is unique to Monte Verde. Although very few of the Monte Verde lithics are recognized as diagnostic artifacts, the two points shown here display affinity to the El Jobo type. El Jobo points are known to be associated with extinct Late Pleistocene animals and have been recovered from the site of Taima-taima on the northeastern Venezuela. (see pages 296-305) Courtesy of Tom Dillehay.

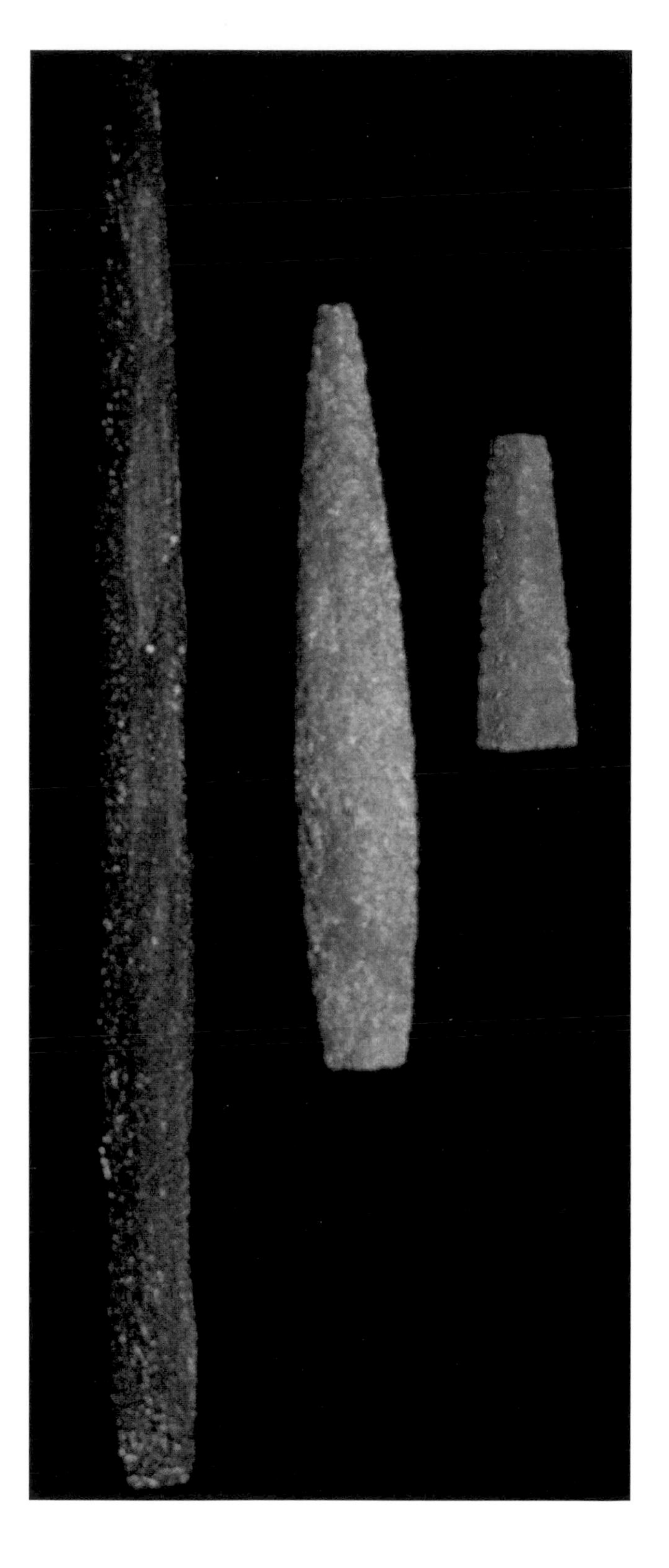

FIGURE 16.11 A bit of burned seaweed attached to a scraper recovered in MV-II, the younger of Monte Verde's deposits. The excavators have identified nine species of chewed or burned seaweed, probably harvested from the coast of the Pacific Ocean, which lay about 90 kilometers west of the site at the time of the site's MV-II occupation. Seaweed is an excellent source of nutrients with medicinal value as an antibiotic and immune system builder. At least two of the species are currently used medicinally by local indigenous people to treat chest and intestinal ailments. This suggest the coastline was an important source of sustenance for the people of Monte Verde, who, in fact, may have entered the area from the Pacific coast itself. Courtesy of Tom Dillehay.

and stems, fruits, berries, pollen, red ochre, and even a human footprint. [Figure 16.11] This unique circumstance owes to the site's apparent history after it was abandoned by the population responsible for MV-II, when the creek's level eventually rose to inundate the site with stagnant water and led to formation of a 15–30 centimeter thick peat layer that covered (and thereby anaerobically preserved) most of the MV-II deposit.

The 2013 field work on Chinchihuapi Creek identified what Dillehay and his colleagues refer to as "multiple, spatially discontinuous, low-density occurrences" of archaeological remains that suggest discrete episodes of ephemeral human activity dating to about 14,500 yr BP and perhaps as early as 19,000 yr BP. The excavations across a 500 meter stretch of Chinchihuapi Creek between site MV-I and the Chinchihuapi sites uncovered twelve small, *in situ* burning episodes in direct association with fragmentary animal bones (both burned and unburned), "economically useful" plant remains, flaked stone artifacts, charcoal, and ash. The lithic artifact assemblage from this portion of the creek valley is not large, numbering only thirty-nine specimens, but these artifacts nonetheless definitively demonstrate the flaking and use of both local and nonlocal lithic material. Dillehay and his colleagues have attributed these newly discovered items to four separate lithic assemblages, two of which are considered to be broadly similar to those from the previously identified MV-I and MV-II site components. The other two assemblages, which span the long gap in time between those components and occur as discrete episodes around 19,000–17,000 yr BP and 16,000–15,000 yr BP, may represent technological precursors to the lithic production techniques employed in the manufacture of the later, abundant MV-II artifacts.

The Monte Verde and Chinchihuapi sites have provided an unprecedented and remarkably full snapshot of human life over an exceptionally long time frame—and perhaps the earliest firmly documented to date—in the Late Pleistocene New World. In their most-ancient incarnations, the sites appear to have witnessed a series of discrete, broadly similar, short-term visits by hunting and gathering peoples who prepared small hearths, produced and then discarded expedient stone tools made from distant raw materials, and then moved on. These peoples were almost certainly highly mobile and probably participated in long-distance exchange or trade networks with other peoples who followed a similar lifeway. These earliest visitors also would have been adapted to a colder climate and more rugged, boreal environmental conditions than the later occupants of MV-II, who faced more-temperate climes and more varied resources with the onset of warmer, wetter conditions between 17,000 and 15,000 yr BP.

While these earlier visitors to Chinchihuapi Creek appear to have followed a lifeway that was similar in some respects to that which had traditionally been

FIGURE 16.12 Archaeological work at Monte Verde was conducted by a team led by Tom Dillehay of the University of Kentucky and the Universidad Austral de Chile from 1977 to 1987. In 1997 the site was inspected by a visiting group of prominent archaeologists sponsored by the Dallas Museum of Natural History. Since 2011 Dillehay has directed biannual excavations. Courtesy of Tom Dillehay.

proposed by the Clovis First hypothesis, the evidence from MV-II instead indicates that its peoples and their activities were quite flexible and far more adaptive than the region's pioneer populations. [Figure 16.12] Thanks to a fortuitously high degree of preservation, the Monte Verde excavators were able to document a far broader range of activities than might previously have been thought possible. The MV-II peoples were engaged in a variety specialized but complementary tasks, ranging from woodworking and plant-fiber and tool production to food processing and medicine, with permanent residents collecting local staple resources and other more-transient visitors venturing year-round to the coast or the mountains to collect exotic items—and even perhaps participating in large-scale exchange and trade networks—all while coalescing to form a loosely organized but interdependent community. According to Tom Dillehay, the kind of diversity on display at Monte Verde is "the hallmark of the first South Americans," and continued research at Monte Verde and other early South American sites will undoubtedly uncover even greater indications of diversity.

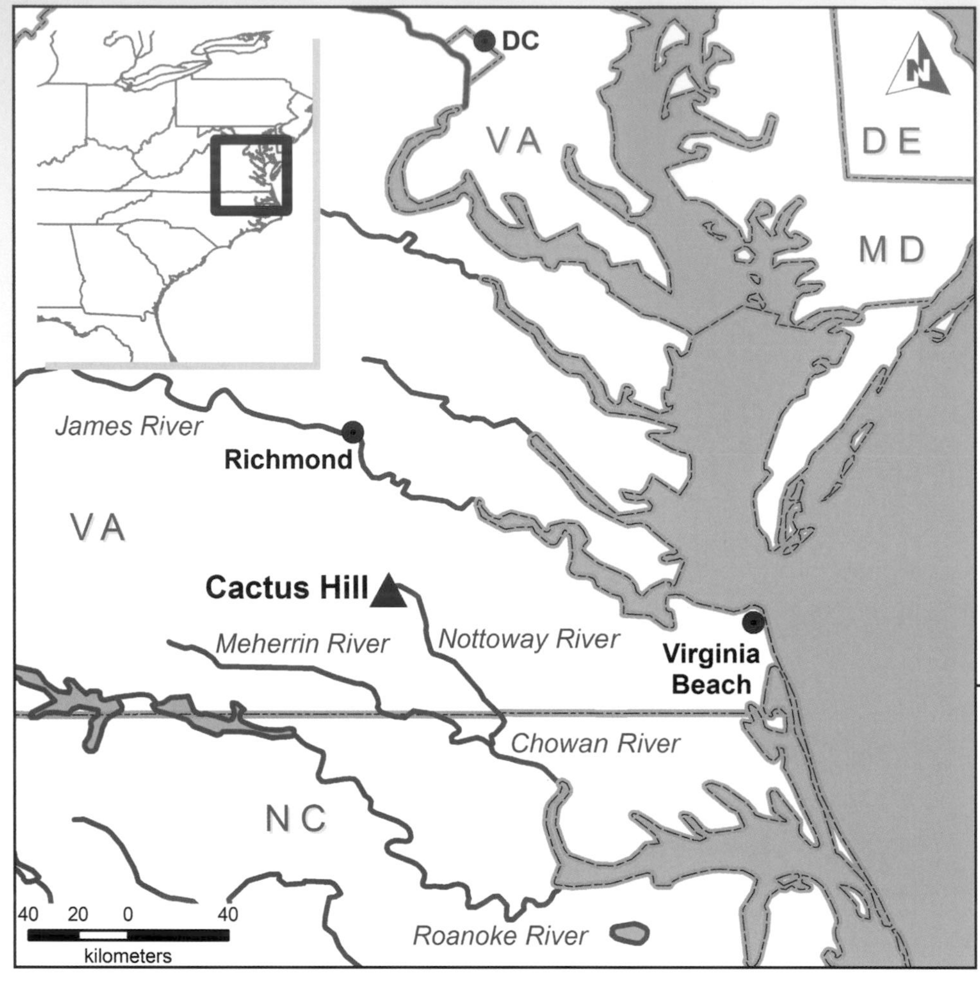

Map by David Pedler

The Cactus Hill archaeological site is located on a wind-deposited terrace adjacent to the Nottoway River in southeastern Virginia. The site gets its name from the prickly pear cacti commonly found growing on the site's sandy soil. Cactus Hill is one of the oldest and most well-dated archaeological sites in the Americas, with the earliest human occupations dating to about 22,000–18,500 yr BP.

17 CACTUS HILL

LOCATION SUSSEX COUNTY, VIRGINIA, UNITED STATES

Coordinates 36°59'15.23"N, 77°19'17.91"W.

Elevation 22 meters above mean sea level.

Discovery Richard Ware in the early1980s.

The Cactus Hill site's deeply stratified archaeological deposits contains a nearly continuous cultural record extending back in time at least 18,000 years. As with three of the other pre-Clovis sites discussed in this book—the Debra L. Friedkin and Gault sites along Buttermilk Creek (see pages 258-271) in central Texas and Paisley Five Mile Point Caves 2 and 5 in south-central Oregon (see pages 236-245)—Cactus Hill's earliest cultural level was discovered beneath an acknowledged Clovis horizon.

Located in Sussex County, Virginia, the Cactus Hill site is about 60 kilometers south of Richmond, 120 kilometers west of Virginia Beach, and 8 kilometers northeast of the of the small town of Stony Creek. The site lies 100 meters east of the Nottoway River within a large northward bend in the midpoint of that river's otherwise northwest–southeast course. The Nottaway River joins the Blackwater River to form the Chowan River on the North Carolina border, 60 kilometers southwest of the site. More than half of the 12,500 square mile Chowan River basin lies within the Piedmont Uplands section of the Appalachian Highlands, and the basin gradually narrows as it flows through the Coastal Plain province to its outlet on the Atlantic Ocean at the Albemarle Sound. The Cactus Hill site is located in a transitional portion of the Coastal Plain about 5 kilometers east of the Piedmont Upland's mapped boundary. The site is named after the prickly pear cactus (*Opuntia*

FIGURE 17.1 General view of the Cactus Hill site from a distance. The site is a migrating loamy sand dune which is episodically stabilized by vegetation. Photograph by Michael Johnson.

humifusa) which commonly grows there. [Figure 17.1]

Although its environs have been modified and fragmented by conventional agriculture, tree-farming, and low-density residential development, the site otherwise lies within the Southeastern Evergreen Forest region near its boundary with the more northerly deciduous Oak–Hickory Forest. In Late Pleistocene times the regional vegetation appears to have been composed of a conifer–hardwood forest that included white pine (presently an upland species that is not native to coastal forests) in combination with beech, hickory, and oak. The immediate area is also thought to have contained fruit trees, smaller shrubs, and numerous riverine plants that would have contributed to a broad diversity of food and material resources. Animal remains are poorly preserved and only scantly represented in the earliest deposits at Cactus Hill, but regional data suggest that caribou (*Rangifer tarandarus*) may have been available for hunting in mountainous areas over 200 kilometers to the east. Paleontological investigations in the Saltville Valley (some 400 kilometers to the east [see pages 286-295) and elsewhere in the region indicate the presence of ground sloth (*Megalonyx* sp.), musk-ox (*Symbos cavifrons, Bootherium* sp.), moose (*Cervalces* sp.), and bison (*Bison* sp.). Numerous small animal species also were undoubtedly abundant in the region, and on the whole the Cactus Hill site's setting would have been extremely attractive to both humans and animals.

The Cactus Hill site lies within a complex of west–east oriented sand hills that have formed along the east bank of the Nottoway River. The site's elevation is about 22 meters above sea level, which places it some 8 meters above the level of the nearby river bed. Resting upon a 3–4 meter thick layer of coarse alluvial sand and clay, the site's upper sediments are composed of a layer of wind-borne sand whose thickness ranges from about 50 centimeters to as much as 1.5–2 meters. Four primary site subareas (labeled Areas A through D) and a transitional zone (labeled Area A-B) were defined in this sand deposit over a 2.4 hectare area that stretches from the headwater of a small Nottaway River tributary on the

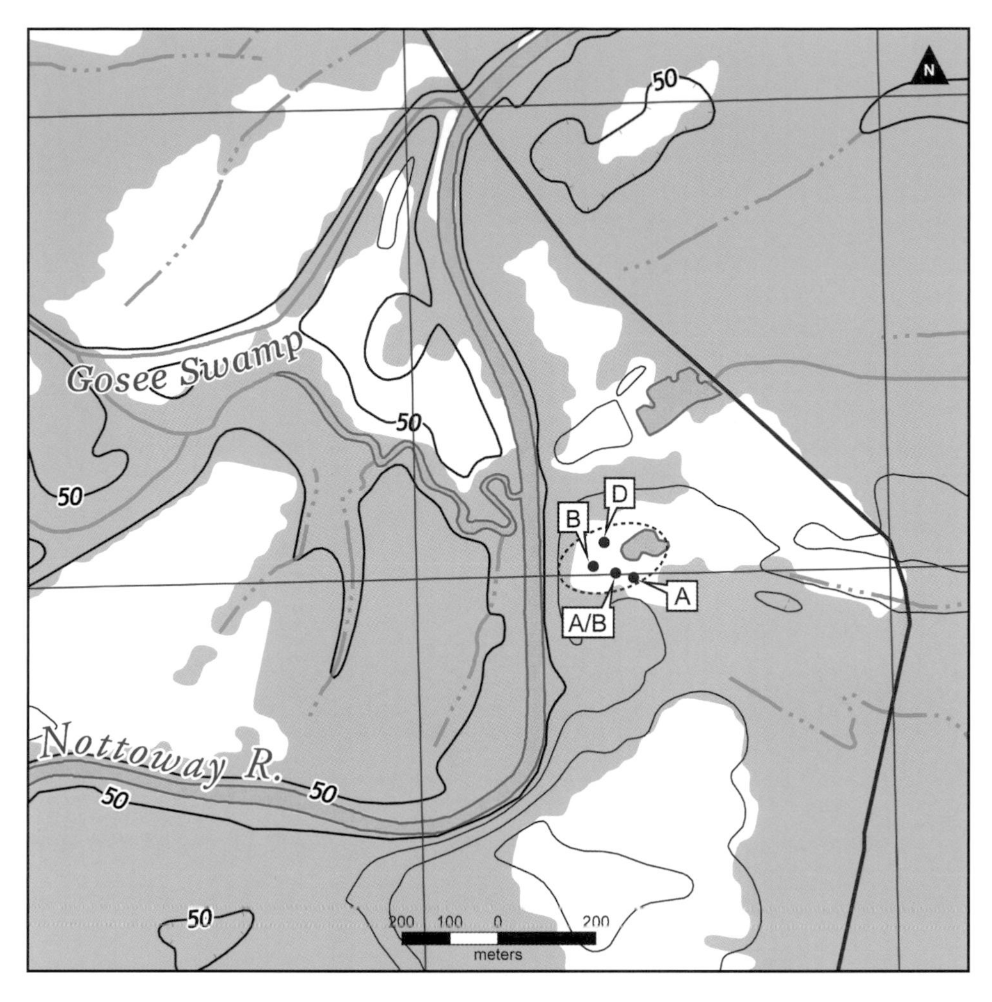

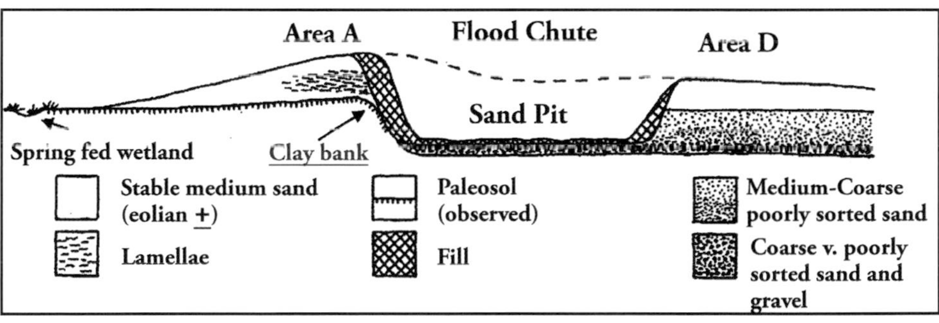

FIGURE 17.2 (above) Topographic map of the Cactus Hill site. Clovis age artifacts recovered from the site have been dated to about 13,300–12,370 yr BP—which is consistent with Clovis-age dates from other parts of North America. Pre-Clovis artifacts have been recovered below Clovis-age deposits in areas A and B. Area B is noteworthy for a hearth dated to 18,530–18,080 yr BP and a concentration of charcoal beneath a cluster of lithic artifacts dated to 21,930–18,490 yr BP. A large cross-mended blade-like flake excavated in Area A/B has been dated to 13,090–12,830 yr BP. Area D has been dated only back to the Early Archaic (11,500–8500 yr BP). Courtesy of Michael Johnson.

FIGURE 17.3 (below) Topographic profile of Cactus Hill areas A and D. Although the entire Cactus Hill shows evidence of human activity, only Area A along with Area B have been dated to Clovis age and earlier. Although Cactus Hills' sandy soil might be thought of as being relatively unstable for the preservation of strata, or layers of materials that archaeologists can date to different periods, its investigators claim evidence of human activity such as stone tools and hearths are stratigraphically consistent within cultural levels. Radiocarbon dates are also consistent with the artifacts and stratigraphy. Courtesy of Michael Johnson.

FIGURE 17.4 The archaeological excavation of cultural materials at Cactus Hill. The work was conducted by professional archaeologists and volunteers under the direction of the private-enterprise Nottaway River Survey, the Fairfax County Parks Authority, and the Archaeological Society of Virginia. Photograph by Michael Johnson.

east to a wetland near the east bank of Nottoway River. [Figures 17.2 and 17.3] The site is bounded to the south and east by swamps, which also generally occupy low areas in all directions within 2 kilometers of the site.

The site was discovered in the walls of a commercial open-pit sand mine by avocational artifact collector Richard Ware of Petersburg, Virginia, in the early 1980s. It was initially tested in 1988 by Joseph McAvoy of the Nottoway River Survey, a commercial archaeological research firm based in Sandston, Virginia. [Figure 17.4] Subsequent intensive archaeological excavations were conducted between 1993 and 2002 by two independent teams of professional and volunteer archaeologists under the direction of McAvoy and Michael Johnson

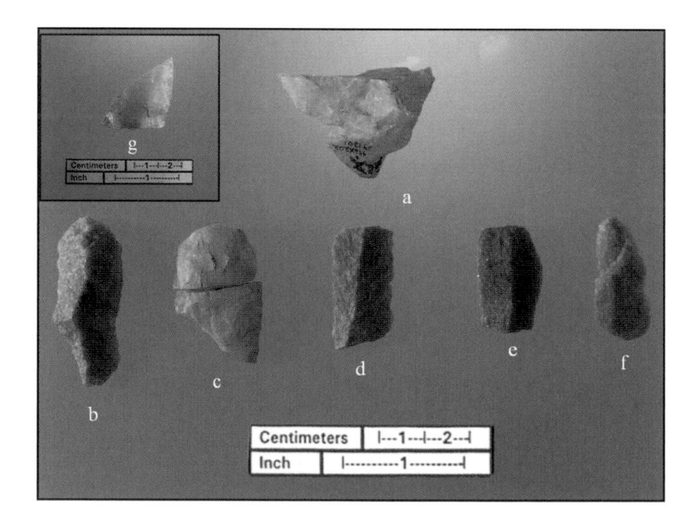

FIGURE 17.5 Pre-Clovis artifacts excavated in Area A at Cactus Hill: (a) unifacial core-like artifact, (b) corner blade-like flake, (c) two pieces of refitted point mid-section, (d–f) blade-like flakes, (g) quartz fluted point preform. (See Figures 17.2 and 17.3.) Photograph by Michael Johnson.

FIGURE 17.6 Stratigraphic profile uncovered by the archaeological excavation at Cactus Hill. The site deposits contained numerous pits, hearths, and artifacts that range in age from the early eighteenth century through the Woodland and Archaic periods to the Paleoindian period. The site's earliest levels contained both Clovis and pre-Clovis materials. Photograph by Michael Johnson.

of the Fairfax County (Virginia) Parks Authority and the Archeological Society of Virginia. The excavations at Cactus Hill were performed in tandem with a multiyear archaeological survey of a 65 kilometer stretch of the Nottaway River and adjacent uplands in the vicinity of the site. Encompassing an 800 square kilometer area in northern Sussex and southern Dinwiddie Counties, this broader research endeavor identified over 100 sites ranging in age from Paleoindian times through the succeeding Archaic, Woodland, and Historic periods. Possible pre-Clovis horizons were identified at two localities, the Chub Sandhill and Blueberry Hill sites, but the only unequivocal, dated pre-Clovis material in the survey area was identified at the Cactus Hill site. [Figure 17.5]

The artifacts recovered from the site were restricted to its uppermost levels, which also contained dozens of variously configured pits, hearths, and clusters of lithic artifacts indicating the presence of activity areas. The youngest deposits appear to have been disturbed by commercial logging at some point in the recent past, but nonetheless include extensive evidence of the site's occupation from the early eighteenth century through the entire Woodland period. The site's occupation throughout Archaic times appears to have been heaviest during the Middle Archaic period (7,400–5,000 yr BP), as evidenced by the recovery of numerous distinctive Morrow Mountain and Guilford projectile points, among lesser quantities of other forms. [Figure 17.6]

FIGURE 17.7 (above) Selection of Clovis tools from Cactus Hill. Proponents of an European "Iberia not Siberia" origin of Native Americans point to the similarly of Cactus Hill artifacts like these to Solutrean tools recovered in France and Spain dated to around 22,000 to 17,000 yr BP. (see page 22) They speculate European seafarers crossed the Atlantic bringing with them their ice-age technology that morphed into Clovis. Skeptics argue the many differences between Solutrean and Clovis are far more significant than the few similarities which can be explained by technological convergence or parallelism. Photograph by Michael Johnson.

FIGURE 17.8 (below) An assemblage of artifacts recovered from pre-Clovis deposits at Cactus Hill. Except for the raw material they bear striking similarity to Miller complex lithic artifacts from Meadowcroft Rockshelter. (see pages 200-215) Photograph by Michael Johnson.

The most well-defined, earliest human use of the site—although still relatively light, as one would expect—appears to have occurred during the Paleoindian period. The site produced fluted Clovis projectile points and slightly later forms, a variety of lithic scrapers and flake tools, gravers (*i.e.*, small cutting, scraping, and/or engraving tools), and blades. For the most part, these tools were produced using locally available chert and quartzite. A hearth in this deposit was dated to 13,300–12,370 yr BP, which is consistent with the ages of other eastern Clovis sites such as Shawnee-Minisink (see pages 78-89), some 500 kilometers to the north. [Figure 17.7]

The Clovis level at Cactus Hill is underlain by an even smaller number of pre-Clovis artifacts in intensively examined Areas A and B, with the larger quantity of the two occurring in Area B. Numbering only 15–20 artifacts from what McAvoy considers to be a "good [archaeological] context," the Area B artifacts include projectile points, blade flakes, blade cores, and flake tools. A hearth identified in this same level was dated to 18,530–18,080 yr BP and a concentration of charcoal beneath a cluster of lithic artifacts produced an age of 21,930–18,490 yr BP. These ages roughly correspond to those obtained from the pre-Clovis deposits identified in Stratum IIa

FIGURE 17.9 Casts of two Cactus Hill points recovered in Area B. (see Figure 17.2) The original points were recovered below Clovis-age levels and support the claim for a pre-Clovis occupation at Cactus Hill. Critics of pre-Clovis claims for these artifacts suggest that through time they were transported downward into the loamy sands at Cactus Hill from the Clovis-age deposits above. Photograph by Michael Johnson.

at Meadowcroft Rockshelter (see pages 200-215), about 450 kilometers to the northwest. [Figure 17.8]

The pre-Clovis artifacts from Cactus Hill indicate the presence of a core and blade technology employing both conical, polyhedral cores and thick, chopper-like cores. These cores appear to have been obtained from locally available quartzite, initially worked and formed off-site, and then brought to the site to be reduced using soft stone hammers or hardwood billets for the production of blades and blade-like flakes. In terms of form and production technique, these artifacts are dramatically different from all succeeding phases of lithic technology represented at Cactus Hill. Perhaps the most distinctive artifacts in the pre-Clovis assemblage are the two relatively small (*i.e.*, about 3.5 centimeters long), thin, triangular lanceolate projectile points recovered from Area B. [Figure 17.9] Together, the pre-Clovis artifacts from Cactus Hill have led the site's investigators to suggest that the early peoples associated with them were generalized foragers who relied on a broad spectrum of plant and animal resources, and not specialized hunters focused on the taking of big-game animals. Little else can be decisively concluded about the lifeway of these peoples, however, as the material footprint they left behind at this apparent hunting camp was only very lightly impressed.

A number of prominent professional archaeologists have cast doubt on the age and validity of Cactus Hill's pre-Clovis artifacts, and in the scholarly press it is frequently overlooked or given scant mention with attendant calls for further work. Critics have cited the very small artifact count, the disturbance and mixing of later artifacts with the site's pre-Clovis level, the apparent absence of similar sites in the vicinity, and even the "misidentification" of later Paleoindian forms such as the distinctive early triangular point. While acknowledging minimal, upper level disturbance in limited portions of the site, the Cactus Hill investigators counter that the site's deep stratigraphic sequence maintains consistent relationships among diagnostic artifacts both within and between discrete cultural levels. They further note that the ages of the site's various overlying levels closely conform to those accepted for similar horizons elsewhere in the Southeast.

The Cactus Hill site has emerged as a key locality in a small but growing number of pre-Clovis sites in eastern North America that appear to share a broadly defined technology involving the manufacture of thin, roughly triangular, unfluted bifaces sometimes accompanied by small blades and distinctive lithic cores. Though by no means explicitly related, the site's pre-Clovis artifacts bear a striking similarity to the earliest lithic artifacts recovered from its contemporary, Meadowcroft Rockshelter (see pages 200-215). A few scholars have claimed an apparent affinity with other extremely early (*i.e.*, 30,000–25,000 yr BP) eastern sites such as Miles Point in coastal Maryland and several others on nearby Chesapeake Bay. Though intriguing, only continued investigation of early sites in eastern North America will determine whether these apparent interrelationships are a mere coincidence or part of a valid pattern.

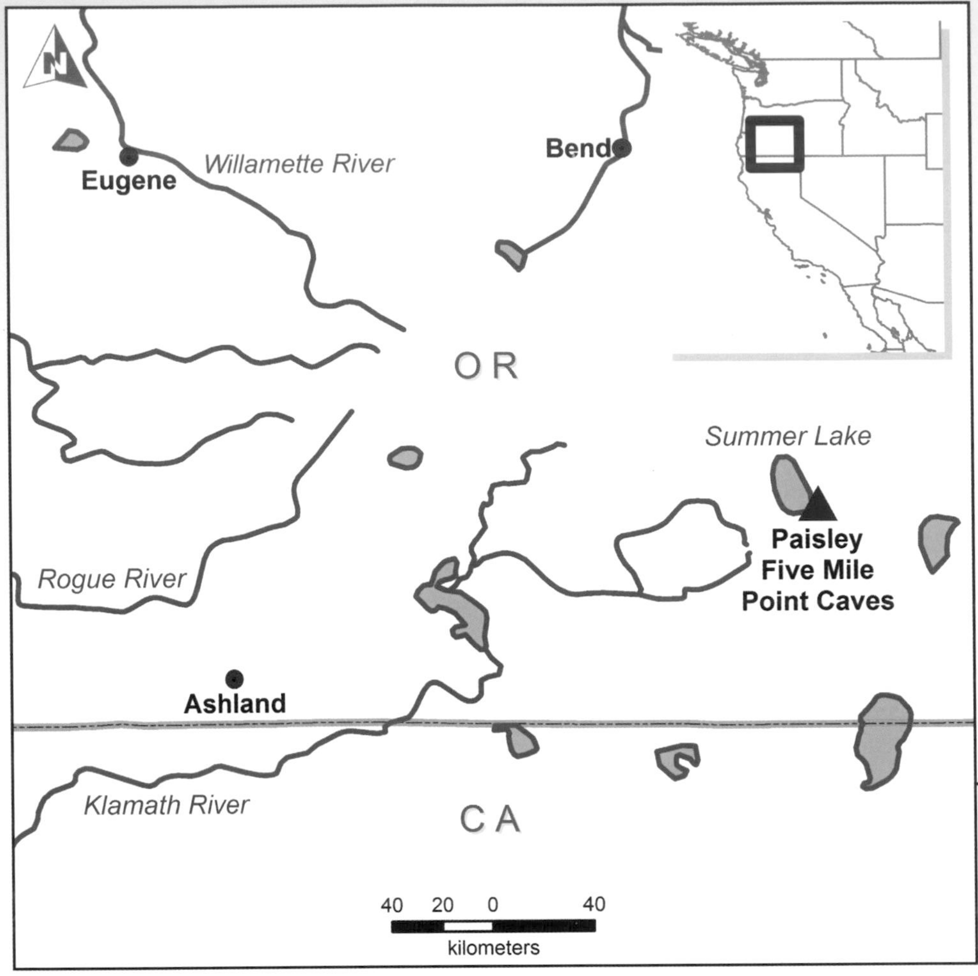

Map by David Pedler

The Paisley Five Mile Point Caves site is comprised of eight caves and rockshelters in an arid, desolate region of south-central Oregon. One of the caves may contain DNA evidence of the oldest definitively-dated human presence in North America. Information about the exact location of the site is restricted to prevent further looting.

18 PAISLEY FIVE MILE POINT CAVES

LOCATION LAKE COUNTY, OREGON, UNITED STATES

Coordinates restricted.

Elevation 1,356 meters above mean sea level.

Discovery Walter Perry in 1937.

The Paisley Five Mile Point Caves in Oregon have produced the Western Hemisphere's oldest directly dated human remains in the form of coprolites (desiccated feces). [Figure 18.1] This definitive proof of an early human presence in the northern Great Basin is bolstered by the recovery of these remains from stable, undisturbed archaeological contexts in direct association with artifacts and the remains of extinct animals, all of which have been radiocarbon-dated to as early as 16,000 yr BP. Furthermore, DNA and floral analyses conducted on these items have provided an unprecedented opportunity to examine the genetic identities and lifeway of the earliest Americans.

The Paisley Five Mile Point Caves are located in the Summer Lake basin of south-central Oregon, practically on the very northwestern edge of the Great Basin, just north of the town of Paisley and about 120 kilometers northeast of Klamath Falls. The site occupies the side of a basalt ridge on a former shoreline of Late Pleistocene pluvial (*i.e.*, rain fed) Lake Chewaucan, a 1,200 square kilometer waterbody that subsumed the present-day basins of Summer Lake, Lake Albert, and the Upper and Lower Chewaucan Marshes until about 18,000 yr BP. [Figure 18.2] Water levels receded dramatically after that time, but between about 14,500 yr BP and 12,500 yr BP the water had again become sufficiently higher

FIGURE 18.1 One of the more than thirty human coprolites recovered from Cave 5. The oldest one dates to approximately 14,500 yr BP. Photography by Brian Lanker.

that the lake stood less than 3 kilometers away from and less than 50 meters below the site. Paisley Caves' proximity to water and the attendant regional diversity of animal and plant resources would have made the site a very attractive place for human habitation during this period. Waterfowl and fish would have been readily available for harvesting, grasslands around the lake would have provided pasture for ungulate animals (such as mammoth, mastodon, llama, horse, bison, deer, elk, and pronghorn), and the uplands to the east would have supported mountain sheep, marmot, and vital plant resources.

The site is composed of eight caves and rockshelters that line a 165 meter southwest-facing stretch of the ridge about 200 meters above the present surface elevation of Summer Lake. Somewhat tightly spaced at a mean distance of about 25 meters from each other, the four caves that have been formally investigated to date are also rather small, ranging in area from about 27 square meters (Cave 1) to about 66 square meters (Cave 5). [Figure 18.3] The periodic exposure of the nearby lakebed and a prevailing southwest wind over several millennia resulted in the caves being filled with layers of sand and silt which became trapped behind piles of rock that had either eroded from or been washed down over the cliff above the site and accumulated at the cave mouths. [Figure 18.4]

FIGURE 18.2 The Paisley Five Mile Point Caves are located on a ridge above Summer Lake, a large, shallow, alkali lake. The marshes around the lake support a wide variety of birds and other wildlife. During the Late Pleistocene the area's currently arid lands were lush, and Summer Lake and nearby Lake Abert subsumed by the much larger 1200 square kilometer Lake Chewaucan. Photograph by fishermansdaughter

Paisley Five Mile Point Caves were first investigated in 1938–1940 under the supervision of Luther Cressman, founder and chair of the University of Oregon's Department of Anthropology and founding director of the Oregon State Museum of Anthropology. Cressman had been made aware of the site by a local resident in 1937, when he made an initial visit to the site in the company of geologist Ernst Antevs. Focusing on Caves 1, 2, and 3, Cressman's excavations revealed that all three of them contained a layer of wind-deposited volcanic material (called "tephra") from the 7700 yr BP eruption of the Cascade Range's Mount Mazama and in one of them, Cave 3, that deposit was underlain by the remains of Late Pleistocene camel, bison, and horse in association with artifacts. In the absence of published details of the site's stratigraphy and recovered materials, however, and without the possibility to confirm his interpretations via modern archaeological dating techniques, Cressman's findings were mostly rejected by a

FIGURE 18.3 Located in the arid, desolate region of south-central Oregon, the eight caves and rockshelters of Paisley Five Mile Point Caves site line a ridge southeast of Summer Lake. Photography by Thomas W. Stafford, Jr.

professional archaeological community that gradually came to regard the Clovis horizon as the hemisphere's pioneer human population.

Cressman's excavations and the surrounding site deposits were left open to destruction by looters for decades, until the renewed investigation of the Paisley Five Mile Point Caves by Dennis Jenkins of the Museum of Natural and Cultural History at the University of Oregon. As part of field investigations and lab analyses beginning in 2002 and continuing to the present, Jenkins and his research team set out to test Cressman's findings by revisiting Caves 1, 2, and 3, and expanding the project to include extensively vandalized Cave 5, which Cressman had not examined. These new excavations were designed to focus on the recovery of bone, coprolites, and cultural material from any *in situ* deposits that had remained undisturbed by Cressman's earlier work and subsequent looting.

As might have been expected, the layer of Mount Mazama tephra formerly identified by Cressman has been re-identified by Jenkins and his team in all four caves, and in each case this layer is overlain by middle and/or late Holocene cultural materials and underlain by older materials extending from the Early Holocene to the Late Pleistocene in age. The earliest evidence of human occupation has been recovered from Caves 2 and 5, where over 180 radiocarbon dates have intensively documented a well-ordered stratigraphic sequence that maps the transition from Early Holocene through Clovis-era to pre-Clovis sediments. [Figure 18.5] Cultural materials recovered from the site include lithic artifacts, cordage (twisted rope or string composed of plant and/or animal materials), wooden artifacts,

FIGURE 18.4 Looking out from the interior of Cave 5 where Clovis-age Western Stemmed projectile points were discovered. Photograph by Brent McGregor.

basketry, textiles, butchered bone, other organic materials such as hair and animal tissue, and hearth charcoal. [Figure 18.6]

About 3,800 lithic artifacts, most of which are morphologically indistinctive forms (such as debitage flakes, bifaces, and cores) were recovered from the older, pre-Mount Mazama tephra strata at Paisley Five Mile Point Caves, and only a relatively small proportion of this admittedly small lithic assemblage was recovered from early Clovis and pre-Clovis contexts. The assemblage nonetheless includes the oldest well-documented examples of the Western Stemmed projectile point, a form long assumed to be descended from Clovis, which was recovered from a context in the southern portion of Cave 5 that dates to 13,430–13,280 yr BP. The recovery of a second Western Stemmed point from the northern portion of Cave 5 may date to 13,050–12,820 yr BP, which makes this locality at the very least a Clovis contemporary. Few of the debitage flakes recovered from these lower levels are associated with these points, however, and most of them represent the diminutive products of tool finishing and refurbishment whose lack of diagnostic attributes precludes their ascription to either Western Stemmed or Clovis technology. [Figure 18.7]

Extensive and diverse plant and animal remains assemblages, in the form of both physical items and residues on artifacts, have also been recovered from the older stratigraphic levels at Paisley Five Mile Point Caves. Notable items among these include a horse maxilla and butchered artiodactyl bone recovered from the same context as a polished and battered handstone

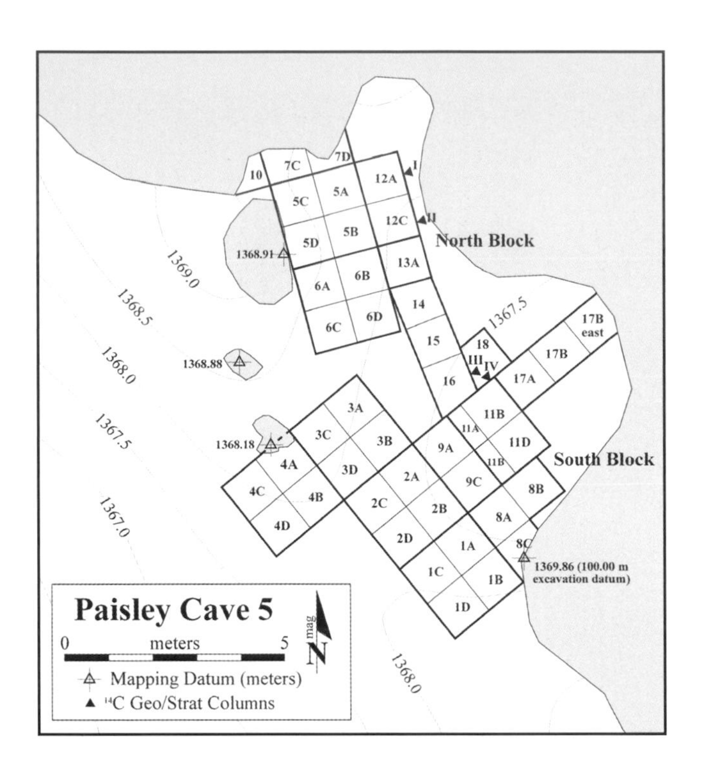

FIGURE 18.5 (left) Floor plan of Cave 5 with excavation pits outlined and numbered. Courtesy of Dennis Jenkins.

FIGURE 18.6 (below) (A) Floor plan showing distribution of radiocarbon dated coprolites and horse and camel bones in Cave 5. (B) Vertical distribution of two Western Stemmed projectile points relative to acceptably dated coprolites and dating column samples in Cave 5. (The ages shown to the right of each profile are uncalibrated radiocarbon dates.) Courtesy by Thomas J. Connolly and Dennis Jenkins.

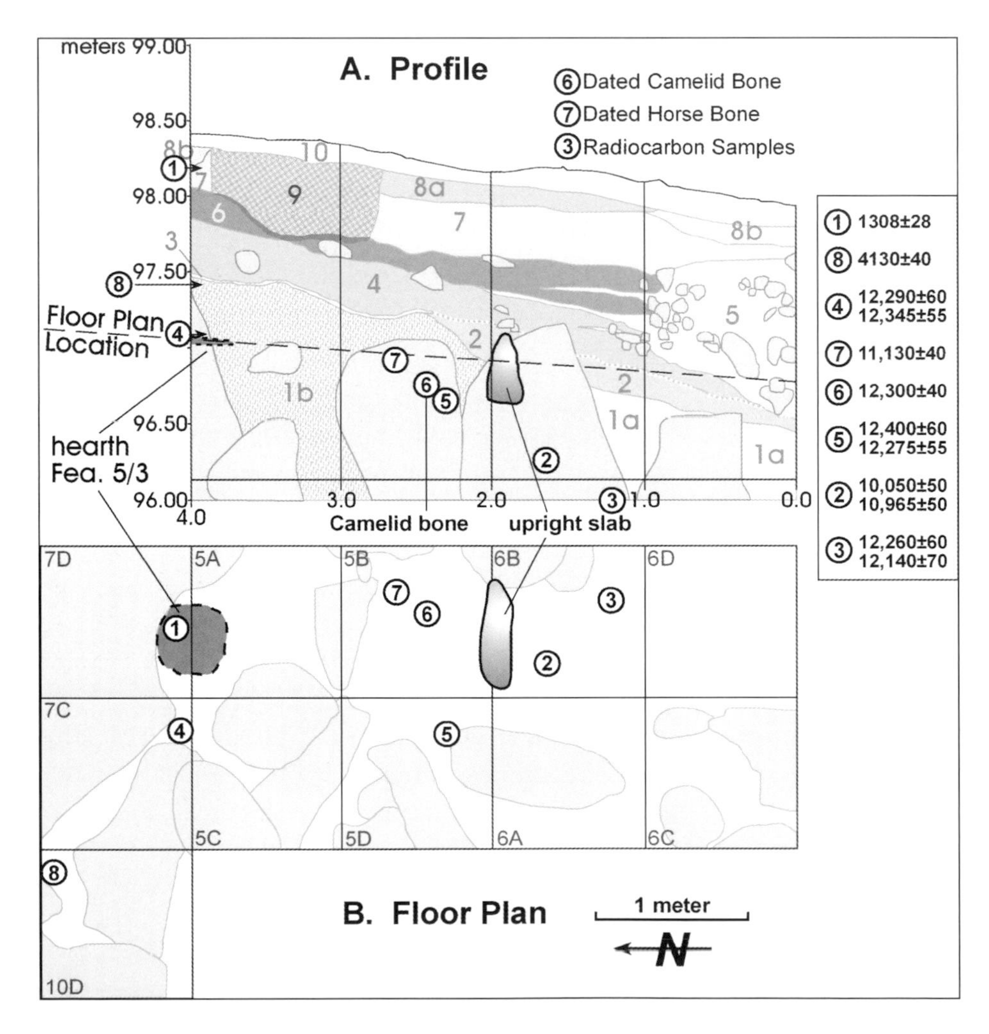

FIGURE 18.7 Western Stemmed projectile point from Cave 5. Western Stemmed projectile points, renowned as the oldest non-fluted projectiles in the far western United States, are generally narrow with sloping shoulders and thick contracting bases. The oldest Western Stemmed point from the site, from the southern block of Cave 5, was in a context dated to 13,430–13,280 yr BP. Courtesy of Loren Davis and Dennis Jenkins.

bearing residues of mammoth/mastodon protein, starches of the carrot plant family and grass seed, and siliceous plant remains, along with an obsidian flake also bearing modified mammoth/mastodon protein. The flake and handstone appear to have been used for the processing of mammoth/mastodon flesh, with the handstone also serving to grind roots and grass seeds, perhaps during a spring or early summer occupation episode of Cave 2 dating to around 14,000–13,500 yr BP. [Figure 18.8] Other items, from Cave 5, include a butchered mountain sheep bone dating 14,850–14,110 yr BP, a modified bear bone dating 14,200–13,980 yr BP, horse bone dating as early as 14,710–14,170 yr BP, camel bone dating as early as 13,740–13,540 yr BP, and numerous smaller animal and plant remains falling roughly into that same age range.

Based on presently available data, most of the cultural features identified in the cave deposits appear to be hearths or fire episodes and trash pits, only a small number of which appear to predate Clovis. The oldest of six radiocarbon assays from a hearth surrounded by a dense concentration of burned bone and obsidian debitage in Cave 2 produced a date of 13,560–13,380 yr BP. Perhaps the best-documented feature at the site is the so-called Bone Pit, which was discovered in Cave 5 beneath a large stone slab that lay adjacent to a second slab that apparently had been propped against a large boulder. Material recovered from the Bone Pit includes bone (camel, horse and mountain sheep), a butchered sheep mandible, human coprolites, human hair, and debitage flakes. At least ten radiocarbon assays provide ages for the Bone Pit. A horse phalanx recovered from immediately above the Bone Pit dates as early as 13,100–12,870 yr BP, and materials recovered from within the Bone Pit proper produced an early age of 14,860–14,140 yr BP, making it about 1,000 years older than Clovis. [Figure 18.9]

The most definitive evidence for a pre-Clovis human presence at Paisley Five Mile Point Caves has also garnered the most controversy and doubt concerning the site's age and depositional integrity. Human coprolites from the Paisley Caves have produced thirty-two radiocarbon dates ranging in age from as early as 14,860–14,140 yr BP to as late as 2350–2320 yr BP, and those dates concur with the ages of the sediments in which the samples occurred. Some of the site's critics, however, have claimed that the coprolites are not human, and that human genetic material from later site occupations—or even the site's excavators—could have contaminated in-place faunal-derived coprolites. Others have contended that the samples do not resemble human coprolites in either morphology or composition, claiming them more likely to have derived from grazing herbivores rather than humans. [Figure 18.10] But multiple lines of evidence examined by multiple laboratories following rigorous, widely accepted analytical protocols have definitively established the coprolites as human, containing only human and plant mtDNA with no other animal species present, and unlikely to be contaminated. Their form and composition also precisely correspond to what one should expect from a prehistoric Great Basin diet.

It is remarkable and perhaps even highly improbable that the pre-Clovis deposits at Paisley Five Mile Point Caves should have been preserved, much less discovered, analyzed, and recognized for what they represent. Like other sites in the Great Basin, the caves were visited numerous times throughout the Late Pleistocene

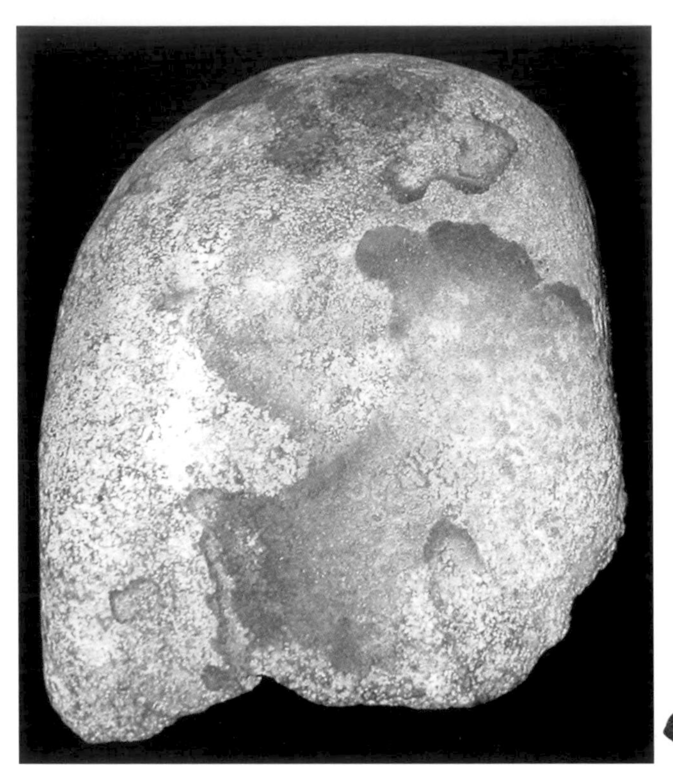

and Early Holocene by bands of hunter-gatherers who typically left a light cultural footprint in the form of relatively few discarded or lost stone tools and perishable items whose integrity otherwise would have been lost due to exposure to the elements, continued deposition and erosion, and displacement by animal and subsequent human activity. The unique circumstances of deposition, preservation, and recovery in the case of Paisley Five Mile Point Caves have made possible a definitive revision of our perspectives on the peopling of the North America and the Western Stemmed Point tradition. Without those unique circumstances, the Paisley Caves pre-Clovis occupation could easily have been missed and attributed to similar later, younger, and less ephemeral cultural episodes. As site investigator Jenkins has noted, "the evidence of pre-Clovis site occupations may be waiting for us to sort out of boxes already on museum shelves."

FIGURE 18.8 (above) Handstone recovered from Cave 2, dated between 13,600–13,460 yr BP (horse maxilla) and 13,820–13,700 yr BP (cut artiodactyl rib). Courtesy of Dennis Jenkins.

FIGURE 18.9 (right, top) Bear bone toothed tool recovered in Cave 5, dated to about 14,200–13,980 yr BP. No other bear bones have been identified at the site. Photography by Brian Lanker.

FIGURE 18.10 (right, bottom) Archaeologists examine sediments in Cave 5 in the area of the earliest human coprolites. Photograph by Thomas W. Stafford, Jr.

Map by David Pedler

Post-glacial lakes in southeastern Wisconsin subsequently became wetlands, which attracted duck hunters until the 1960s. Gradually, farmers extended their cultivation into these low, wet areas of muck soil by ditching and installing drain tile. Their ditching machines chanced upon the mammoth fossils at the Schaefer and Hebior sites.

19 SCHAEFER AND HEBIOR MAMMOTH

LOCATION KENOSHA COUNTY, WISCONSIN, UNITED STATES

Coordinates Schaefer, 42°38'38.69"N, 87°59'31.45"W; Hebior, 42°38'7.28"N, 87°59'54.49"W.

Elevation Schaefer, 238 meters above mean sea level; Hebior, 236 meters above mean sea level.

Discovery Schaefer, Franklin Schaefer and Phil Sander in 1964; Hebior, John Hebior in 1979.

The Schaefer and Hebior Mammoth sites have sometimes been associated with the Chesrow complex, a Late Pleistocene archaeological cultural complex composed of at least six habitation sites that roughly fall within a 200 square kilometer area on the western shore of Lake Michigan in southeastern Wisconsin, immediately adjacent to the Wisconsin–Illinois border. These sites are believed to share a broadly defined technological affinity that is exemplified at the complex's namesake locality, the Chesrow site. Though their ascription to the Chesrow complex has been rigorously questioned by a number of archaeologists, the Schaefer and Hebior sites are nonetheless unique and important localities as they demonstrate the presence of butchered mammoth remains and associated lithic artifacts in valid archaeological contexts which pre-date the Clovis horizon. [Figures 19.1 and 19.2] Schaefer and Hebior are also the northernmost—and thus the closest to glacial-margin environmental conditions—of any of the localities within the contiguous United States that are described in Part Two of this book.

The Schaefer and Hebior sites are located in Kenosha County, Wisconsin, about 18 kilometers southwest of Racine, 45 kilometers south of Milwaukee, and 90 kilometers north-northwest of Chicago, Illinois. The sites

SCHAEFER SITE
KENOSHA CO
FEMUR &
MAXILLA

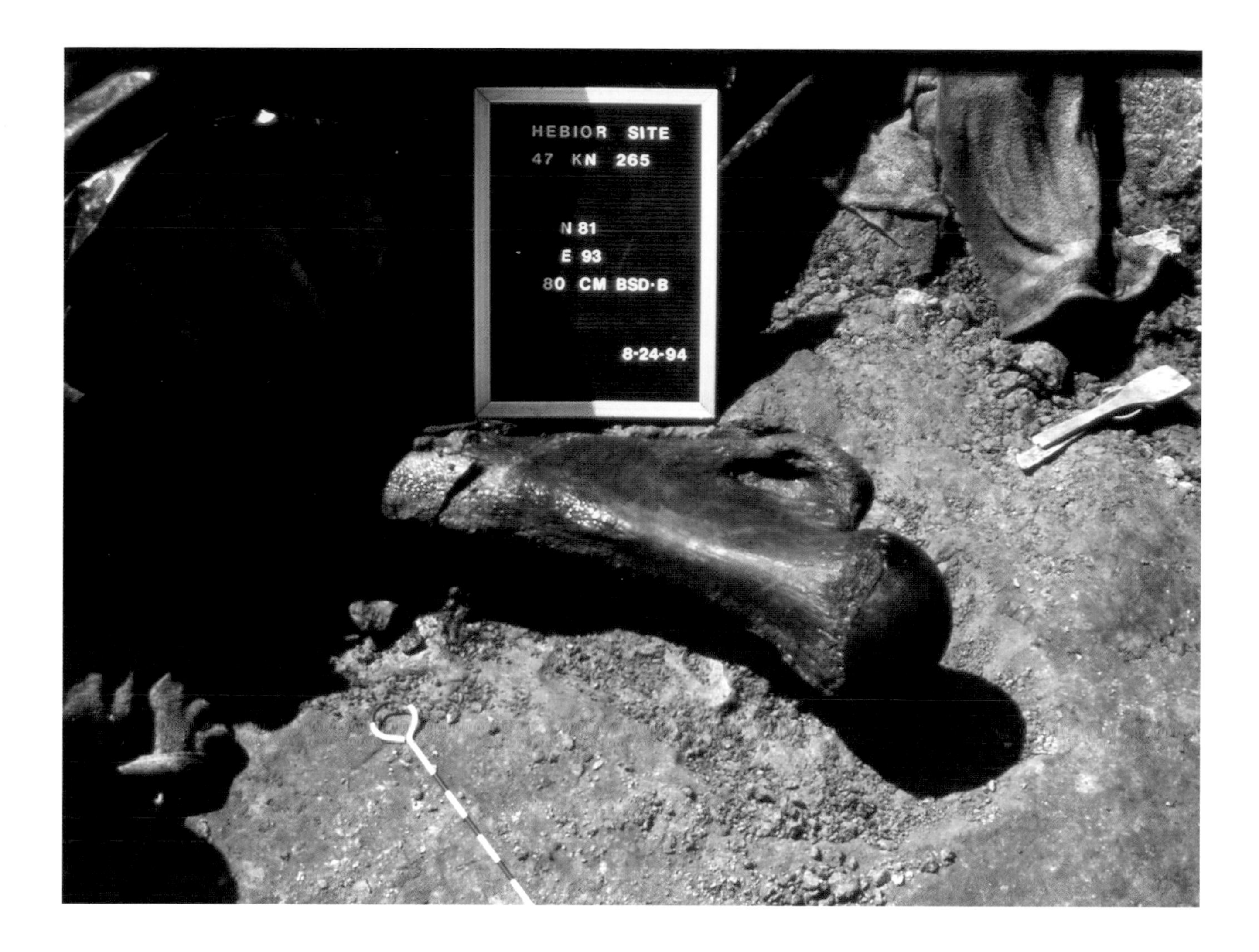

FIGURE 19.1 (left) Early in the excavation at Schaefer, a mammoth femur was exposed. It lay next to the cranium, which is on its left side. The upper molar shows parallel lines which make up its many plates. Courtesy Kenosha Public Museums.

FIGURE 19.2 (above) An area of approximately 25 square meters of the Hebior site was excavated in 1994. Like the Schaefer mammoth, the Hebior mammoth was an adult male. They are among the very few large extinct mammals found east of the Mississippi River which bear the marks of butchering by humans. Courtesy of the Milwaukee Public Museum.

lie 15 kilometers west of Lake Michigan's western shore and virtually straddle the boundary between the Great Lakes and upper Mississippi River drainage basins. The local watershed in the upper Mississippi River basin is that of the Des Plaines River, which reaches its confluence with the Illinois River 70 kilometers to the southwest, which in turn reaches its confluence with the Mississippi River about 460 kilometers southwest of the Schaefer and Hebior sites.

The Lake Michigan lobe of the Laurentide glacier would have completely covered the Schaefer and Hebior sites and extended more than 500 kilometers to the south at the Last Glacial Maximum (26,500–19,000 yr

BP), but by the time of the sites' occupation the glacier had receded to the present-day upper Lake Michigan basin, over 200 kilometers to the north. The glacial retreat from the region began around 17,000 yr BP and left behind a series of parallel, north-south oriented glacial moraines (*i.e.*, accumulations of soil and rock that form at the edges of glaciers) interspersed with shallow inter-morainal valleys. As a result of further glacial retreat and high-volume water drainage from the north, a dense network of small lakes, ponds, and low-flow drainages appeared in the region's lower elevations. Schaefer and Hebior formed around these glacial features, and with continued glacial drainage they were covered—at first, rapidly—by a succession of Late Pleistocene and middle Holocene sediments. At the time of the sites' brief visits by Paleoindians, the local vegetation appears to have been a mosaic of open spruce parkland and sedge wetlands.

The Schaefer and Hebior sites are located within about 1 kilometer of each other on neighboring farms in the same valley, and both were discovered during agricultural mechanical drainage-tile excavations designed to remove excess water from the ground surface to promote cultivation. The Schaefer site, initially discovered in 1964, was not formally investigated until a 1992–1993 project directed by archaeologist Daniel Joyce of the Kenosha Public Museums in Kenosha in cooperation with the Great Lakes Archaeological Research Center in Milwaukee. [Figure 19.3] While working at Schaefer, the archaeological research team was alerted to the similar find at Hebior, which was subsequently excavated in 1994 by the Great Lakes Archaeological Research Center. As Schaefer is the more extensively and comprehensively documented of the two localities to date, it best serves as the primary focus of this discussion.

The Schaefer site is located about 20 meters from the edge of a ponded drainage on top of Stratum 5a, a layer of fine silty clay containing numerous organic fragments and the remains of molluscan species which indicate the presence of a more or less still-water environment at the time of the site's formation. The Schaeffer mammoth bones were deposited on this layer and shortly thereafter encased and covered by Stratum 3e, a thick layer of pond sediment in which organic material and shells are less common than in the lower level. This pond sediment layer was in turn covered by a relatively thin layer of peat (Stratum 2b), and the site was then capped by the thick layer of organic muck soil (Stratum 2a) that comprises the present-day ground surface. The stratigraphy observed at Hebior is roughly similar to that observed at Schaefer, but includes additional layers of sand and gravel resulting from its formation in a non-pond environment.

Both the Schaefer and Hebior sites contain mammoth bone beds, each composed of the remains of a single male Jefferson's mammoth (*Mammuthus jeffersoni*), as reported by Daniel Joyce. The Schaefer mammoth bones were concentrated in a 9 square meter area, completely disarticulated (*i.e.*, separated at the joints) and tightly arranged in a configuration that scholars have compared to that of Mammoths 1 and 2 at Blackwater Draw (see pages 46-57). [Figures 19.4 and 19.5] With the exception of a few foot bones and a single humerus, the Schaefer mammoth's forelimbs appear to have been moved elsewhere at the time of the butchering event and thus remain unrecovered. [Figure 19.6] Ten of the bones identified in the Schaefer bone bed show a total of thirty human-made cut marks and wedge marks (*i.e.*, indications on the bone surface that a wedge was used to pry apart the butchered mammoth), and it appears that the butchering operation commenced with the animal's legs before its body was disarticulated and the pieces were stacked. Only one of the mammoth's tusks was recovered, which led the excavators to surmise that it was removed to permit turning over the skull during butchering. [Figure 19.7] Archaeologist Daniel Joyce, who has extensively researched and written about the remains from Schaefer and Hebior, has concluded that the mammoth was freshly butchered and probably hunted.

Unlike the Schaefer mammoth bone bed, which was more or less only peripherally affected by the excavation of the 46 centimeter wide drainage-tile trench, the trench excavation at Hebior traversed the central portion of the deposit. The bones were nonetheless concentrated, like those at Schaefer, but not as fully disarticulated and appear to have undergone only minor movement following the site's deposition. The Hebior mammoth shows nine cut marks, most of which were identified on the bones of its feet. The configuration

FIGURE 19.3 Chris Foos takes "aerial" photographs from a dredge bucket, 10 October 1992, of the exposed bed of mammoth bones at Schaefer. Landowner Franklin Schaefer watches on the left. Courtesy Kenosha Public Museums.

FIGURE 19.4 (above) Overhead view of the exposed Schaefer bone bed. Multiple lines of evidence indicate that the bones are in their primary depositional position. The "bar" running across the lower portion of this photograph is the 46 centimeter wide 1964 tiling machine trench. The femur in the center has a pelvis below and above. To the right of the femur is the cranium. Ribs and vertebrae are scattered throughout. (see Figures 19.5 and 19.12) Photograph by Chris Foos.

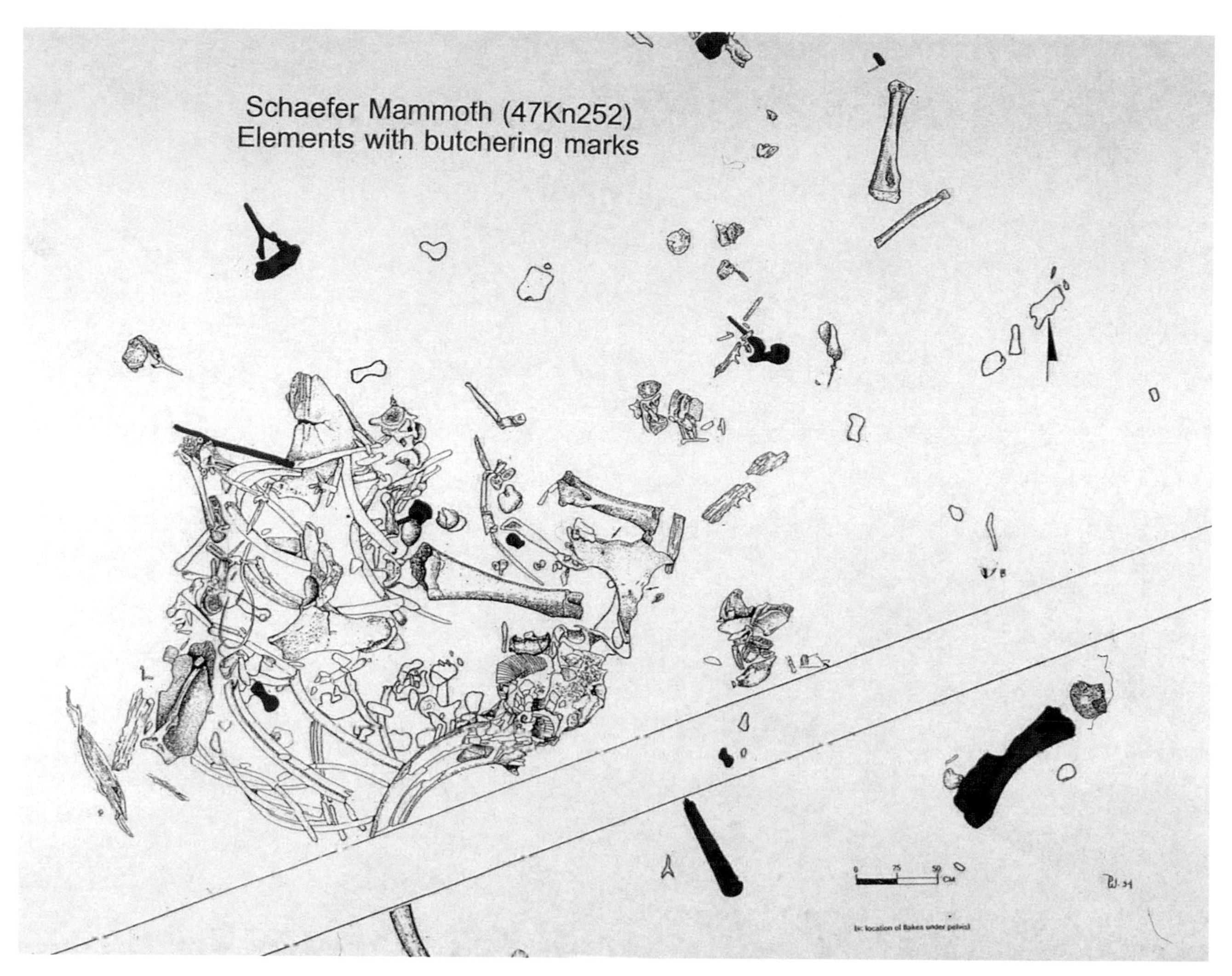

FIGURE 19.5 (above) Plan map of the Schaefer bone bed pictured in Figure 19.4. Bones with cut marks in red. Courtesy Kenosha Public Museums.

FIGURE 19.6 (left) Mammoth tusk excavated at Schaefer. As with all proboscideans, mammoth tusks are enlarged incisor teeth. Mammoths had particularly long tusks, which were more curved than those of modern elephants. Courtesy Kenosha Public Museums.

FIGURE 19.7 (right) Detail of Schaefer mammoth bone with cut mark. Courtesy Kenosha Public Museums.

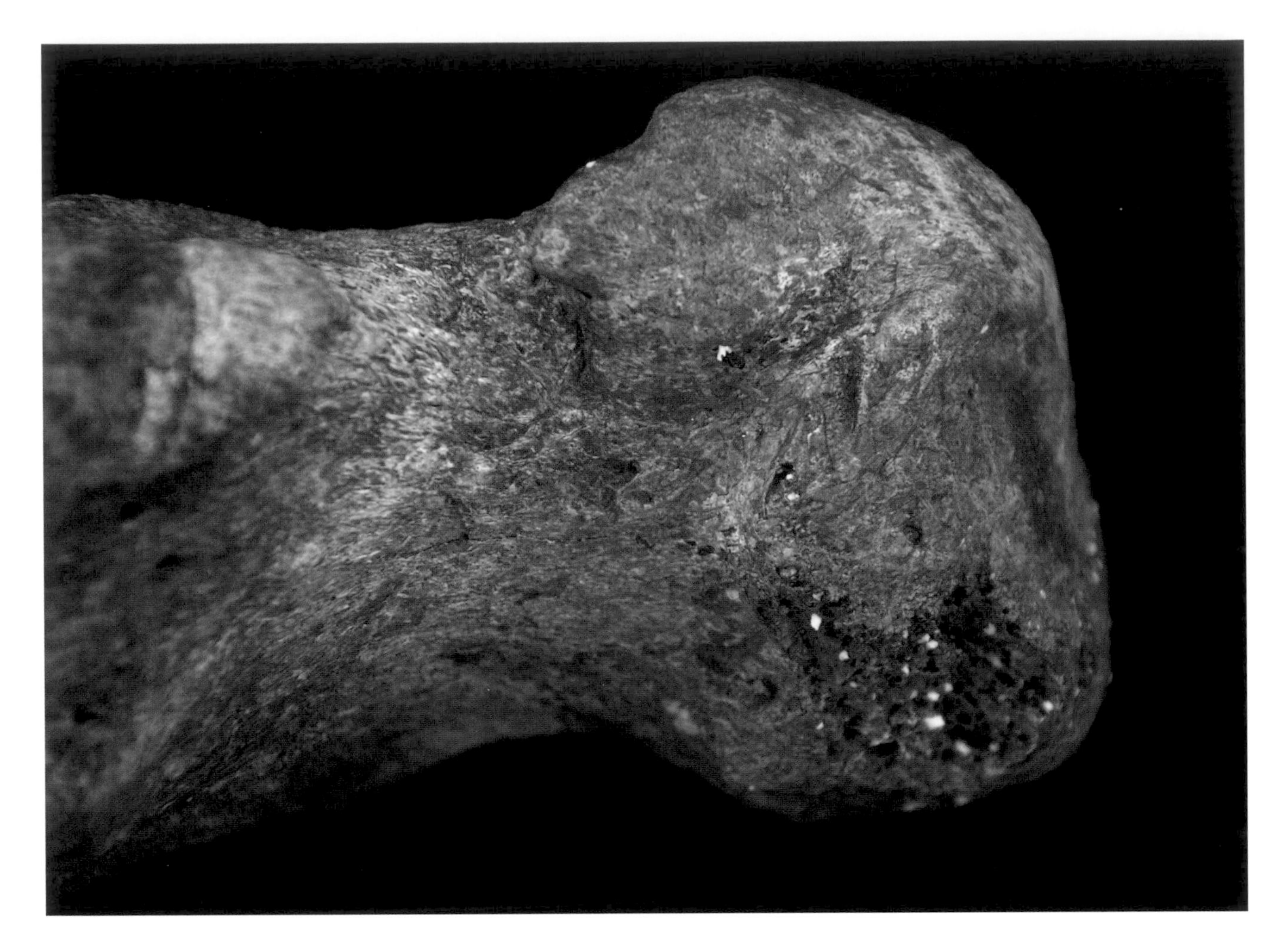

FIGURE 19.8 Detail of mammoth bone from Hebior bearing V-shaped cut marks characteristic of those made by Paleoindians during butchering. Courtesy of the Milwaukee Public Museum.

and location of the cuts suggest that the animal was butchered when fresh. [Figure 19.8] In contrast to the Schaefer mammoth bone bed, which appears to have been quickly buried following the butchering event, the Hebior remains are weathered and thus appear to have been exposed to the elements for a lengthy period of time prior to their burial. [Figure 19.9]

Both sites produced only very small numbers of lithic artifacts, but all were in direct association with the mammoth remains. The two artifacts from Schaefer, found underneath the mammoth's pelvis, include a small blade-shaped flake (produced from a bifacial core) and a larger blade that appears to have been heat-treated. [Figure 19.10] Neither of the artifacts show use wear, and both are made of locally occurring chert. The Hebior site yielded four artifacts, including a chert flake, two chert bifaces, and a dolomite chopper. [Figure 19.11] The bifaces were recovered in association with vertebrae, and both show use wear resulting from the processing of meat and/or hide. No artifacts were recovered from the sites that could be attributed to a particular cultural period, although the presence of the two blades at Schaefer suggests the beginnings of a core and blade industry lacking the more-sophisticated technology witnessed at later Clovis sites.

FIGURE 19.9 (above) The Hebior Jefferson's mammoth fossil skeleton on exhibit at the Milwaukee Public Museum. This extinct North American elephant reached 4 meters at it shoulders and weighed as much 10 tonnes. This mammoth ranged across the United States and Mexico and as far south as Costa Rica. It preferred open spruce parkland, and ted on sedge and grass, both available at Schaefer and Hebior once glaciers retreated northward. Courtesy of the Milwaukee Public Museum.

FIGURE 19.10 (left) Two lithic artifacts excavated during the second season at the Schaefer site. They were located beneath and in contact with the pelvis. Courtesy Dan Joyce, Kenosha Public Museums.

FIGURE 19.11 Lithic artifacts found in close association with the Hebior Mammoth, a flake, two bifaces and a dolomite chopper. The chopper on the left is seen *in situ* in Figure 19.12. Courtesy of the Milwaukee Public Museum.

Unfortunately, the limited artifact assemblages from both sites and the absence of cultural features (*e.g.*, fire pits, living floors, *etc.*) apart from the bone beds preclude any direct observations about subsistence of the humans who visited Schaefer and Hebior, other than that they butchered mammoths at these localities. In very general terms it may be assumed that, given the sites' geographic position, their visitors would have exploited the broad variety of plants and animals that were available in a transitional zone mosaic of open boreal forest, tundra, mixed coniferous-deciduous forest, and pine woodlands.

The mammoth remains from Schaefer were comprehensively dated via fifteen AMS assays on purified bone collagen that range in age from as early as 15,150–14,630 yr BP to as late as 14,630–14,030 yr BP. [Figure 19.12] The three AMS assays available for Hebior, also run on purified bone collagen, range from as early 15,170–14,670 yr BP to as late as 15,040–14,270 yr BP. The age ranges for both sites indicate that they are at least 1,000 years older than the Clovis horizon and perhaps as many as 2,000 years older than Clovis.

The Chesrow complex was first proposed, tentatively and preliminarily, by former Marquette University archaeologist David Overstreet to explain an apparent late Paleoindian phenomenon in the western Lake Michigan drainage basin whose lithic artifact forms bore broad similarities to those from a diverse variety of Clovis-era contemporaries and later Paleoindian analogs in the Great Lakes basin. In contrast to those sites, however, the lithic industries at Chesrow localities

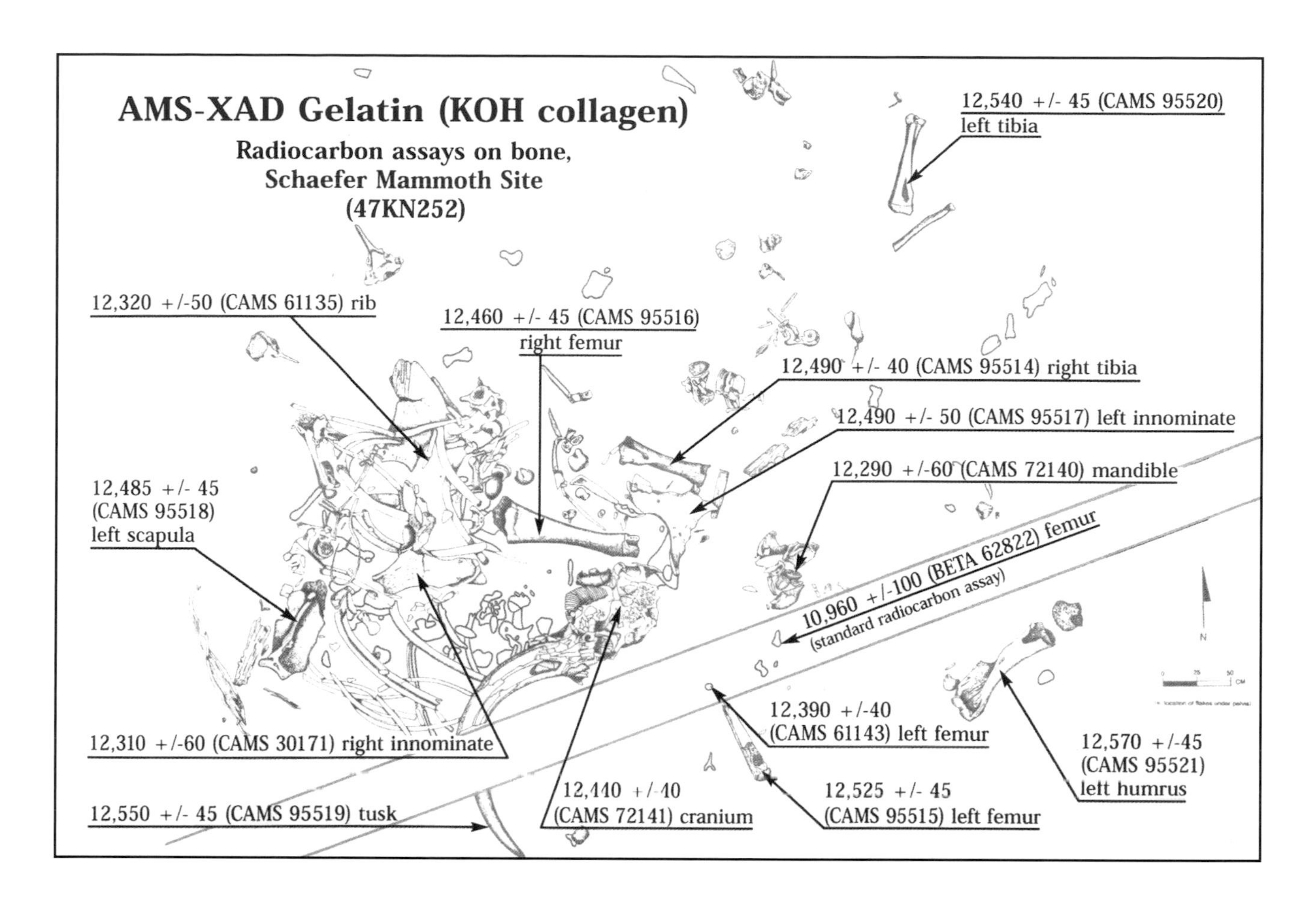

FIGURE 19.12 The mammoth remains from Schaefer were comprehensively dated via fourteen AMS (accelerator mass spectrometry) assays on purified bone collagen that range in age from as early as 15,150–14,630 yr BP to as late as 14,630–14,030 yr BP. (The ages shown in the diagram are presented in uncalibrated radiocarbon years.) Courtesy Kenosha Public Museums.

focused on locally available, poor-quality raw materials rather than exotic tool stone from far-flung quarries, and that apparent preference led to a cruder technology that sometimes involved the recycling of artifacts from earlier archaeological components.

The discoveries of mammoth remains at Schaefer and Hebior in direct association with artifacts, and the apparent similarity between Chesrow lithic technology and that witnessed at the two sites, suggested to Overstreet that despite obvious problems with Chesrow chronology (the sites are undated), "Chesrow remains the most viable candidate for human-mammoth interaction in the southeastern Wisconsin." Archaeologist Daniel Joyce, on the other hand, argues for a decoupling of Schaefer and Hebior from the Chesrow complex, reasonably asserting that "until we recover a Chesrow point associated with a butchered mammoth or mastodon in the Western Great Lakes, or Chesrow projectile points *in situ* below intact Clovis deposits, we have no evidence of a Chesrow/mammoth link." Whatever their ultimate cultural affiliation, the Schaefer and Hebior sites provide convincing evidence of human-mammoth interaction that pre-dates the Clovis horizon. Work at the sites and in the region is ongoing, and holds great promise to refine our understanding of Late Pleistocene lifeways in near-glacial environments which had long been thought to be uninhabitable.

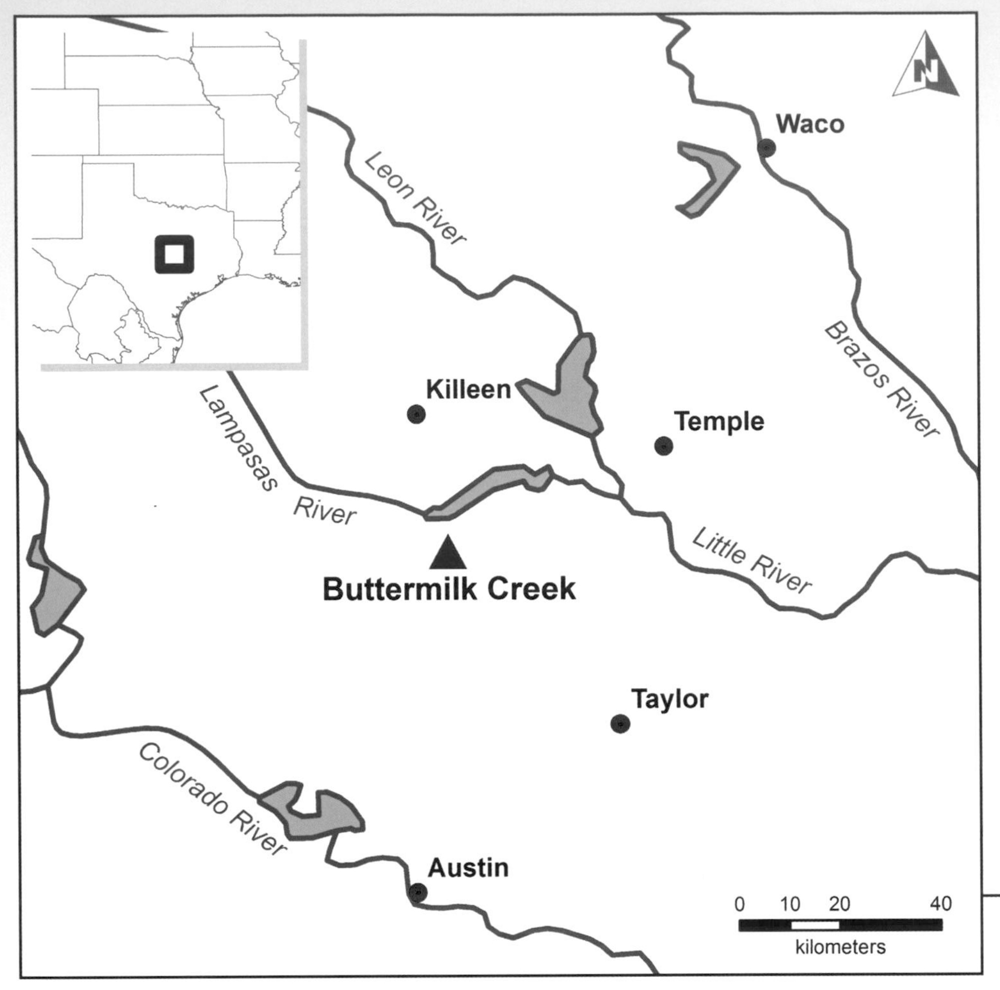

Map by David Pedler

Debra L. Friedkin and Gault—two of the most important Paleoindian sites in North America—are located along Buttermilk Creek in a small wooded valley just at the point where three springs come together to form Buttermilk Creek, which has never gone dry in historic times. The creek flows west to east for 13 kilometers to its confluence with the Salado Creek, which eventually joins the lower half of the Brazos River, empting into the Gulf of Mexico.

20 BUTTERMILK CREEK (DEBRA L. FRIEDKIN AND GAULT)

LOCATION BELL COUNTY, TEXAS, UNITED STATES

Coordinates 30°53'32.00"N, 97°42'35.00"W.

Elevation 275 meters above mean sea level.

Discovery Debra L. Friedkin and Michael R Waters in 1961; Gault, James E. Pearce in 1929.

In contrast to their North American pre-Clovis contemporaries, the Buttermilk Creek sites of central Texas are quite far removed from the Atlantic Coast and the southern extent of the Laurentide ice sheet during the Last Glacial Maximum (26,500–19,000 yr BP). Composed of two sites known as Debra L. Friedkin and Gault, the Buttermilk Creek localities are further distinguished by the presence of a well-defined (and in the case of Gault, very large) Clovis horizon that overlays the earlier materials. [Figures 20.1 and 20.2] Collectively ascribed to a cultural complex named after the stream that runs immediately to the south, the pre-Clovis levels at these two sites share technological characteristics with their later Clovis-age levels. Interestingly, however, the Buttermilk Creek complex also shows some affinity with the stone technologies witnessed at other far more northerly pre-Clovis sites, none of which contained overlying Clovis materials. Separated by a distance of about 250 meters, it is probable that Friedkin and Gault represent a more or less continuous, single site composed of a series of seasonal residential occupations. [Figure 20.3]

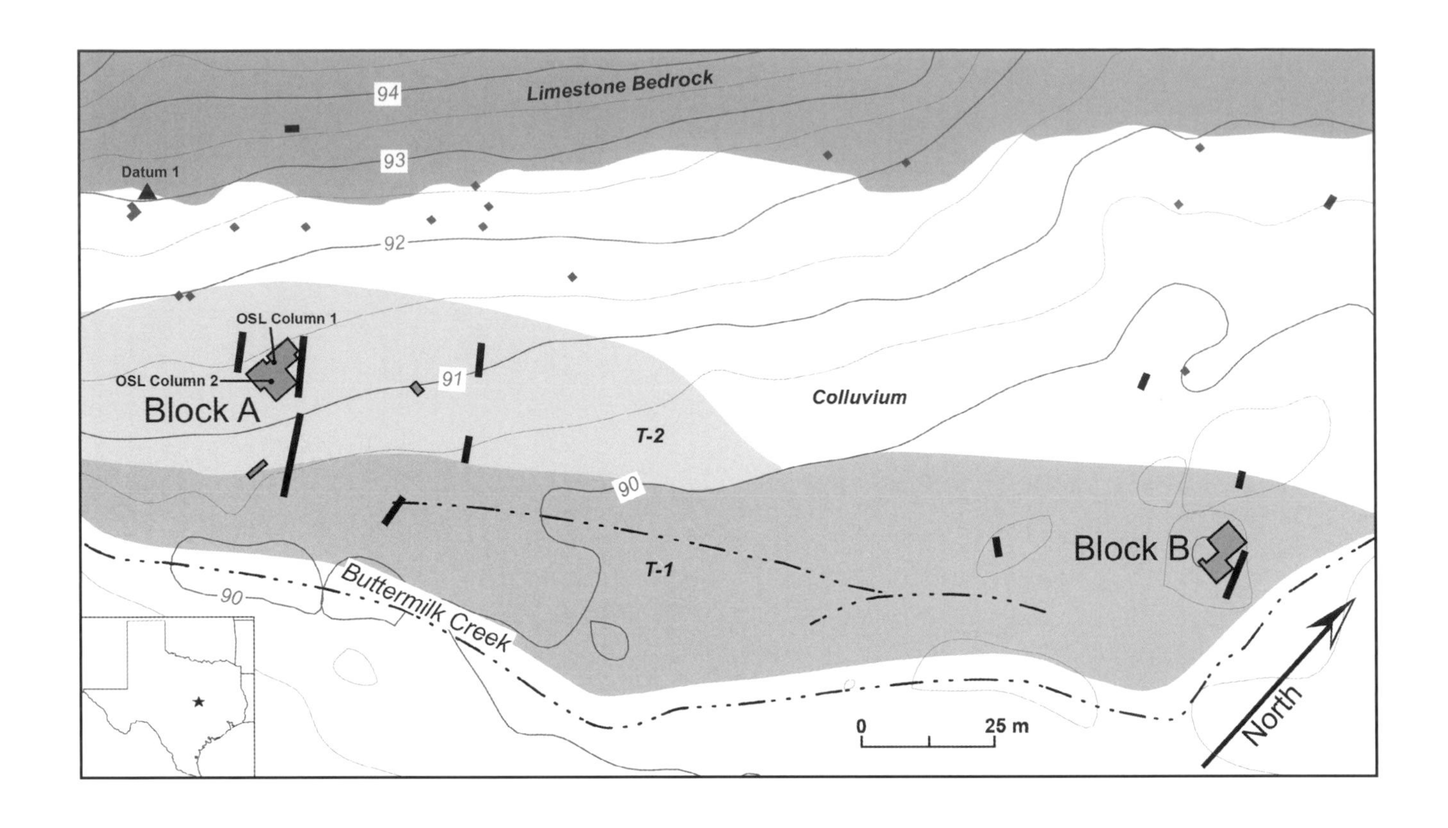
Limestone Bedrock
94
93
92
91
90
Datum 1
OSL Column 1
OSL Column 2
Block A
Colluvium
T-2
T-1
Block B
Buttermilk Creek
0
25 m
North

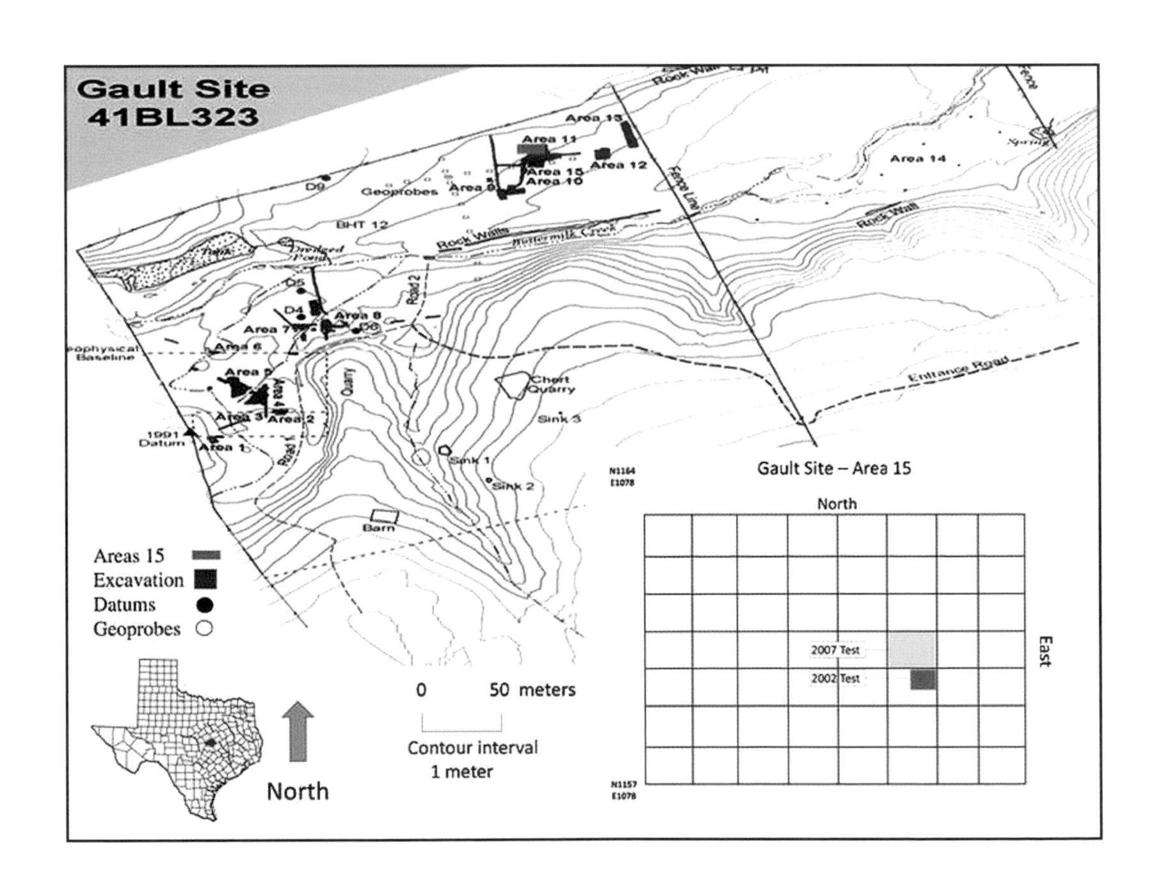
Gault Site
41BL323
Area 13
Area 11
Area 12
Area 15
Area 10
Area 9
Geoprobes
D9
BHT 12
Rock Walls
Buttermilk Creek
Fence Line
Area 14
Rock Wall
Road 2
Area 8
Area 7
Area 6
Area 5
Area 3
Area 2
Area 1
1991 Datum
Quarry
Chert Quarry
Sink 3
Sink 1
Sink 2
Entrance Road
Barn
Road 1
Areas 15
Excavation
Datums
Geoprobes
North
0 50 meters
Contour interval
1 meter
Gault Site – Area 15
North
East
2007 Test
2002 Test

FIGURE 20.1 (left, top) Friedkin site map with excavation areas and trenches (black rectangles and squares). Courtesy Center for the Study of the First Americans.

FIGURE 20.2 (left, bottom) The Gault site sprawls over 16 hectares along Buttermilk Creek. Courtesy of Gault School of Archaeological Research.

FIGURE 20.3 (above) In this aerial view of Buttermilk Creek, the two white tents (center) are the Gault excavation structures. To the right of the two tents, there is a gravel driveway leading from two houses that crosses the valley. On the right of this (on the right margin of this photograph) there is a small rectangular field and single house. The field and house are on the Friedkin site. Courtesy of Gault School of Archaeological Research.

Friedkin and Gault are located in southwestern Bell County, Texas, about 70 kilometers north of Austin and about 330 kilometers northwest of Galveston on the Texas Gulf Coast. In physiographic terms, the sites lie within a portion of the Edwards Plateau called the Lampasas Cut Plain, a landform characterized by a mesa-type topography interrupted by wide lowland valleys between the mesa uplands and, in the vicinity of the sites, closely bounded by the adjacent Live Oak-Mesquite Savanna and Blackland Prairie. This unique setting occupies a resource-rich transitional zone between limestone uplands and the more distant coastal plains, making the area what Gault's principal investigator Michael Collins refers to as "a special place" that has been so "for a very long time." In hydrological terms, the region lies within the Brazos River basin, an expansive 115,000 square kilometer watershed that originates in eastern New Mexico and stretches over 1,000 kilometers to its mouth on the Gulf of Mexico

FIGURE 20.4 View of the 1930 excavations at Gault under the direction of James E. Pearce. Pearce focused on one of the most obvious archeological features at the site—deposits of domestic waste (midden)—the thick rocky layer extending from the surface to below the workmen's waists. Pearce and his colleagues recovered abundant Late Prehistoric (1,300 yr BP to historic times) and Archaic (8,000–1,300 yr BP) materials and, although not recognized at the time, a small assemblage of Clovis artifacts. Courtesy of Gault School of Archaeological Research.

south of Houston. Friedkin and Gault occupy the first and second stream terraces of the alluvial flood plain of Buttermilk Creek, a minor tributary of the Lampasas River that joins the Brazos about 100 kilometers east of the sites.

Presently, the Buttermilk Creek sites have ready access to potable water and are surrounded by grassland and open woodland composed of burr oak, walnut, pecan, ash, elm, bois d'arc, willow, and cottonwood. Though the species compositions have doubtlessly changed, these conditions appear to have been roughly similar during the sites' earliest human occupations. Judging from the faunal remains recovered, particularly from Gault, the local environment supported a diverse array of now-extinct larger animals such as horse, bison, and mammoth along with smaller mammals, frogs, and birds. Interestingly, the remains of mammoth, horse, and bison are limited to the lower cultural deposits at Gault and are overlain only by bison, which suggests to the excavators that the Clovis cultural interval at Gault appears to have spanned the regional extinction of horse and mammoth. Together, the Buttermilk Creek complex sites encompass an area of about 17 hectares. The larger and longer running of the two excavations has been conducted at Gault.

Initial work at Gault was conducted in 1929–1930 by University of Texas archaeologist James E. Pearce, who was granted access to the property by landowner and farmer Henry Gault. [Figure 20.4] Apparently focusing their investigation on an Archaic period (8,000–1,300 yr BP) layer of domestic waste (called a midden), Pearce and his colleagues also recovered a small assemblage of Clovis artifacts along with the far more abundant materials from the Archaic and Late Prehistoric (1,300 yr BP to historic times). The site lay unprotected after Pearce's investigation, and as the property's ownership changed hands over the course of several decades Gault was thereafter subjected to extensive, large-scale artifact collection and looting, apparently by very large groups of collectors who sometimes resorted to the use of heavy machinery. A commercial "pay to dig" operation continued at the site into the 1990s, charging collectors as much as 25 dollars per day to further destroy the site.

Limited excavations at the site in 1991 under the auspices of the University of Texas at Austin and the Texas Historical Commission confirmed that intact Paleoindian deposits lay beneath Gault's irreparably damaged Archaic period deposits, but the looting continued. A change in the site's ownership ultimately led to the its protection and the commencement of professional investigations in 1998 by archaeologists Michael Collins and Tom Hester along with University of Texas paleontologist Ernest Lundelius. A subsequent 1999–2002 excavation under the direction of Collins recovered more than 1.2 million artifacts—about half of which are of Clovis age—and continuous professional research has been conducted since the transfer of the property to the Gault School of Archaeological Research in 2007. The adjacent Friedkin site, on the other hand, has been spared such depredation, and professional archaeological investigations began there under the direction of Michael R. Waters of Texas A&M University in 2006.

Both Buttermilk Creek sites have been demonstrated to contain lengthy occupational sequences encompassing every prehistoric period from Late Prehistoric through Archaic and Paleoindian times. At Gault, whose entire prehistoric deposit encompasses an extraordinarily large area of about 16 hectares, the 3 hectare Clovis occupation (dating 13,100–12,750 yr BP) is underlain by a thin, sterile (*i.e.*, artifact-free) layer which is in turn underlain by pre-Clovis deposits referred to by Collins as "Older–than-Clovis." [Figure 20.5] Gault's Paleoindian material has, to date, been far more extensively examined than the deeper, more difficultly accessed pre-Clovis deposit. Gault's abundant Clovis material assemblage numbers in the hundreds of thousands and includes stone tool chipping debris, a large and diverse array of Clovis points and other stone tools (*i.e.*, adzes, bifacial knives, endscrapers on blades, gravers on blades, and serrated blades), bone tools from several faunal species, and a large number of associated pieces of bone, ivory, and teeth.

The apparently smaller, underlying pre-Clovis deposit at Gault displays broad technological similarities to Clovis technology in the form of prismatic blades and, somewhat more vaguely, to Clovis's blade tools and bifaces. This site level also contains distinctly unique biface and projectile points (including fragments of two very small, expanding stem unfluted projectile points and at least one unfluted lanceolate biface) that are quite different from Clovis. [Figures 20.6 and 20.7] The pre-Clovis level at Gault also contains a distinct and rare cultural feature for the time frame: a 4 square meter stone pavement composed of naturally shaped rocks that includes two arc-shaped artifact concentrations, one composed of the bones of large mammals and the other composed of flaked stone chipping debris. [Figure 20.8] Several of the flaked stone artifacts have been heat-treated and the snapped blade artifacts identified in the assemblage show evidence of use wear, all of which suggest the presence of a food processing activity area or even a domestic structure. [Figure 20.9]

At Friedkin, more than 15,000 artifacts have been recovered from the pre-Clovis levels in one of two areas excavated at the site. This Buttermilk Creek complex assemblage at the site is overwhelmingly dominated by stone chipping debris but also includes twelve bifaces, one core, twenty-three edge-modified flake tools, five blade fragments, fourteen bladelets, and one piece of polished hematite. The excavators have identified three stone tool production technologies for the site: preform, chopper/adze, and core production. Of the bifaces from

FIGURE 20.5 (above) The cultural levels at Gault. Artifacts from the Late Prehistoric period are minimal at Gault, consisting of a few projectile points and pottery sherds. Most of the Archaic period deposits have been completely disrupted by previous unregulated looting. Mixed in with scattered fragmentary dart points are cans, bottles, and other modern trash. Below this are soils containing modest number of projectile points from the Late Paleoindian period (11,600–9000 yr BP in Texas). Folsom age (12,900–11,000 yr BP in Texas) artifacts are also sparse as Gault. Underlying these four upper deposits, are Clovis age (13,300–12,700 yr BP in Texas) artifacts numbering in the hundreds of thousands. The Clovis occupation zone in underlain by a thin sterile (*i.e.,* artifact-free) layer which is in turn underlain by pre-Clovis deposits, including bifaces and projectile points that are quite different from Clovis. Courtesy of Gault School of Archaeological Research.

FIGURE 20.6 (right, top) Comparison of Clovis and Older than Clovis flake tools and cores from Area 15 at Gault: (a–f) Clovis and (g–l) Older than Clovis. Courtesy of Gault School of Archaeological Research.

FIGURE 20.7 (right, bottom) Comparison of Clovis and Older than Clovis biface technologies from Gault Area 15: (a–l) Clovis and (m–x) Older than Clovis. Courtesy of Gault School of Archaeological Research.

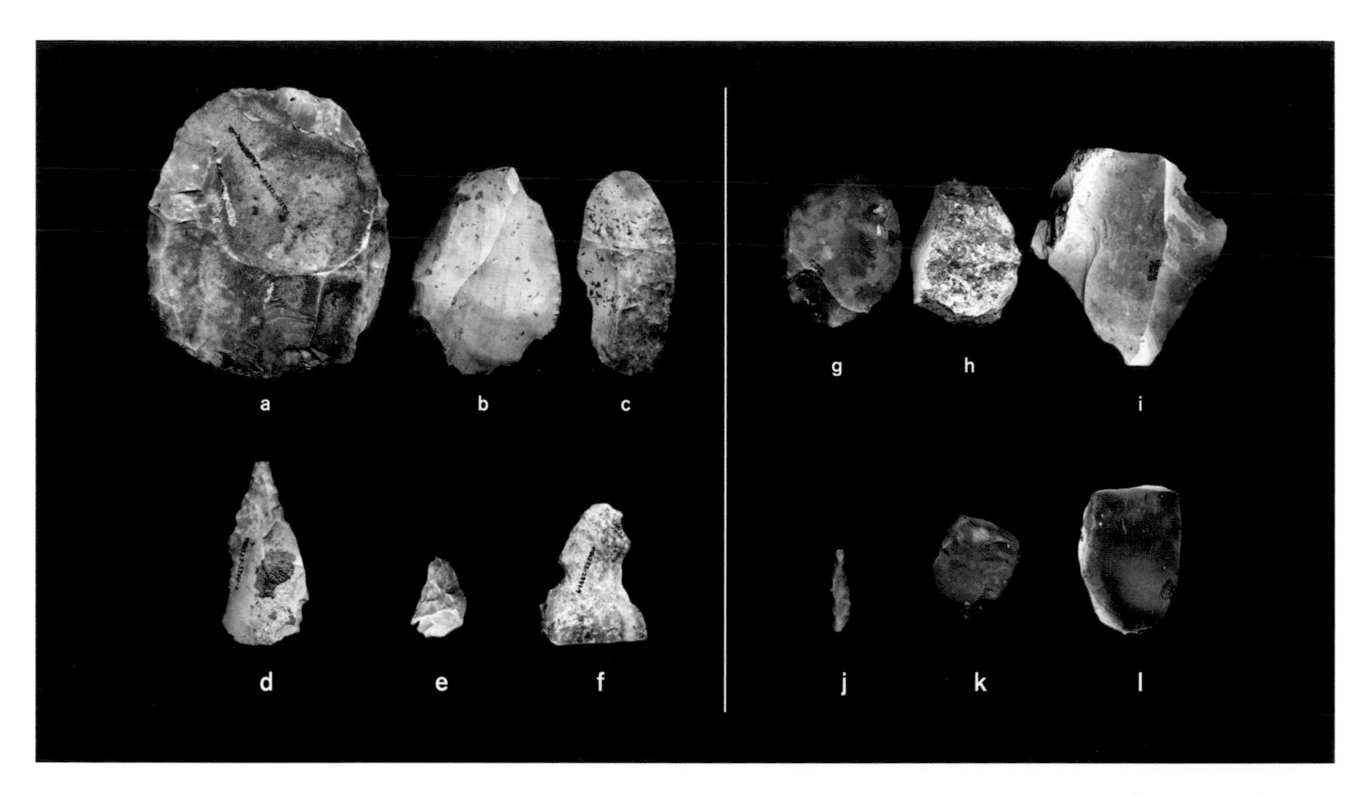
a
b
c
d
e
f
g
h
i
j
k
l

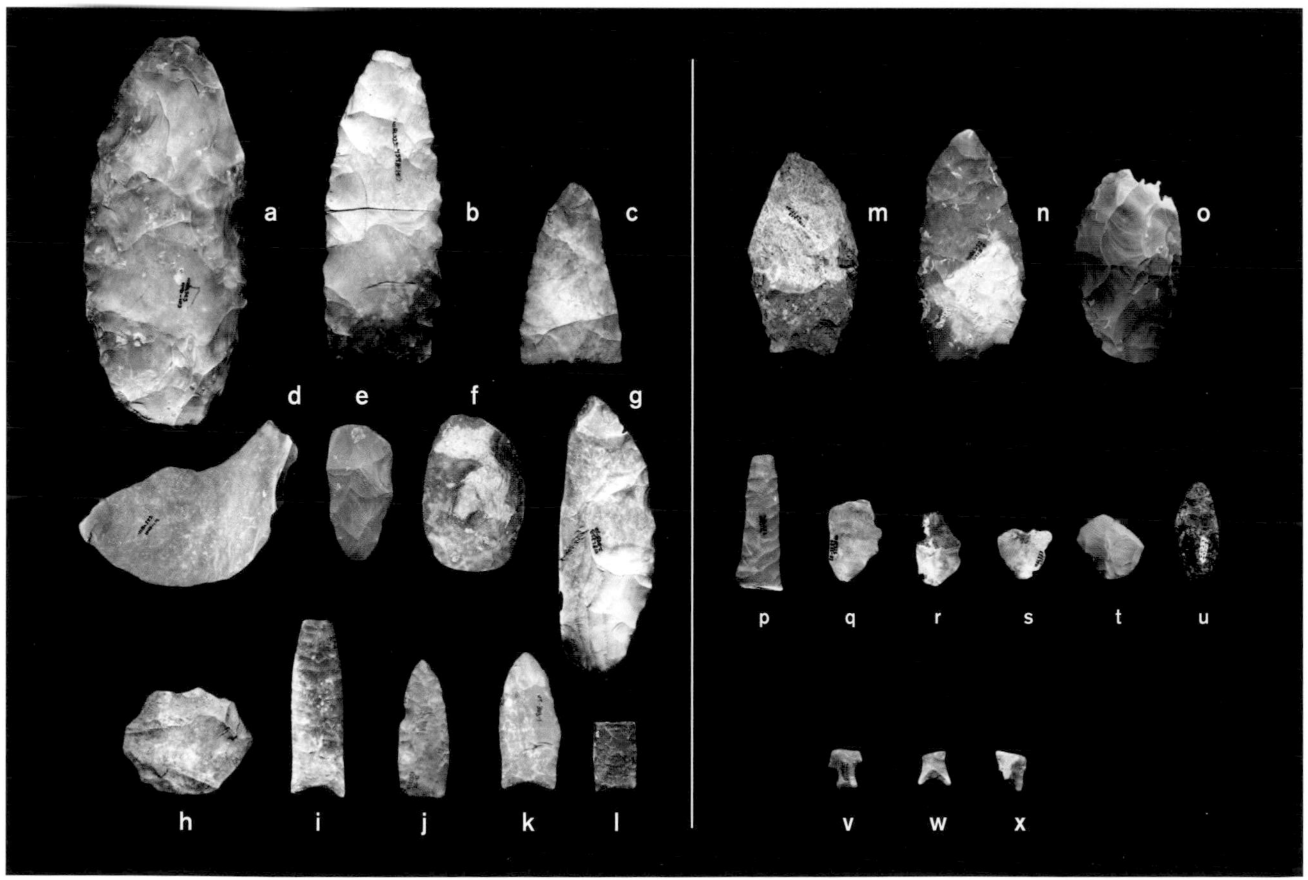
a
b
c
m
n
o
d
e
f
g
p
q
r
s
t
u
h
i
j
k
l
v
w
x

FIGURE 20.8 (above) One of the most the most unusual finds at Gault is a 4 square meter stone pavement aligned to the four cardinal directions. Courtesy of Gault School of Archaeological Research.

FIGURE 20.9 (below) During the 1991 excavations at Gault, six engraved limestone rocks were recovered. Subsequently, an additional twenty-one Clovis-age engraved stones have been earthed at the site, representing the earliest examples of representational art in North America. In this example, parallel lines intersect other lines at various angles to form rectilinear or diamond-shaped girds, a motif common among these incised artifacts. A Clovis-age engraved stone has been recovered at Shawnee-Minisink (see page 85) and a younger Folsom-age one at Blackwater Draw. Courtesy of Gault School of Archaeological Research.

the site, ten have been classified as late-stage fragments, and microscopic analysis of the blades and bladelets indicates the presence of use wear. [Figures 20.10 and 20.11]

Because organic preservation was poor in Friedkin's artifact-bearing floodplain clay deposits, which precluded conventional radiocarbon dating, forty-nine optically stimulated luminescence (OSL) ages were obtained from two stratigraphic columns excavated on the Buttermilk Creek floodplain's second terrace. The temporal sequence indicated by diagnostic stone artifacts from Friedkin closely corresponds to that identified at Gault (though with perhaps fewer late Paleoindian artifacts from the latter site), and the forms recovered from those temporal horizons occur in correct stratigraphic order and in close correlation with the OSL ages obtained from the site deposits. The OSL dates from Friedkin suggest that its pre-Clovis occupation dates between 15,500 yr BP and 13,200 yr BP, which is consistent with the age range obtained from corresponding levels at Gault. [Figure 20.12]

The Buttermilk Creek sites occupy a unique position in pre-Clovis archaeological research, in that they both contain the unprecedented combination of prolific Clovis and pre-Clovis levels in their exceptionally long-lived archaeological records. The Buttermilk Creek complex stone tool kit, which is generally lightweight and small in size, was procured almost exclusively from extensive local outcrops of Edwards Plateau chert. In the opinion of Friedkin site investigator Michael Waters, this tool kit "also provides an ancestral assemblage from which the biface- and blade-dominated Clovis tool kit could have evolved." No organic materials have been identified at either site to date, but use wear noted on stone tools indicates they were employed in the processing of organic materials. These sites have also been instrumental in refining the archaeological understanding of Clovis. In stark contrast to the once-held notion of Clovis as a highly mobile social unit focused on the hunting of large game, the excavators believe that Gault and Friedkin were frequently re-used during Clovis times by visitors who followed a more broadly based subsistence lifestyle in terms of dietary choice (*e.g.*, frogs, birds, small mammals, plants, *etc.*), woodworking, and plant processing. [Figure 20.13]

FIGURE 20.10 (next spread, left) Friedkin artifacts. Late Prehistoric (a) Perdiz; Archaic (b) Castroville, (c) Edgewood, (d) Ensor, (e) Wells, (f) Gary, and (g and h) Angostura; Paleoindian (i and j) Golondrina, (k) Dalton, (l, m, and n) Folsom, (o) Clovis blade, (p and q) Clovis flakes, (r) Clovis point midsection, and (s) Clovis biface tip. Courtesy Center for the Study of the First Americans.

FIGURE 20.11 (next spread, right) "Older than Clovis" Buttermilk complex artifacts recovered at Friedkin: (a) lanceolate point, (b) chopper/adze, (c) disk-shaped flake core, (d) broken flake with notch, (e) engraving tool, (f, g, and h) flake tools, (i) polished hematite [iron oxide], (j) biface flake, (k) broken flake, (l) broken biface, (m and n) blade midsections, and (o–s) microliths–bladelets. Courtesy Center for the Study of the First Americans.

a
b
c
d
e
f
g
h
i
j
k
l
m
n
o
p
q
r
s

a
b
c
d
e
f
g
h
i
j
k
l
m
n
o
p
q
r
s
t

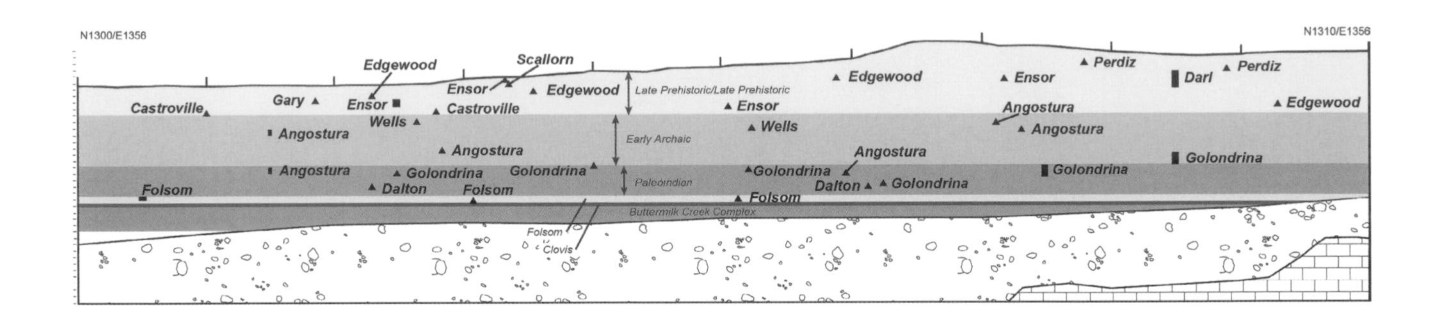

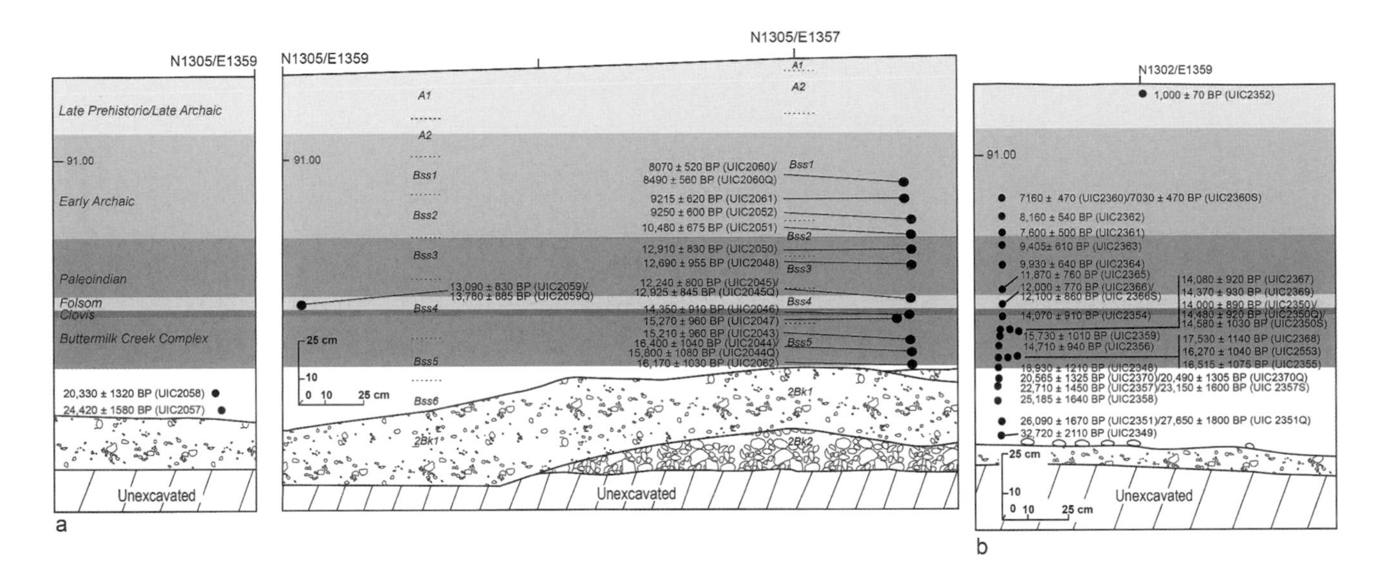

FIGURE 20.12 (above) North to south stratigraphic profile of Friedkin Block A showing the location of lithic complexes seen in Figures 20.10 and 20.11. (below) Block A stratigraphy, archaeological complexes, and optically stimulated luminescence (OSL) ages from column 1 (A) and column 2 (B). BP indicates years before the present. Courtesy Center for the Study of the First Americans.

FIGURE 20.13 Humans have been present along Buttermilk Creek for over thirteen millennia. In fact, they are still present there in the form of the teams of professionals and students who continue to investigate its prehistory at Gault and Friedkin. With its favorable climate, abundant food resources, readily available water from the creek and local springs, and large quantities of Edwards chert for tool making, Buttermilk Creek is what Michael Collins refers to as "a special place" that has been so "for a very long time." Courtesy of Gault School of Archaeological Research.

Pleistocene artifacts from Topper, South Carolina.

CONTROVERSIAL PRE-CLOVIS SITES

The validity of any archaeological site rises or falls on its ability to satisfy the three evaluative criteria (see page 23) that were developed during the course of the historic Glacial Man debate of the nineteenth century and which have guided the method and practice of archaeological science ever since. The following sites—separated by a span of over 8,000 kilometers and ranging from the Arctic Circle to a high plateau in the Colombian Andes Mountains—have thus far failed to do so to the satisfaction of many in the professional archaeological community. Claims of a pre-Last Glacial Maximum temporal provenance for one of these localities, Bluefish Caves (see pages 306-312) in Canada's western Yukon, if ultimately proved to be valid, would radically re-order our understanding of early human life at the entryway to the hemisphere at a time when eastern Beringia was thought to be uninhabited.

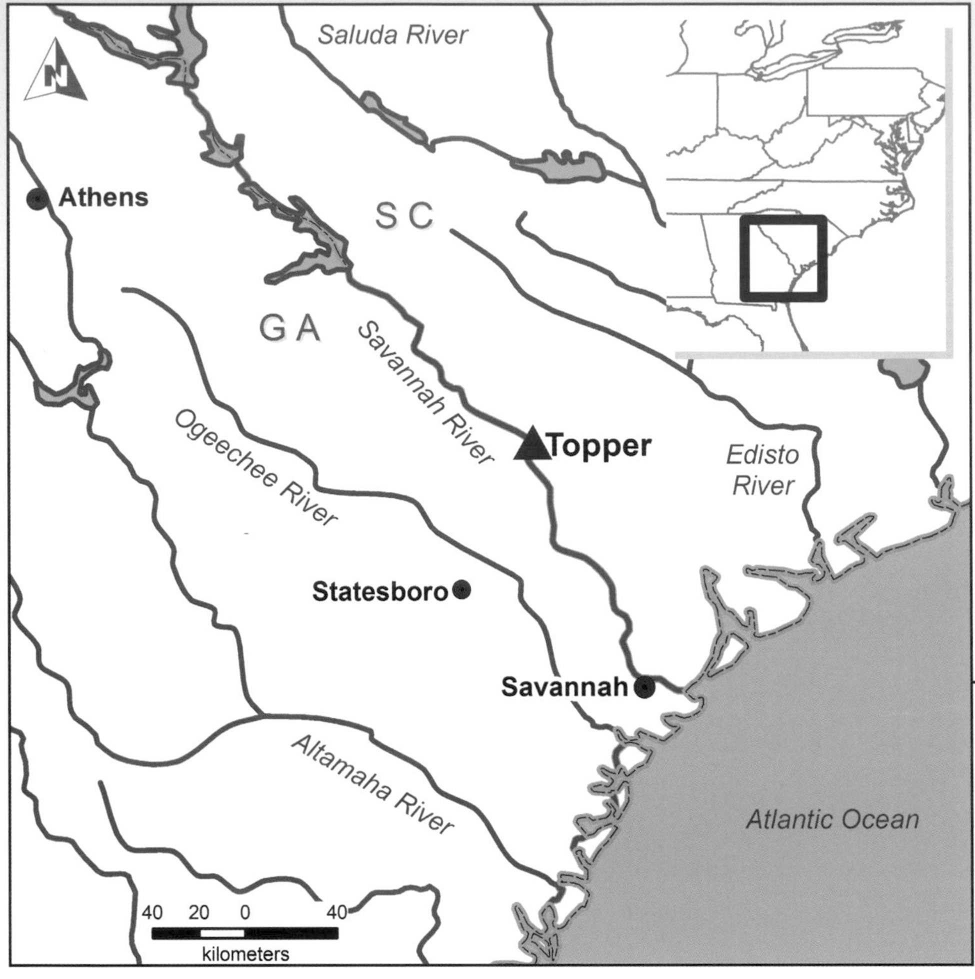

Map by David Pedler

The Topper site is one of a series of prehistoric quarries located on the east bank of the Savannah River, approximately halfway between its headwaters in the Blue Ridge Mountains and where it debouches into the Atlantic Ocean below Savannah, Georgia. It is noted for controversial artifacts believed by some archaeologists to evidence human habitation of the New World earlier than the Clovis culture by as much as 35,000 years.

21 TOPPER

LOCATION ALLLENDALE COUNTY, SOUTH CAROLINA, UNITED STATES

Coordinates restricted.

Elevation 27 meters above mean sea level.

Discovery Albert C. Goodyear in 1998.

The Topper site is an exceptional archaeological locality. It is at once a rare example of a deeply stratified, buried archaeological site on the Atlantic Ocean's southern coastal plain and the only excavated and recorded Clovis site on the coastal plain in Georgia and the Carolinas. The renown of its massive Clovis deposit is very well deserved. The site has also endured as a controversial Paleoindian locality due to claims for a deep antiquity whose earliest reaches are on par with that proposed for the Paleoamerican levels at Pendejo Cave (see pages 162-171).

The Topper site is located in Allendale County, South Carolina, about 70 kilometers southeast of Augusta, Georgia, 110 kilometers north of Savannah, and 17 kilometers west of Allendale, the county seat. The site occupies a terrace on the east bank of the Savannah River, which served as an important inland transportation route in the early Euro-American settlement of the American Southeast. Forming the boundary between Georgia and South Carolina, the river traverses three physiographic regions between its headwaters in the mountains of North Carolina and its outlet on the Atlantic Ocean, flowing from the Blue Ridge province through the Piedmont province to the Coastal Plain province on the east. The 28,500 square kilometer Savannah River basin encompasses an impressive diversity of ecosystems, including upland forests, bottomland hardwood forests, swamps, freshwater and marine marshes, free-flowing tributary streams, a coastal estuary, and a variety of modern developments such as agricultural systems, commercial pine plantations, and dammed reservoirs. Topper is situated within the Coastal Plain, roughly midway between the Atlantic Coast and the eastern foothills of the Appalachian Mountains.

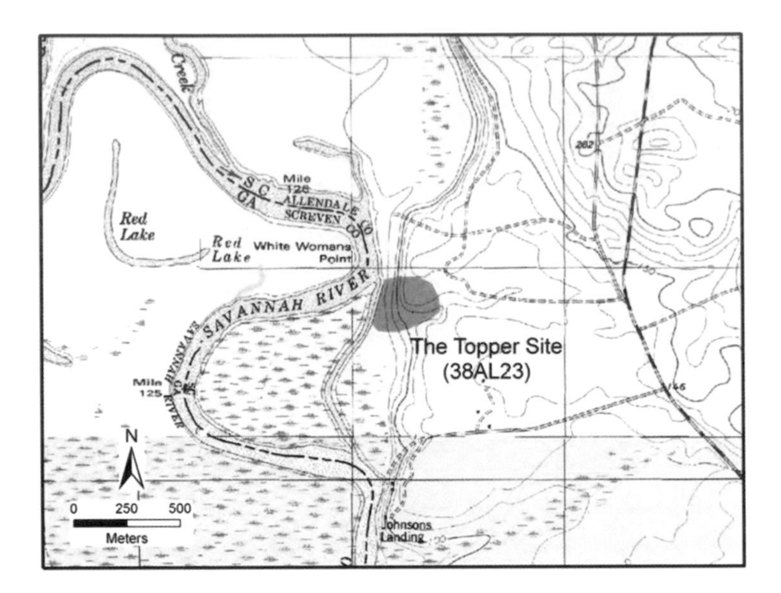

FIGURE 21.1 Throughout its history, Topper has been situated in the species-diverse boundary between upland hardwood forests and coastal plains pines. Courtesy of Shane Miller.

The lightly populated environs of the Topper site are dominated by closed canopy evergreen forest interspersed with sporadic small patches of mixed (*i.e.*, both hardwood and evergreen) woodland and, to the west along the meandering Savannah River channel, bottomland floodplain hardwood forest. These woodlands and their associated wetlands are among the most species-diverse ecological communities in North America, with large, rich populations of both plants and animals. (The Southeastern Coastal Plain's vascular plant species alone account for nearly 25 percent of all such species found in North America.) At the peak of the most recent Ice Age, the Last Glacial Maximum, the region's vegetation was composed of temperate deciduous species such as oaks. But with the gradual warming of the Late Pleistocene and Early Holocene, mixed hardwood species and spruce became common in interior and upland regions while the Coastal Plain province became dominated by southern pines. The Topper site appears to have been positioned on the boundary between these two resource-rich ecosystems. [Figure 21.1]

The Topper site lies at an elevation of 27 meters above sea level on a hilltop overlooking the Savannah River about 130 meters from its east bank and 7 meters above the river. The archaeological deposit is composed of four principal zones: a lower zone on the terrace, an upper hilltop zone elevated about 3–4 meters above the terrace, a third zone on the gently sloping hillside adjacent to the hilltop, and a centrally located chert quarry at the steeply sloping foot of the hill. [Figure 21.2] The site measures about 150 meters from east to west and about 130 meters from north to south. The site's quarry zone is covered by what the Topper research team had first thought to be naturally occurring scree (*i.e.*, downslope-migrating gravel) but which was subsequently discovered to have resulted from prehistoric quarrying of the toolstone that crops out there.

The Topper site was initially identified in 1981 by University of South Carolina archaeologist Albert Goodyear, who was led to the site by forester John Topper, the site's namesake. Goodyear's subsequent 1983–1984 survey of the central Savannah River watershed led him to designate the site as one of a handful of quarries for the Allendale Coastal Plain chert, then a known Archaic period toolstone occurring in beds and as cobbles along the Savannah River. Excavations conducted at Topper by Goodyear in 1986 identified an *in situ* Archaic occupation, which was discovered to be underlain by an intact and extensive Clovis occupation in 1998. In 2004, a major Clovis deposit identified on the Topper hillside was excavated by Mississippi State University archaeologist Shane Miller. In order

FIGURE 21.2 (top) Schematic profile diagram of sediments at Topper. The upper stratum (yellow) of silty sand washed down from the adjacent hillside contains relatively undisturbed and consistently sequenced cultural materials dating from the eighteenth century to Clovis times. The Clovis artifacts found at the base of these sediments have been dated to 13,200 yr. BP. Below these sands are alluvial (river-derived) Pleistocene sands (white) that extend to 2.2 meters below the surface. These have been dated to 15,200–14,800 yr. BP. Chert clusters found at the base of this layer are the basis for the site's pre-Clovis claims. Beneath these sands is a scoured, gray silty clay terrace (red). Similar chert fragments have been found in this soil all the way to its base dated in excess of 50,000 yr BP. Courtesy of Albert Goodyear.

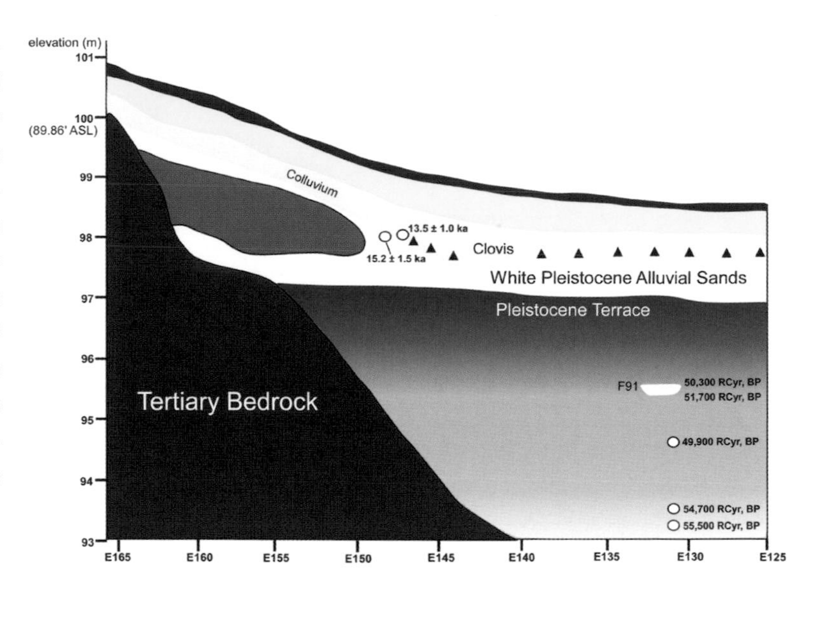

FIGURE 21.3 (bottom) Profile of the Topper stratigraphy. The upper layer of sand was formed by slope wash and the lower Pleistocene sands were deposited by the river atop an ancient river terrace. The upper layer (shown in the background) contained artifacts spanning historic times down to Clovis (upon which the excavator is standing). The Topper Assemblage artifacts were recovered from the Pleistocene terrace. Courtesy of Albert Goodyear.

FIGURE 21.4 A relatively dense accumulation of Clovis points uncovered during the 2005 field season. This is one of a number of clusters of Clovis artifacts found at the base of the upper horizon (Stratum 3b) at Topper. Courtesy of Shane Miller.

to provide an independent assessment of the site's chronology and stratigraphy, Texas A&M University archaeologist Michael Waters and his colleagues conducted archaeological and geologic investigations beginning in 1999 and concluding with the publication of their results in 2009. As of 2013, almost 600 square meters have been excavated at Topper.

The site sediments are composed of at least three discrete geologic horizons. The upper horizon (Stratum 3b) is 1–1.4 meter thick layer of silty sand that washed down the slope and contains a relatively undisturbed and consistent sequence of diagnostic cultural materials dating from the eighteenth century to Clovis times. The Clovis artifacts cluster at the base of this stratum, to which Waters attributes an age of 13,200 yr BP based on an optically stimulated luminescence date obtained from soil. (This technique was employed because *in situ* charcoal, wood, and plant remains suitable for conventional radiocarbon analysis are reported to be rare at the site). [Figure 21.3]

The site's Clovis level is underlain by a 2 meter thick layer of Pleistocene-aged alluvial sand (Stratum 2b) which is in turn underlain by a 2 meter thick layer of silty clay (Stratum 1) that appears to be a buried Pleistocene river terrace. The site's most controversial artifacts, known to some as the "Topper Assemblage," were discovered as far as 2 meters below the Clovis horizon in the lower reaches of Stratum 2b and the upper portion of Stratum 1. Waters ascribes a minimum age of 15,200–14,800 yr BP to the former, again based on optically stimulated luminescence dating of soil. The deeper of these two ancient strata is radiocarbon-dated to an age of >54,700 yr BP, based on the analysis of a hickory (*Carya* sp.) nutshell.

The most distinctive artifacts from the Topper site are undoubtedly those comprising the exceptionally prolific Clovis deposit that has been identified in the terrace and hillside zones of the site, which essentially surround the quarry. [Figure 21.4] Excavations have recovered 40,000 artifacts to date, including numerous and diverse bifaces in all stages of manufacture, fluted-point preforms (*i.e.*, initially worked but unfinished points), fluted points, unifacial tools, blades, denticulates (*i.e.*, cutting tools with multiple notches resembling saw teeth), an unusually broad variety of scrapers, flake tools, cores, and countless pieces of debitage. [Figure 21.5] Several researchers, including Mississippi State University archaeologist Shane Miller, have concluded that the Clovis flintknappers at Topper employed distinctive manufacturing techniques that are commonly associated with Clovis peoples elsewhere. Both scholars note, however, that biface manufacture at the site shows far more variability that is commonly witnessed at Clovis sites, particularly in respect to biface and blade size. Apparently adapting to limitations imposed by initial lithic raw material size and variations in quality—and perhaps even subsistence needs unique to Topper—the site's Clovis stoneworkers demonstrated an uncommon technological flexibility. [Figure 21.6]

In marked contrast, the pre-Clovis artifacts from Topper are rather indistinct and initially were somewhat tentatively described by Goodyear as being "unusual." [Figures 21.7 and 21.8] The crude technology proposed for these artifacts is based on the primary reduction of toolstone and the initial production of tools via the smashing together of chert cobbles. Goodyear and his research team have identified evidence of this activity in lower Stratum 2b in the form of discrete concentrations of broken chert containing numerous small flakes. [Figure 21.9] A large quantity of debitage has been reported among the site's 13,000 pre-Clovis artifacts, which also includes blades, scrapers, a possible microblade core, and hundreds of "burin-like" tools that were apparently produced via the so called "bend-break" technique. (This technique, which uses broadly applied pressure to an entire artifact rather than percussion or pressure flaking at a specific location on the artifact, has been documented in Paleolithic and Paleoindian assemblages.) Unlike the overlying Clovis assemblage—as well as other presumably contemporary pre-Clovis assemblages recovered elsewhere in the eastern United States (see Meadowcroft Rockshelter, pages 200-215, and Cactus Hill, pages 228-235)—definitive evidence of bifacial and unifacial lithic technology is conspicuously absent from the Topper pre-Clovis assemblage. [Figures 21.10 and 21.11]

Despite an initial openness, and even enthusiasm in some quarters, very few professional archaeologists presently accept the validity of the Topper site's pre-Clovis artifacts. The prevailing opinion appears to hold that a vast majority of the assemblage is non-cultural, and that any bona fide artifacts recovered from lower Stratum 2b originated in the acknowledged cultural deposits above, then migrated downward through the site's sand sediments. Similar claims have been made for the pre-Clovis levels at Cactus Hill, but the indisputable human manufacture and distinctiveness of the assemblage at that location has mollified at least some of that site's critics.

The criticism that the Topper site has lacked a thorough, published analysis of its pre-Clovis occupation has very recently been addressed by University of

FIGURE 21.5 (left) This broken 15 centimeter long Clovis biface is called "Alaina's Sword" in honor of the Topper unit supervisor Alaina Williams, who found it. It is one of the Clovis artifacts unearthed in the site's upper strata of a 1–1.4 meter thick layer of silty sand. If it hadn't broken during knapping, the finished point could have been hafted onto a shaft some 13,000 years ago. Courtesy of Albert Goodyear.

FIGURE 21.6 (below) Topper Clovis-age prismatic blades, which are generally similar to blades found at other Clovis sites. Photo by Daryl P. Miller.

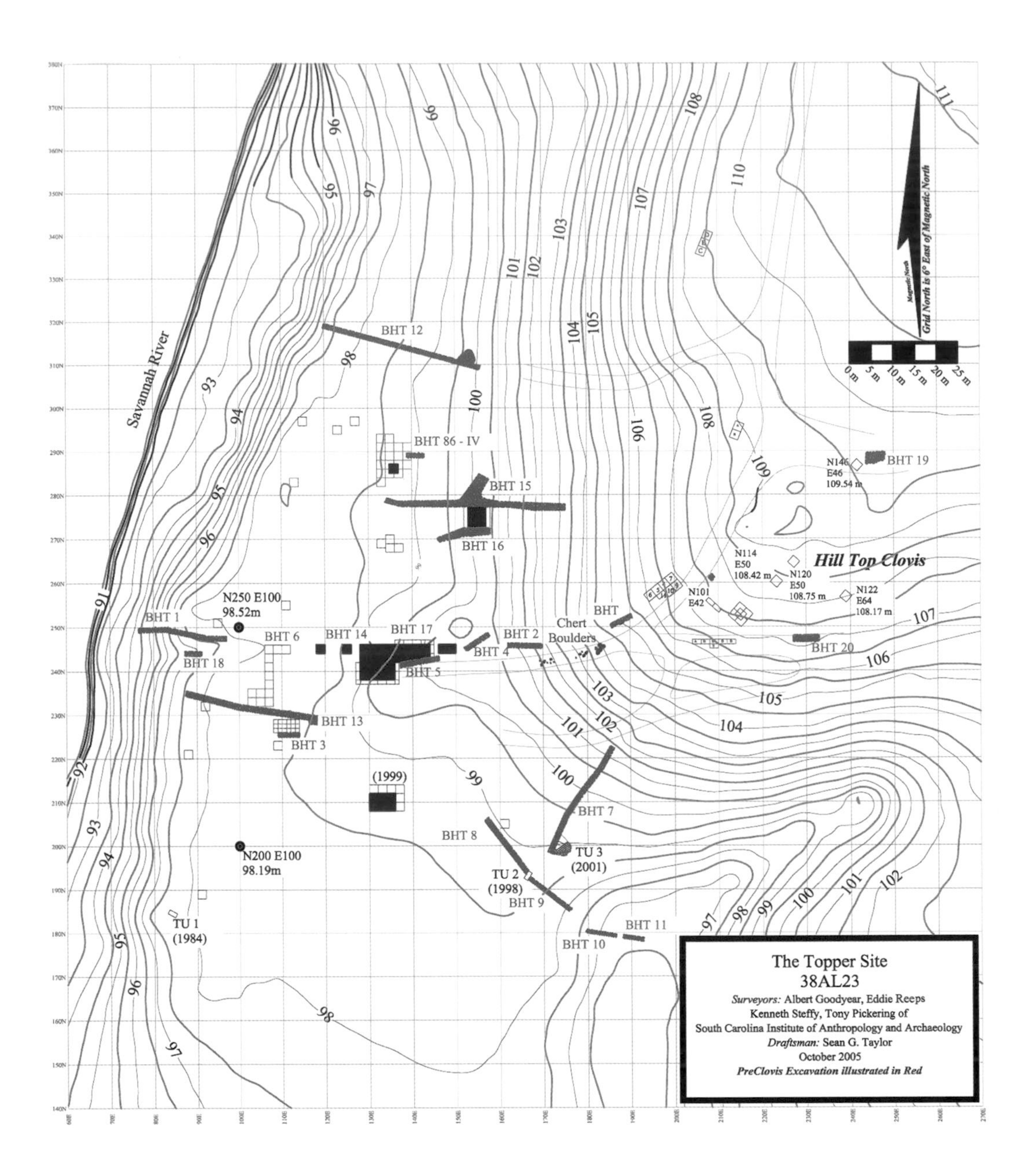

FIGURE 21.7 Topographic site map of Topper showing pre-Clovis excavations and backhoe trenches on the terrace, chert outcrops at the escarpment, and the extensive Clovis occupation on the hillside. Courtesy of Sean Taylor.

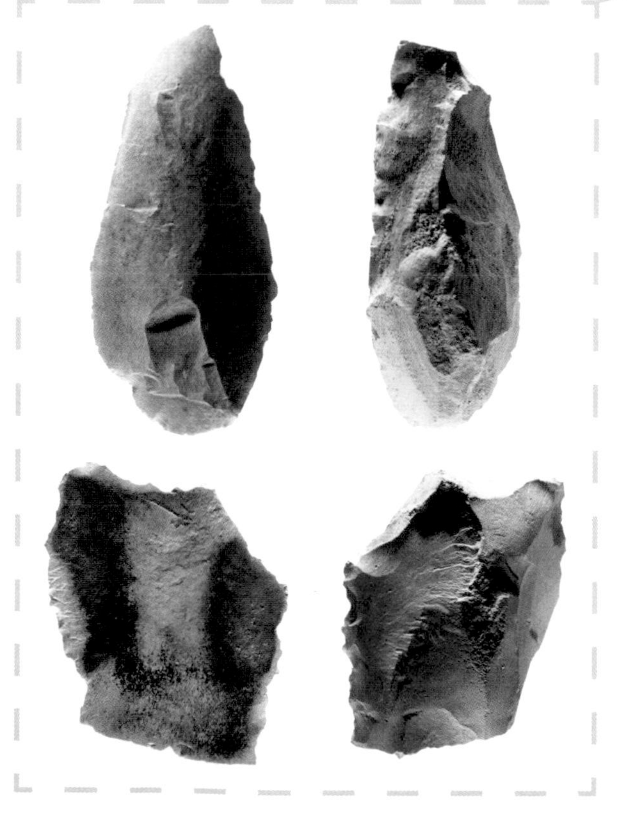

FIGURE 21.8 (top) In 1997 Goodyear and his team resumed work at Topper. After digging 30–50 centimeters below the base of the Clovis level, they encountered what they proposed were small flake tools in the Pleistocene alluvial sands below. After further excavating down about 1.8 meters, the team found this grouping of rocks, including a large stone which they identified as an anvil. Courtesy of Albert Goodyear.

FIGURE 21.9 (bottom) Examples of the flake tools associated with the boulder-sized anvil. Top, uniface flake (obverse and reverse); and bottom, unifacial sidescraper (obverse and reverse). Photo by Daryl P. Miller.

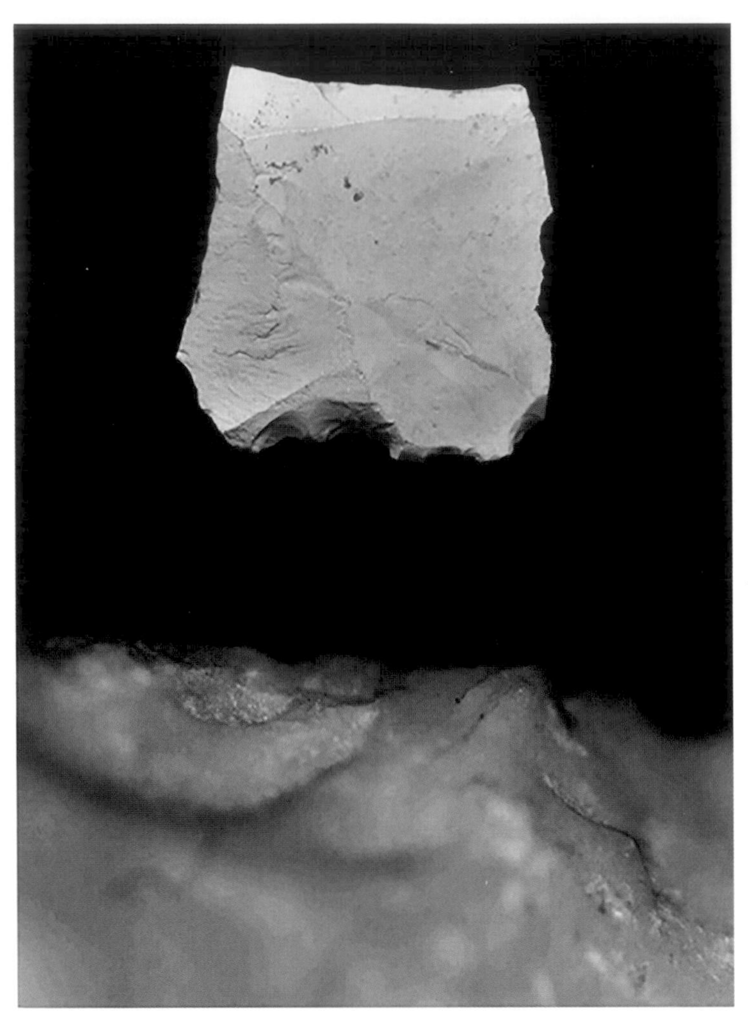

FIGURE 21.10 (top) A unifacially retouched scraper from the pre-Clovis levels with microscopic evidence of use wear. Photograph courtesy of Douglas Sain.

FIGURE 21.11 (bottom) Bend-break flake tools from the Pleistocene sands. Upper five are transversely broken creating approximately 90 degree angles; bottom four may be triangular break forms. Over 1,000 bend-break flakes have been recovered from all pre-Clovis levels at Topper, making them one of the predominant artifact types from the site. Courtesy of Albert Goodyear.

Tennessee graduate student Douglas Sain in his 2015 doctoral dissertation. Sain has concluded that the pre-Clovis Topper Assemblage artifacts are indeed genuine, and that a small sample of the tools show microscopic evidence of human use in the form of edge polish (the smoothing of sharp edges via repetitive action), striations (fracture lines resulting from contact with another object), residue (plant or animal material adhering to the artifact), and edge damage (chipping of the artifact's edge through use). [Figure 21.12] His research has also concluded that the Topper Assemblage artifacts are indeed *in situ*, and hence did not migrate downward into the deposit from overlying archaeological levels. It remains to be seen what the professional archaeological community will make of Sain's findings, but if the Topper Assemblage finds widespread acceptance a radical reworking of our understanding of pre-Clovis stone technology will be in order.

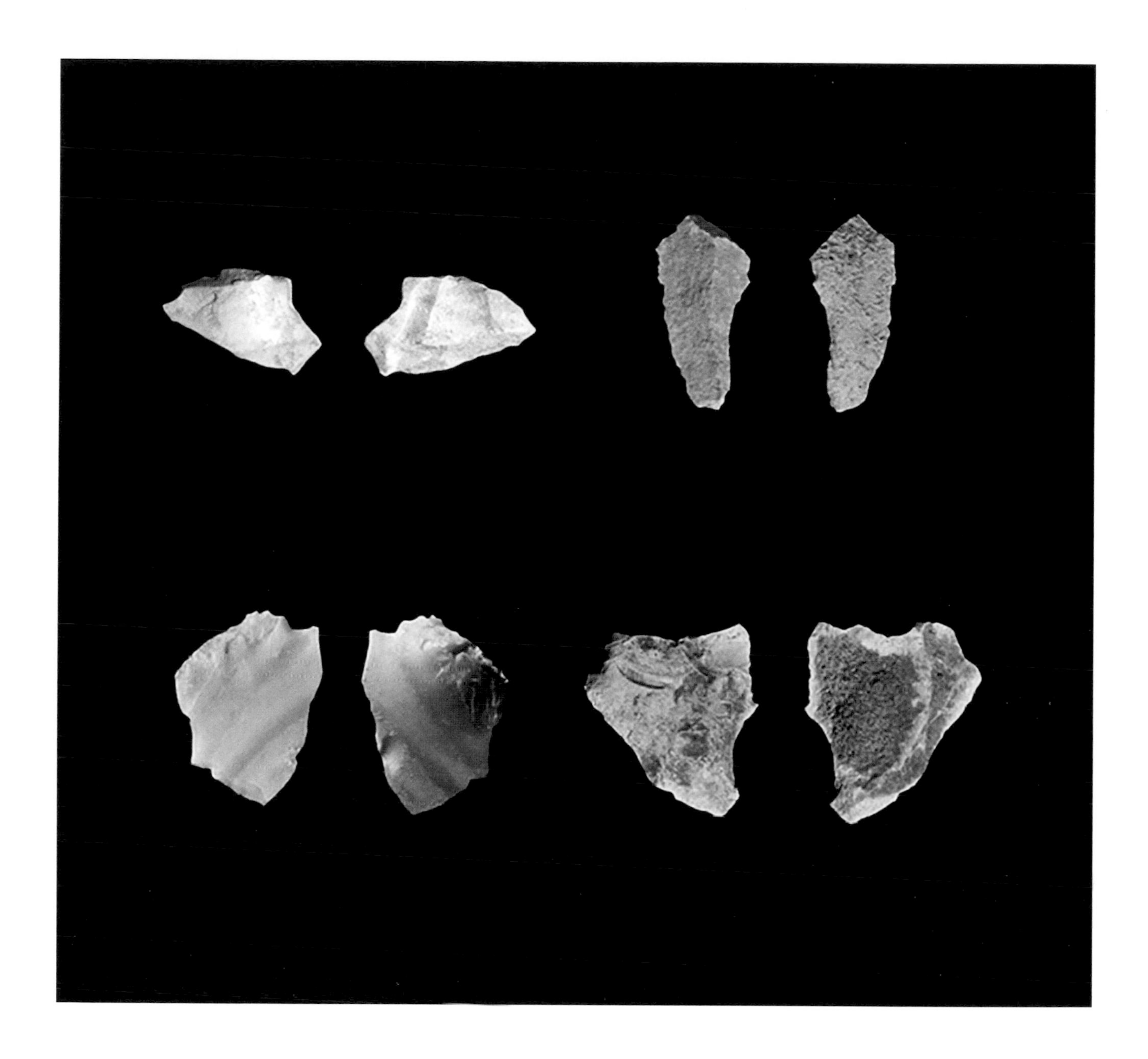

FIGURE 21.12 A 1999 independent assessment of these fragments suggested that they could just as likely been the result of forest fire, freeze-thaw cycles, or fracturing through stream transport. It was also noted that these same fragments are found throughout the Pleistocene sands dated from 15,000 yr BP to in excess of 50,000 yr BP, raising the question of why there was no lithic technological change for at least 35,000 years. Courtesy of Douglas Sain.

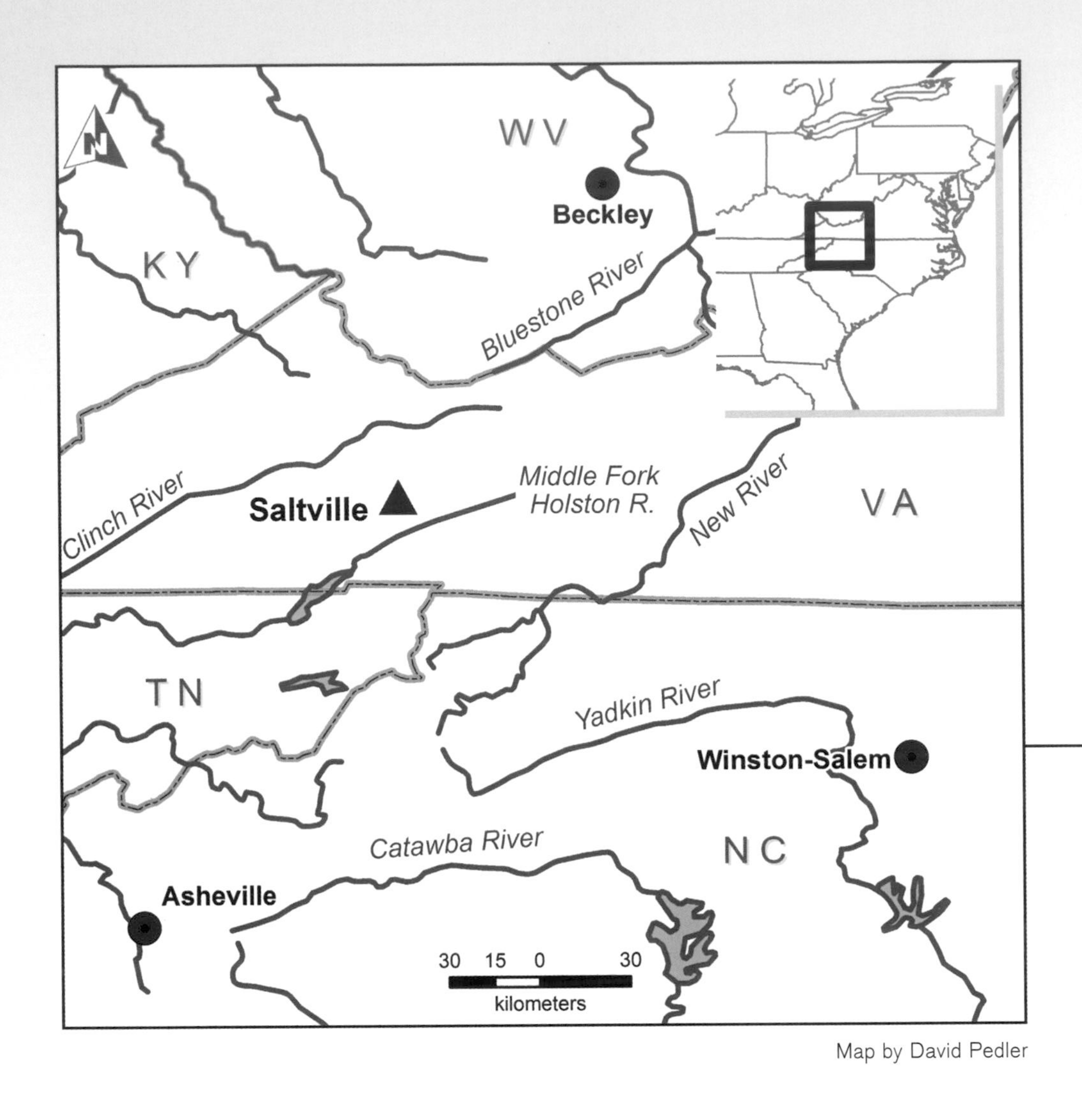

Map by David Pedler

During the Late Pleistocene, Saltville was located in the midst of pine-spruce parkland, interspersed with valley-bottom marshes. The earliest archaeological reference regarding the site is from Thomas Jefferson's journal in which he notes a mastodon tooth from Saltville given to him by Arthur Campbell in 1782.

22 SALTVILLE

LOCATION SMYTH COUNTY, VIRGINIA, UNITED STATES

Coordinates 36°52'18.90"N, 81°46'23.91"W.

Elevation 525 meters above sea level.

Discovery Arthur Campbell in 1782.

The Saltville site and the surrounding Saltville Valley have been studied for over two centuries, but remarkably little has been published that directly bears on the archaeological record. To date, only a relatively short archaeological monograph and brief scholarly publications have appeared, and any other treatments of the site have been limited to presentations at professional conferences and mentions in secondary regional or continental archaeological syntheses. The site nonetheless remains a potentially crucial locality for our understanding of pre-Clovis in eastern North America as the depth of its proposed antiquity would make it the oldest archaeological locality in the Appalachian Ridge and Valley region.

The Saltville site is located in western Smyth County, western Virginia, within the town limits of Saltville, about 110 kilometers southwest of Beckley, West Virginia, 160 kilometers northeast of Asheville, North Carolina, and 160 kilometers northwest of Winston-Salem, North Carolina. The site lies within the Saltville Valley about 1.3 kilometers south of the North Branch of the Holston River, which together with that river's North and Middle Forks compose one of the major drainage systems of southwestern Virginia, eastern Tennessee, and the upper Tennessee River basin. An artificial reservoir created in 1964 by a commercial salt extraction enterprise lies just to the south of the site, and mechanical excavations associated with this operation apparently resulted in the exposure of river and lake sediments in areas surrounding a former saltmarsh that had been covered by overlying layers of mud and gravel. The history of salt mining and extraction in the area dates back to at least to the Civil War, and may well have occurred in prehistoric times.

The Saltville Valley is surrounded on all sides by relatively steep slopes and mountains ranging about 600–800 meters above sea level, which is typical of the region's topography. In Late Pleistocene times, the now-extinct Saltville River flowed north through the valley, joining the North Fork of the Holston River almost 2 kilometers northeast of the site. The upper portion of the Saltville River is thought to have been captured into the Holston River by an adjacent watercourse (McHenry Creek, presently a minor tributary stream) around 16,000 yr BP, and soon thereafter its lower reaches formed upland Lake Totten after the river's blockage downstream from the site immediately northwest of the present-day town of Saltville. The lake appears to have relatively quickly reached a depth of at least 5 meters and the lake bed deposits thickened with sediments derived from the surrounding unstable valley slopes. [Figure 22.1]

Ongoing paleoecological investigations of the Saltville Valley since the early 1980s have identified almost twenty discrete archaeological localities, but the most intensive work has been conducted at two sites called SV-1 and SV-2, the latter of which has yielded Pleistocene faunal remains and purported pre-Clovis artifacts. [Figure 22.2] Site SV-2 is located in the southwestern portion of the Saltville Valley on the north bank of the above-mentioned reservoir. [Figure 22.3] The site's elevation on the Saltville Valley floor is about 525 meters above sea level. Local relief varies from 75 meters to the ridgetops on the valley's northern side to over 200 meters on the ridgetops to the south. In the vicinity of site SV-2 the valley floor is about 400 meters wide. The site measures about 20 meters from east to west and about 10 meters from north to south, and its long orientation roughly follows the former course of the extinct Saltville River. [Figure 22.4]

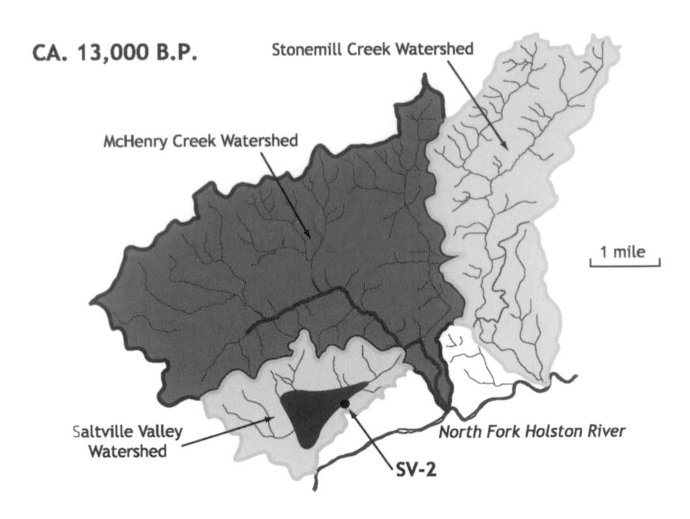

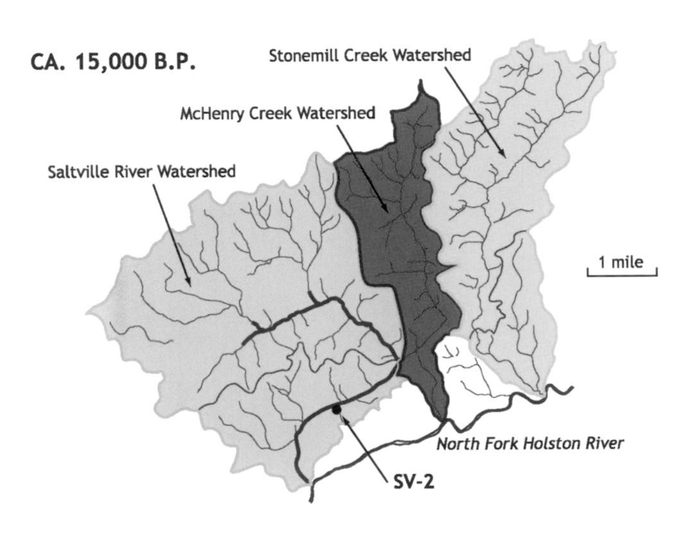

FIGURE 22.1 (above) Sometime before 16,000 yr BP the river that ran through the Saltville Valley was blocked and formed a lake. Parts of the flooded valley floor became covered and sealed with newly deposited mud which proved to be an ideal environment for the preservation of plant and animal remains. Courtesy of Jerry N. McDonald.

FIGURE 22.2 (right, top) Saltville Valley lies astride highly variable layers of shales, siltstones, sandstones, and limestones. The various distinct colors of the muds deposited during the Late Pleistocene have assisted geologists in distinguishing the stratigraphy at Saltville SV-2. Courtesy of Jerry N. McDonald.

FIGURE 22.3 (right, bottom) SV-2 site map. The oldest deposits at this site that have yielded putative evidence of human activity are dated to around 17,920–17,460 yr BP. They contain evidence of extensive utilization of a single mastodon (*Mammut americanum*) carcass, reportedly including the butchering of the animal, possible working of its hide, shattering of much or most of the skeleton, extraction of grease from the shattered bone, and sectioning of the tusks for the probable production of ivory shafts. Courtesy of Jerry N. McDonald.

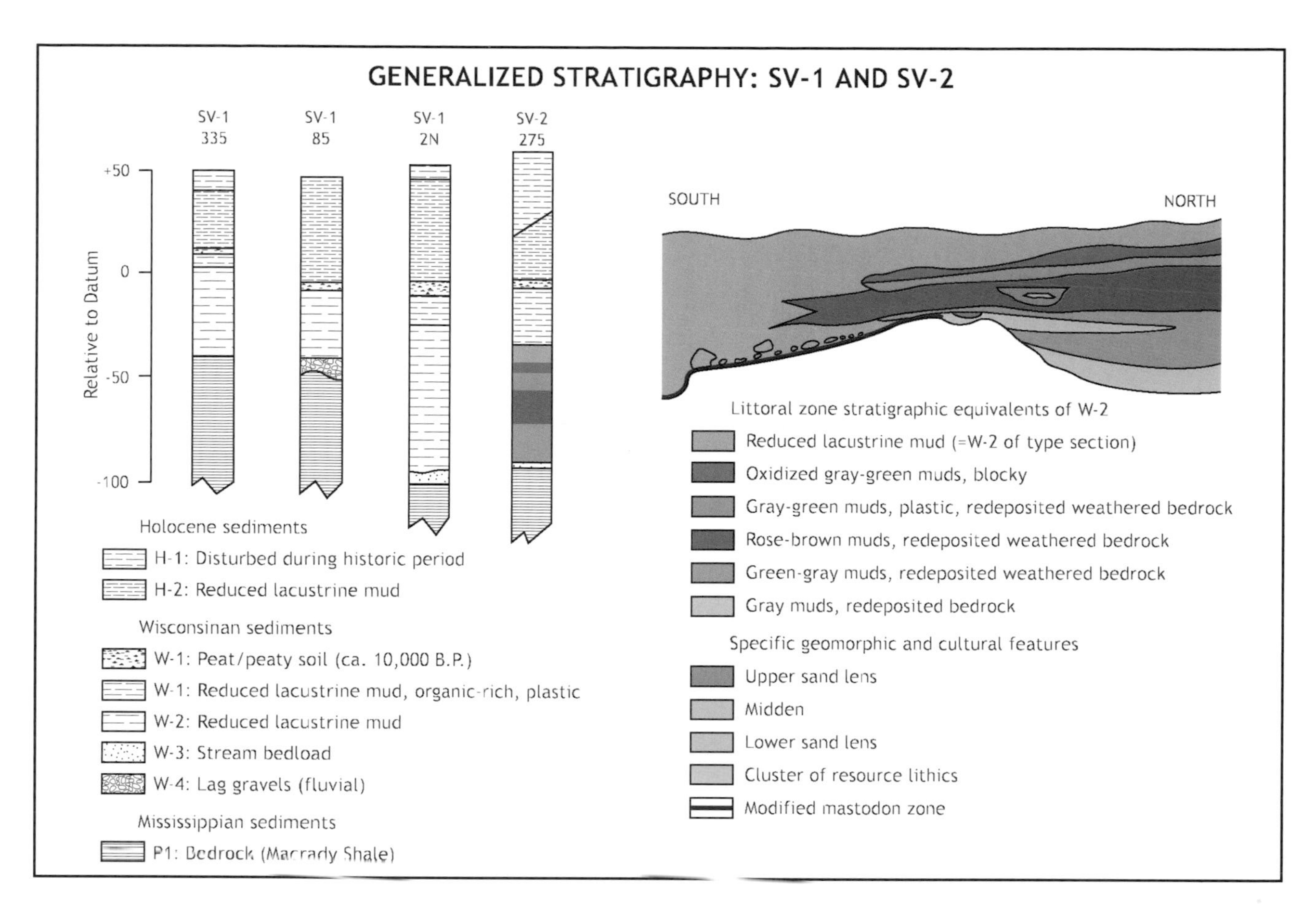

GENERALIZED STRATIGRAPHY: SV-1 AND SV-2
SV-1 335
SV-1 85
SV-1 2N
SV-2 275
+50
0
-50
-100
Relative to Datum
SOUTH
NORTH
Holocene sediments
H-1: Disturbed during historic period
H-2: Reduced lacustrine mud
Wisconsinan sediments
W-1: Peat/peaty soil (ca. 10,000 B.P.)
W-1: Reduced lacustrine mud, organic-rich, plastic
W-2: Reduced lacustrine mud
W-3: Stream bedload
W-4: Lag gravels (fluvial)
Mississippian sediments
P1: Bedrock (Macrady Shale)
Littoral zone stratigraphic equivalents of W-2
Reduced lacustrine mud (=W-2 of type section)
Oxidized gray-green muds, blocky
Gray-green muds, plastic, redeposited weathered bedrock
Rose-brown muds, redeposited weathered bedrock
Green-gray muds, redeposited weathered bedrock
Gray muds, redeposited bedrock
Specific geomorphic and cultural features
Upper sand lens
Midden
Lower sand lens
Cluster of resource lithics
Modified mastodon zone

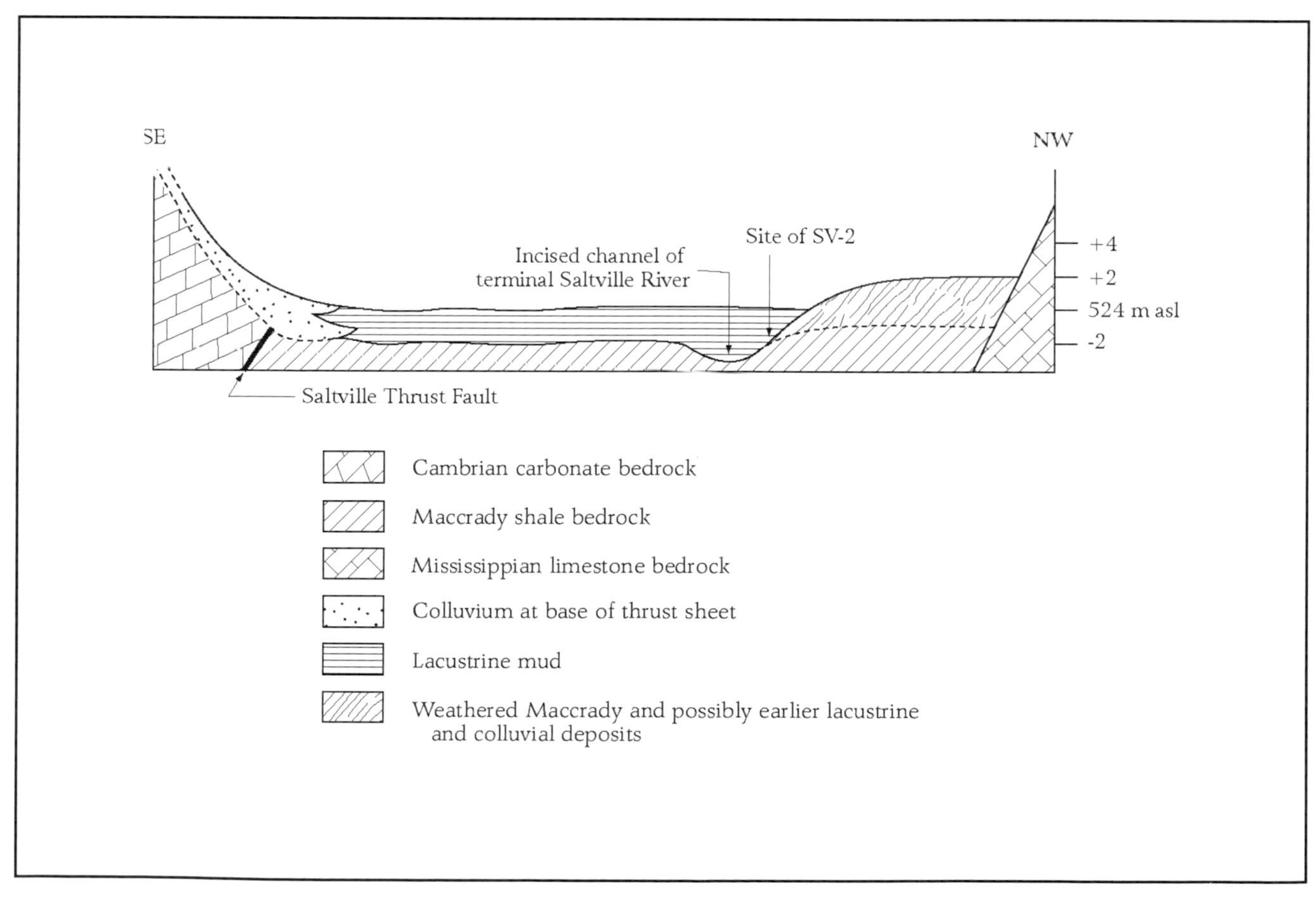

SE
NW
Incised channel of
terminal Saltville River
Site of SV-2
+4
+2
524 m asl
-2
Saltville Thrust Fault
Cambrian carbonate bedrock
Maccrady shale bedrock
Mississippian limestone bedrock
Colluvium at base of thrust sheet
Lacustrine mud
Weathered Maccrady and possibly earlier lacustrine and colluvial deposits

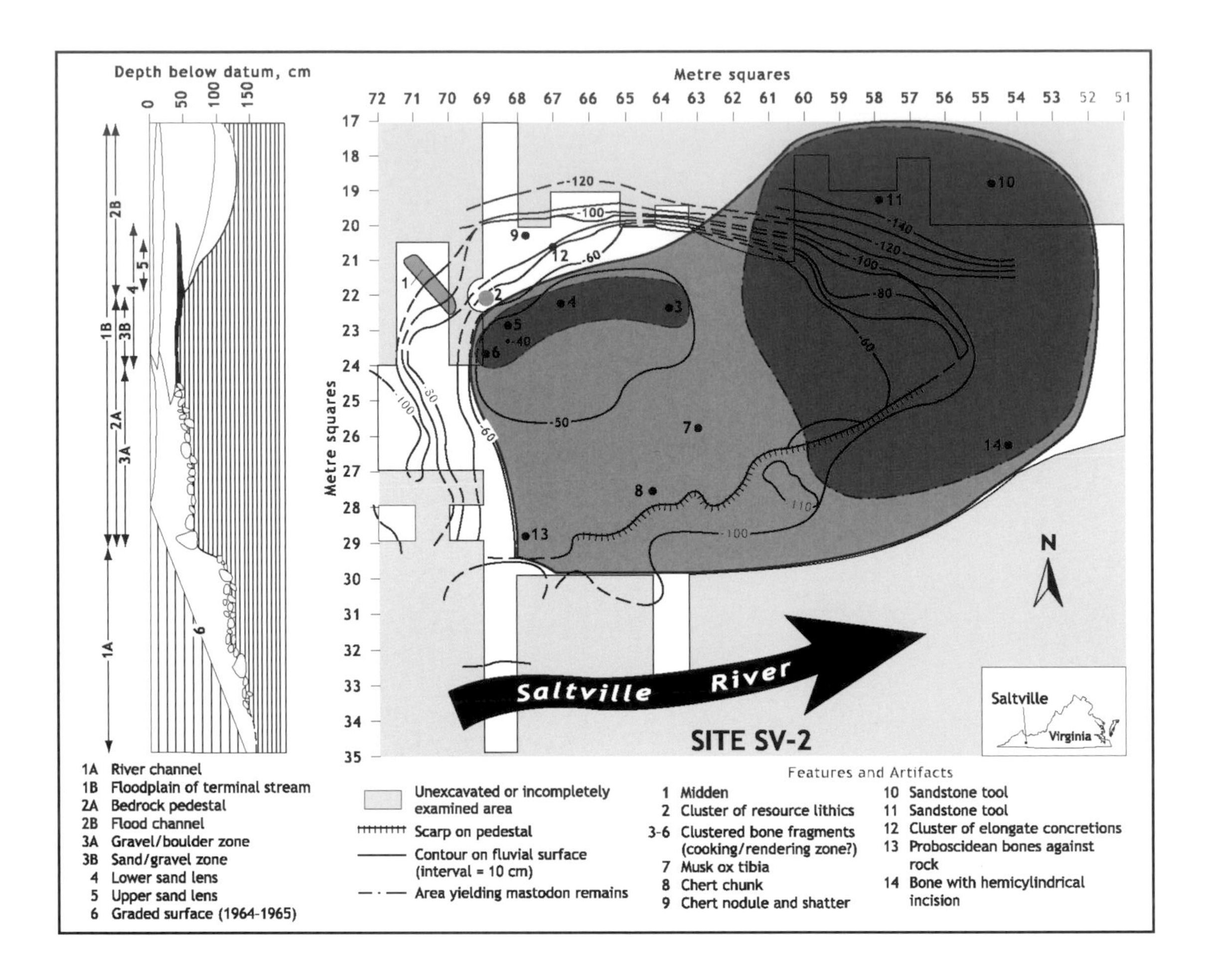

FIGURE 22.4 During the Late Pleistocene, the Saltville River rose in uplands east of Saltville Valley, then entered the valley at its southwestern end, and incised a channel along the northwestern edge of the valley. Both sites SV-1 and SV-2 are located in sediments that filled, overlaid, or abutted parts of this ancient channel. Courtesy of Jerry N. McDonald.

The Saltville Valley was first investigated as a paleontological locality at least as early as 1782, when Colonel Arthur Campbell mentioned the recovery of "bones of an uncommon size" in a letter written to Thomas Jefferson which was apparently accompanied by a mastodon tooth specimen. Accidental discoveries resulting from mining activity continued throughout the nineteenth and twentieth centuries, and formal scientific investigations were sporadically undertaken between 1917 and 1980 by various research organizations including the Carnegie Museum, Virginia Polytechnic Institute, and the Smithsonian Institution.

Although fluted projectile points had been known from other portions of the Saltville Valley, archaeological excavations on site SV-2 proper were not undertaken until the 1992 recovery of a putative stone tool in association with the ribs of a single mastodon. [Figure 22.5] Conducted from 1992 to 1997 under the direction of paleogeographer Jerry McDonald of the Virginia Museum of Natural History, the investigation ultimately excavated almost 200 square meters of the SV-2 sediments. The excavators identified three cultural layers (labelled "horizons") within a 1 meter thick

FIGURE 22.5 This flat, hand-sized sandstone object, both sides of which are shown here, was unearthed in the putative pre-Clovis deposits at Saltville. It is thought to have been modified and used as a hand axe or knife for butchering the associated mastodon (*Mammut americanum*) remains. Courtesy of Jerry N. McDonald.

layer of sediment, the oldest of which dates to about 17,920–17,460 yr BP. The so-called middle horizon dates to about 17,180–16,610 yr BP and the uppermost youngest horizon is presumed to date to the estimated time of Lake Totten's formation at about 16,000 yr BP. These radiocarbon ages indicate that the sediments containing the site's oldest archaeological horizon pre-date the regional Clovis horizon by almost 5,000 years.

The cultural materials reported from SV-2's oldest horizon are quite few in number, amounting to only seven specimens. The five identified lithic artifacts include two sandstone knives, a triangular stone described as a "hammer- or hand axe-sized object found in association with sectioned tusk fragments," a putative chert tool, and a single chert flake. Both of the chert specimens are of a lithic raw material that does not appear to occur in the Saltville River watershed. The two remaining specimens are composed of bone, both of which are reported to exhibit wear patterns resulting from their prehistoric use. Finally, the site's oldest layer also contained a "concretion" (putatively bound together by congealed animal fat), a single alleged cultural feature in the form of what the excavators call a "mastodon butcher zone," and two arrangements of unmodified bone that are thought to be associated with cultural activity. [Figures 22.6 and 22.7]

The overlying middle and youngest horizons at SV-2 are also rather ephemeral. The middle horizon yielded three apparent bifacial thinning flakes and microdebitage, a series of concretions similar to the one identified in the oldest horizon, a cluster of large pebbles and cobbles of non-local "resource lithics" (none of which appear to show unequivocal signs of human modification), and a concentration of material created by "presumed human agency" containing fish bones, charcoal, and apparent lithic artifacts. The youngest horizon yielded only a single concentration of material containing a diverse assortment of animal remains, 125 pieces of "possible microdebitage," and three bifacial reduction flakes. [Figure 22.8]

The most distinctive object from SV-2 is the bone tool recovered from the top of a gravel surface in the oldest horizon on the site's west-central portion about 9 meters southwest of the mastodon butcher zone. [Figure 22.9] Described as having the "size, shape, and edges of a knife or scraper," this specimen represents

INCH
CM

FIGURE 22.6 (left, top) Seven concretions of weathered bedrock arranged in an unusual orientation putatively indicating human placement were found in the middle pre-Clovis deposits at Saltville dated to about 17,180–16,610 yr BP. Courtesy of Jerry N. McDonald.

FIGURE 22.7 (left, bottom) A brightly polished tooth of a mammoth (*Mammuthus primigenius*) (lower right) from the late river gravels and an unabraided femur of a musk ox (*Bootherium bombifrons*) from the early lacustrine muds of SV-1 at Saltville dated to about 17,180–16,610 yr BP. Remains of large mammals of the Late Pleistocene have been discovered and reported from Saltville for more than 200 hundred years. Courtesy of Jerry N. McDonald.

FIGURE 22.8 (below) Four relatively intact mastodon (*Mammut americanum*) tusk segments, straight and approximately 25–30 cm in length, have been recovered (above), along with a longer, curved segment about 40 cm in length (below). Courtesy of Jerry N. McDonald.

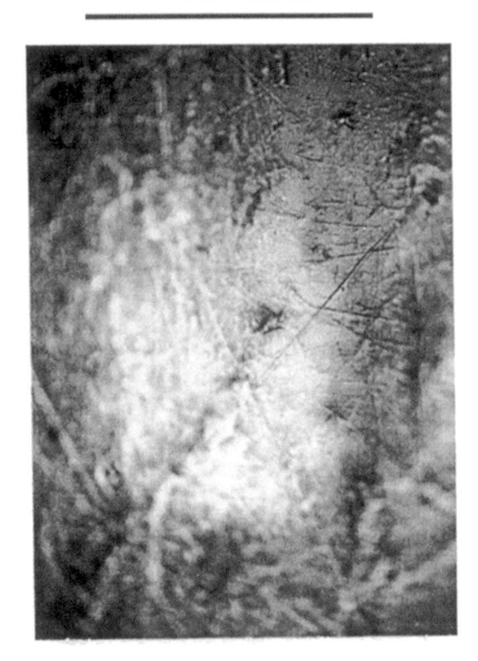

FIGURE 22.9 This portion of a musk ox (*Bootherium bombifrons*) tibia from SV-2 deposits at Saltville shows putative evidence of butchering and use that left polishing on both ends. Courtesy of Jerry N. McDonald.

FIGURE 22.10 What is considered to be a bone tool also was recovered from the oldest pre-Clovis deposits at Saltville dated to about 17,920–17,460 yr BP. The bone is probably a musk ox (*Bootherium bombifrons*) tibia. An analysis by Marvin Kay, University of Arkansas archaeologist and lead investigator at the Kimmswick Bone Bed (see pages 90-101), concluded, "The tibia was intentionally fractured, was hand-held, and was used as a mattock hide beaming tool." Courtesy of Jerry N. McDonald.

a right tibia fragment of what appears to be Harlan's musk ox (*Bootherium bombifrons*), a species which became extinct at the close of the Pleistocene. (A radiocarbon sample obtained from this object produced the age of 17,920–17,460 yr BP recorded for SV-2's oldest horizon.) It measures almost 20 centimeters in length, ranges 36–49 millimeters in breadth, and displays two fractures that appear to have been produced when the bone was still green or fresh. University of Arkansas archaeologist Marvin Kay, a prehistoric tool analyst who specializes in the identification of micro-wear and who was also a member of the Monte Verde site (see pages 216-227) research team, has indicated that this object was "intentionally fractured," "hand-held," and "used as a mattock hide beaming tool."

Unfortunately, because of the relative paucity of archaeological data from the SV-2 locality, very little can be said concerning the lifeway of the Saltville Valley's first human visitors beyond the fact that they appear to have been involved in the predation or scavenging and processing of a single proboscidean (probably mastodon) and, toward that purpose, had fashioned a tool from the tibia of a single Harlan's musk ox. [Figure 22.10] Nor can much be said concerning the lithic technology that might have operated at the site, as many of the SV-2 lithic artifacts have been only tentatively identified as such, the results of their technological analysis do not appear to have been published in detail, and the lithic artifact illustrations available to date are of insufficient quality and detail to permit an independent assessment of their character and validity. Perhaps as a result of these uncertainties and other potential ambiguities, recent major scholarly studies have tended to overlook SV-2 as a valid pre-Clovis locality, but considerable potential remains for further reporting and investigations on SV-2 and elsewhere is the Saltville Valley.

Map by David Pedler

The lithic technology uncovered at the northwestern South American sites of Taima-taima and Tibitó is quite different from, and apparently older than, Clovis technology.

23 TAIMA-TAIMA AND TIBITÓ

LOCATION TAIMA-TAIMA, FALCÓN STATE, VENEZUELA; TIBITÓ, DEPARTMENT OF CUNDINAMARCA, COLOMBIA

Coordinates Taima-taima, 11°29'57.12"N, 69°31'19.91"W; Tibitó, 4°59'5.13"N, 73°58'56.63"W.

Elevation Taima-taima, 40 meters above mean sea level; Tibitó, 2,500 meters above sea level.

Discovery Taima-taima, José María Cruxent and Alex Krieger in 1962; Tibitó, Gonzalo Correal in 1979.

The Late Pleistocene archaeology of northwestern South America, with a few crucial exceptions, has been largely dismissed or just simply ignored by North American archaeologists until relatively recently. Traditionally convinced that the colonization of the hemisphere occurred exclusively on dry land via the ice-free corridor (see page 30), they considered the region to be too far south for the presence of significantly early sites, and that any sites discovered would be Clovis descendants. Taima-taima and Tibitó are two of the region's longest standing pre-Clovis candidates, yet they frequently receive only passing mention—if any at all—in archaeological syntheses of the peopling of the New World.

The Taima-taima site is located on the Caribbean Sea in Falcón State, northwestern Venezuela, about 95 kilometers southwest of the island nation of Curaçao and 320 kilometers northwest of the Venezuelan capital, Caracas. Falcón's capital city, Coro, lies about 20 kilometers to the southwest. (Coro is recognized as an outstanding UNESCO world heritage site and considered to be a remarkably conserved early Spanish colonial era historic town.) The Caribbean shoreline is about 500 meters north of the site, and the Paraguaná Peninsula and Gulf of Venezuela are 15–20 kilometers to the west. The low rolling hills of the Falcón coastal plain surround the site, and the mountainous Venezuelan

FIGURE 23.1 Taima-taima in 1980, view to the south. There is little hard evidence about the paleoenvironment around the time when Taima-taima was allegedly used as a kill site, some 13,000–10,500 yr BP. Floral data from the site has led one of site's principle investigators, Claudio Ochsenius, to suggest that that the site was covered in semiarid vegetation, similar to that of today. The artesian spring at Taima-taima was frequented by megafauna as well as smaller species such as deer and large rodents in Late Pleistocene times. Courtesy of José Oliver.

Coastal Range lies 30 kilometers to the south. The Coro region is quite dry, and many of its streams flow only seasonally. The modestly sized Tocuyo River flows through an interior valley of the Coastal Range about 90 kilometers south of the site. Northern Venezuela's principal drainage basin, the 880,000 square kilometer Orinoco River watershed, lies well over 200 kilometers to the southeast.

The boundary shared by two important ecological regions—desert and xeric (*i.e.*, very dry) shrubland on the west, and tropical dry broadleaf forest on the east—falls immediately to the west of Taima-taima, essentially placing the site in a transitional zone that more closely resembles the former. [Figure 23.1] The region's most prominent natural feature is the Médanos de Coro Desert, a barren expanse that surrounds the Gulf of Venezuela and occupies much of the Paraguaná Peninsula. This hot, dry area is covered in a mosaic of sand dunes, coastal grassland, open scrub, cacti, briars, and fresh-water marsh habitats that provide a refuge for native plant species and a rich variety of native and migratory birds. Immediately east of the desert lie a number of ancient springs which in Late Pleistocene times appear to have attracted an exceptionally broad range of megafauna, horses, llamas, smaller carnivores, and reptiles. In the Late Pleistocene, the region would have been similarly arid and dotted with open scrub and herbaceous savannah vegetation.

Measuring only 40 meters north–south by 20 meters east–west, the Taima-taima site is centered on the edge of a basin-shaped, artesian spring-fed waterhole whose source is an aquifer at the base of the Sierra de San Luis, a low-altitude mountain range located 10 kilometers to the south. A series of intersecting, low-relief coastal hills and ridges surround the site, which is elevated 40 meters above sea level. Isolated artifacts have been recovered in the general vicinity of Taima-taima, but the site deposit does not appear to expand beyond the limits of the excavation. A similarly configured paleontological locality, Muaco, lies about 3 kilometers to the southwest.

Taima-taima was discovered by Spanish-born Venezuelan archaeologist José María Cruxent and American archaeologist Alex Krieger in 1962. Cruxent and his research team had excavated 150 square meters of the site by the mid-1970s, when vandals destroyed most of the exposed faunal remains, essentially ending their identification and analysis. In an attempt to refine Cruxent's tentative results and resolve scholarly disputes that had arisen about the site's archaeological validity, a final excavation season was undertaken at Taima-taima in collaboration with University of Alberta archaeologists Ruth Gruhn and Alan Bryan in 1976. The results of their 80 square meter 1976 excavation, along with whatever could be gleaned from the prior work at the site, were published in 1979.

The 1976 excavation identified four discrete strata at the waterhole. At the base of the archaeological deposit lies a "pavement" of limestone cobbles that is embedded in sand and covered with the broken remains of Pleistocene megafauna. [Figure 23.2] The basal layer is overlain by a 1 meter thick layer of sand (Unit I) containing waterlogged, well-preserved bone, organic material, and undisputable artifacts in its lower reaches. Unit I is dated by fifteen radiocarbon ages ranging from 16,510–15,740 yr BP to 15,310–14,180 yr BP. The middle and upper reaches of Unit 1 also yielded bone and organic material, but the unit's contact with overlying lower Unit II has not been directly dated. Unit II lacks organic material for age assessment, but the upper surface of Unit III produced radiocarbon ages ranging from 12,400–11,770 yr BP to 11,220–10,740 yr BP. Uppermost Unit IV contains no organic remains and therefore remains undated. [Figure 23.3]

The principal feature identified at Taima-taima is the scatter of Pleistocene faunal remains, which number in the hundreds and include examples of mastodon, glyptodon (a car-sized mammal roughly resembling a modern armadillo), bear, giant ground sloth, native horse, and various smaller mammals. Only the mastodon remains in the assemblage, which mostly included fragments of varying age and only one identifiable individual, appear to have been butchered on the site. [Figure 23.4] Bones representing the partial remains of a juvenile mastodon (probably *Haplomastodon waringi*), were recovered from lower Unit I in association with indisputable flaked stone artifacts and a minimum age of 15,810–15,240 yr BP, based on radiocarbon analysis of the animal's apparent organic stomach contents. [Figure 23.5] Conspicuous cut marks were identified on the animal's left humerus and at least two of its ribs, and the artifacts included a quartzite El Jobo projectile point fragment associated with the pelvis and a chert flake next to the left ulna.

The artifact assemblage from Taima-taima is relatively small and mostly composed of expedient stone and bone tools, which in most cases appear to have been only minimally modified. The lithic artifacts are dominated by crude scrapers, several of which appear to have been hafted to some form of shaft or handle, followed by lesser numbers of hand axes and anvils. The bone scrapers and cutting tools from the site are somewhat ambiguous as bona fide artifacts owing to their poor preservation and possible modification by natural rather than human agency, as is acknowledged by Cruxent.

Of indisputable human manufacture, on the other hand, are the chert debitage flakes and two El Jobo projectile point fragments from the site—one of which, as noted, was associated with the butchered mastodon in Unit I. [Figures 23.6 and 23.7] The distinctive El Jobo type is a bifacial, bipointed, narrow projectile that can attain a length of up to 12–18 centimeters. Recognized as a Late Pleistocene diagnostic artifact type, El Jobo

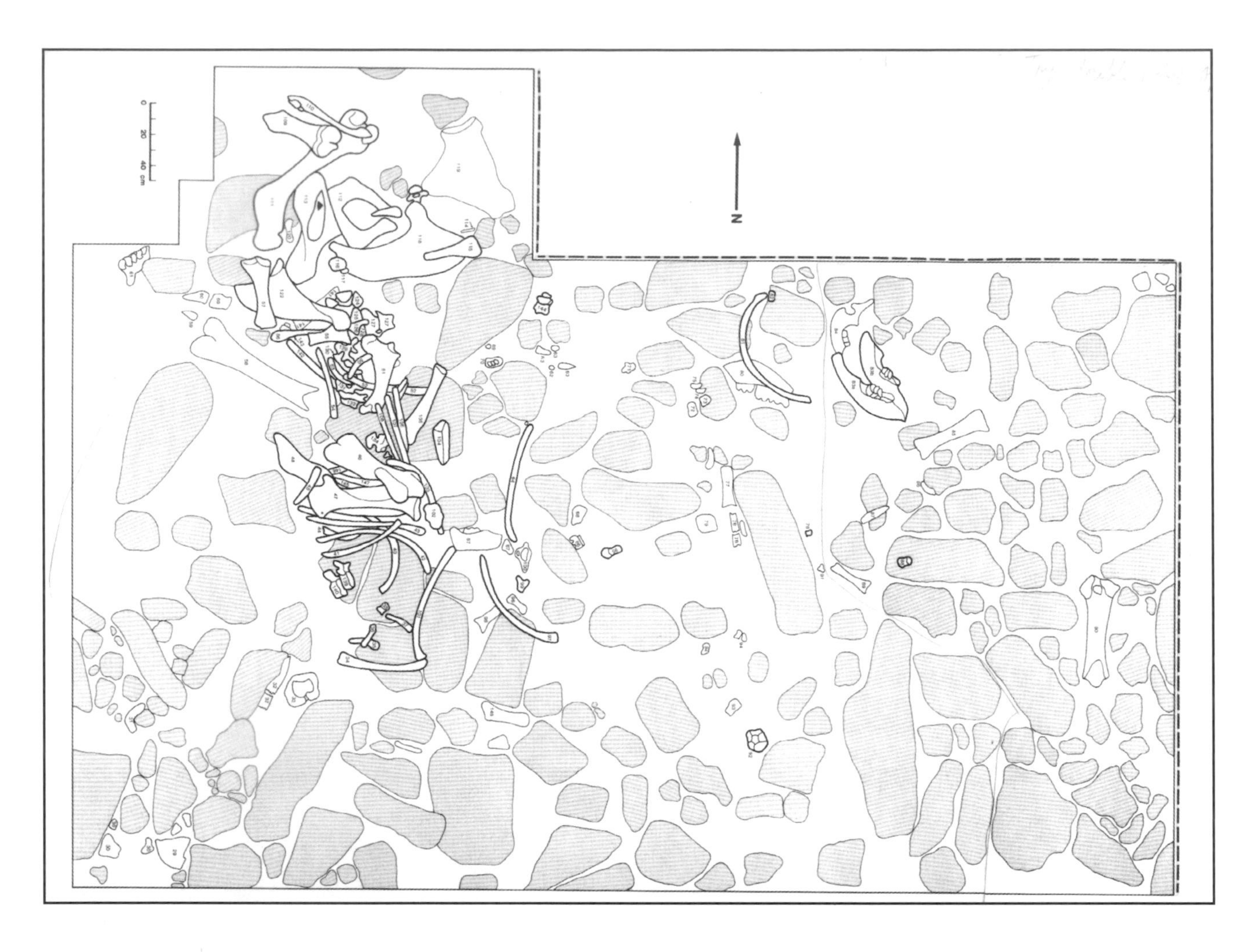
0 20 40 cm
N

FIGURE 23.2 (left, top) Map of the Unit 1 (deepest) bone bed. Courtesy of Ruth Gruhn.

FIGURE 23.3 (left, bottom) Excavation block area of Taima-taima in 1976. Between 1962 and 1975 close to 150 square meters of deposits were excavated at Taima-taima. In 1976 this 160 square meter excavation block was exposed. The remains of a glyptodon carapace were discovered in the upper Late Pleistocene layer. Glyptodons were large, armored, distant relatives of modern armadillos. The carapace provided protection similar to a turtle's shell except that it was composed of more than one thousand bony plates. Unlike turtles, glyptodons could not withdraw their heads into their carapace. Glyptodons survived until the Early Holocene (which began around 11,700 yr BP) and would have coexisted with humans at Taima-taima. Courtesy of José Oliver.

FIGURE 23.4 (right, top) View of the Unit 1 (deepest) bone bed. Three fragments of El Jobo points and scrapers, together with crudely modified stones that are thought to have been used as tools, were found among this confusion of mastodon and glyptodon bones. Courtesy of Ruth Gruhn.

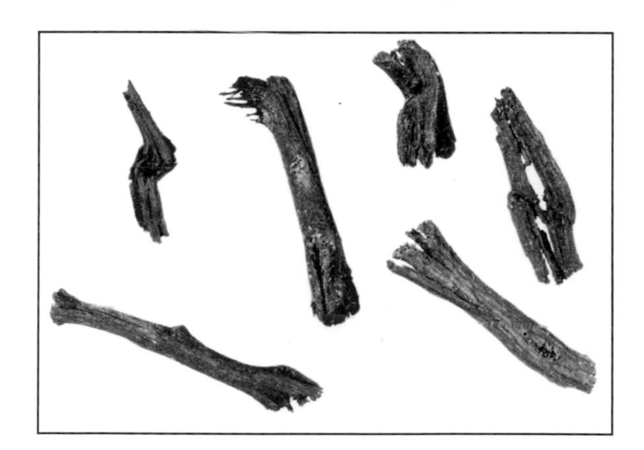

FIGURE 23.5 (right, bottom) Sheared wood twig fragments found abundantly in close association with the butchered mastodon and hypothesized to be stomach contents preserved in saturated gray clayey sand. A sample of this material has been dated to 15,810–15,240 yr B.P. Photograph from British Museum (Natural History), courtesy of A. Sutcliffe.

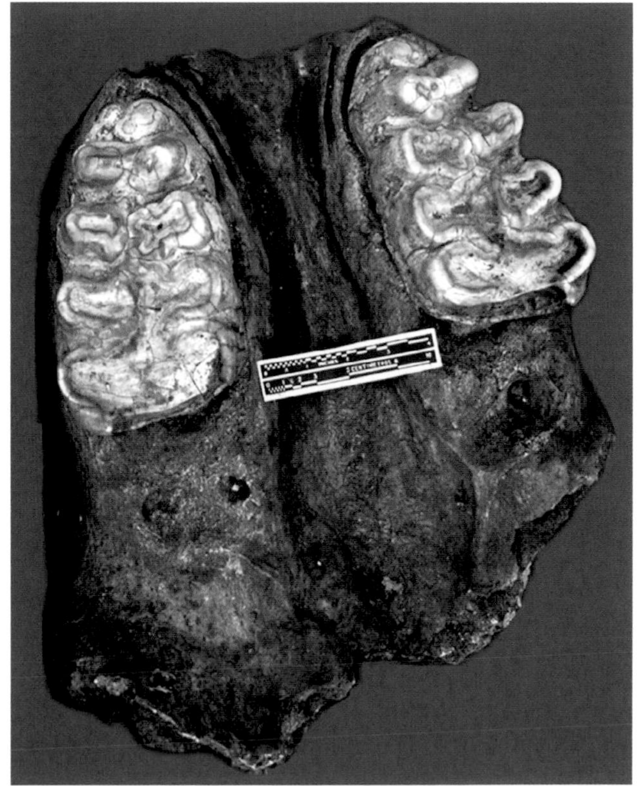

FIGURE 23.6 (above) An El Jobo projectile point rests next to the tibia of a mastodon (*Haplomastodon sp.*) at Taima-taima. (see Fibure 23.4) The minimum age of this El Jobo mastodon kill is 15,240 yr BP, demonstrating that a technological tradition existed in northern Venezuela at least a millennium and a half earlier than the Clovis complex of North America. Courtesy of José Oliver.

FIGURE 23.7 (left) The mandible of a butchered juvenile South American mastodon (*Haplomastodon* sp.) recovered at Taima-taima. The molars had a complex pattern of ridges and knobby protrusions, seen here, giving the animal a large chewing surface that enabled it to feed on grass. This elephant-like Ice Age animal had two tusks on either side of its trunk and ranged from Venezuela southward to Argentina. Courtesy of José Oliver.

FIGURE 23.8 El Jobo projectile points from Taima-taima. Typically, they are thick bipoints made of fine-grained quartztic sandstone abundantly available throughout northwestern Venezuela. If the data from Taima-taima are correct, El Jobo lithics date as early as 15,500 yr BP. El Jobo projectile points are found across northern South America, and similar points have been found at Monte Verde, Chile. (see pages 216-235) Courtesy of José Oliver.

FIGURE 23.9 A Las Casitas straight-stemmed triangular projectile point from Taima-taima. Similar projectile points are found in a wide geographic distribution throughout South America after 12,000 yr B.P. (see page 27) Courtesy of José Oliver.

points have been recovered elsewhere in northwestern Venezuela and—perhaps coincidentally—strongly resemble points recovered from Monte Verde (see pages 216-227) in southern coastal Chile, almost 6,000 kilometers to the south. [Figures 23.8 and 23.9]

The Tibitó site is located in the central reaches of the Department (or state) of Cundinamarca, Colombia, 330 kilometers northeast of Cali, 30 kilometers north of Bogotá, and 8 kilometers west of the municipality of Tocancipá. The Bogotá River is the Tibitó region's principal drainage, flowing 150 kilometers from its headwaters in the foothills of the Colombian Andes' eastern range to its confluence with the Magdalena River in Girardot, about 120 kilometers southwest of Tibitó. Flanked on both sides by the eastern and

FIGURE 23.10 Burned and unburned deer and horse bones and stone tools are exposed around a boulder near the edge of a dried up bog at Tibitó. Courtesy of Tom Dillehay.

central ranges of the Colombian Andes, the Magdalena River is the largest river system in Colombia. Its 260,000 square kilometer drainage basin covers almost one-quarter of the Columbian land mass and enters the Caribbean Sea at the major city of Barranquilla, almost 1,000 kilometers from the river's headwaters in Colombian Massif region. The region's climate is seasonally wet, and its montane forests are very rich in animal and plant diversity.

Tibitó is situated in a marshy area on the east bank of the Bogotá River between two parallel, low-relief ridges on the Bogotá plain at an elevation of 2500 meters above sea level. The site was excavated in 1979 by National University of Colombia archaeologist Gonzalo Correal, who identified three discrete site areas. One of these areas yielded lithic and bone artifacts in association with the remains of extinct Pleistocene fauna including mastodon (*Haplomastodon waringi*) gomphothere (*Cuvieronius hyodon*), and horse (reported as *Equus* [*Amerhippus*] *lasallei*) along with deer and fox. [Figure 23.10] Some of the animal remains showed evidence of human modification in the form of cutting, breaking, and burning—perhaps resulting from marrow extraction. The apparent selection of high-value cuts such as ribs and limbs suggests that the animals were initially dismembered elsewhere in the valley and transported to Tibitó for processing. A single radiocarbon assay suggests an age of 13,780–13,340 yr BP.

The site's artifact assemblage is composed mostly of lithic flake-tool cutting implements and scrapers, a small number of cores, and a few bone artifacts which Correal interprets to represent knives and drills. Tom Dillehay has noted two of the region's early flaked stone industries—Tequendamiense and Abriense—among the Tibitó artifacts. The former industry is named after the deeply stratified rockshelter site of Tequendama, located about 30 kilometers southwest of Bogotá, which yielded bifacial artifacts made from imported stone in contexts dating to as early as 13,310–12,140 yr BP. At Tequendama, this earlier deposit is overlain by artifacts of the apparently somewhat later Abriense industry. First identified at and named after the rockshelter site of El Abra, which lies about 5 kilometers northeast of Tibitó, the Abriense industry is characterized by so-called "edge-trimmed" unifacial tools made from local materials. While the Tequendamiense industry is thought to have lasted for no more than 3000 years, the Abriense endured until historic times.

Taima-taima and Tibitó appear to represent Late Pleistocene mastodon scavenging or kill sites that are (in the case of Tibitó) at least slightly older than Clovis. Taima-taima probably witnessed repeated visits by hunting or foraging parties, whereas it is likely that single-use Tibitó served as a short-term camp. The sites were doubtlessly selected for their proximity to one or more kill or scavenging sites, and the material culture left behind—though sparse—suggests the presence of a lifeway that had adapted and evolved in place over a considerable length of time. And though they are situated by 900 kilometers on profoundly different landforms, the sites' geographic positions point to the early, vital importance of coastal travel and access to major river systems with prominent outlets on those coasts.

Since the publication of their respective investigations, the sites have garnered significant criticism. The legitimacy of Tibitó has justifiably been called into question due its lone radiocarbon date. Clearly, the definition of a well-dated and thoroughly documented stratigraphic sequence would clarify the relationship between the site's cultural and faunal materials. But archaeologist Tom Lynch has also questioned Tibitó's lack of diagnostic projectile points and the validity of some of its artifacts (notably the scrapers), while also suggesting the artifacts might simply be a later Abriense occurrence that had intermixed with the bone deposit. Tom Dillehay has dismissed this possibility as unlikely, instead suggesting that Tibitó might preserve ephemeral evidence of multiple visits through time, the earliest occurring during early Clovis times or perhaps slightly before.

Broader criticisms of Taima-taima have been issued by a greater number of scholars, who have variously claimed that the waterhole spring's flow has comingled the site's faunal and cultural materials, that earlier artifacts have migrated downward from overlying sediments, and that the presumed mastodon stomach contents are intrusive and therefore unassociated. Dillehay has opined that Taima-taima's context appears to be valid, at least by the standards applied to North American Clovis localities, but that the site's age "still remains an open question." The professional consensus holds that further work is required to validate the age of Taima-taima. The site remains open to the public. [Figure 23.11]

FIGURE 23.11 In early 2000 the Taimia-taima site was enclosed and opened to the public. Photograph by Humberto Arias.

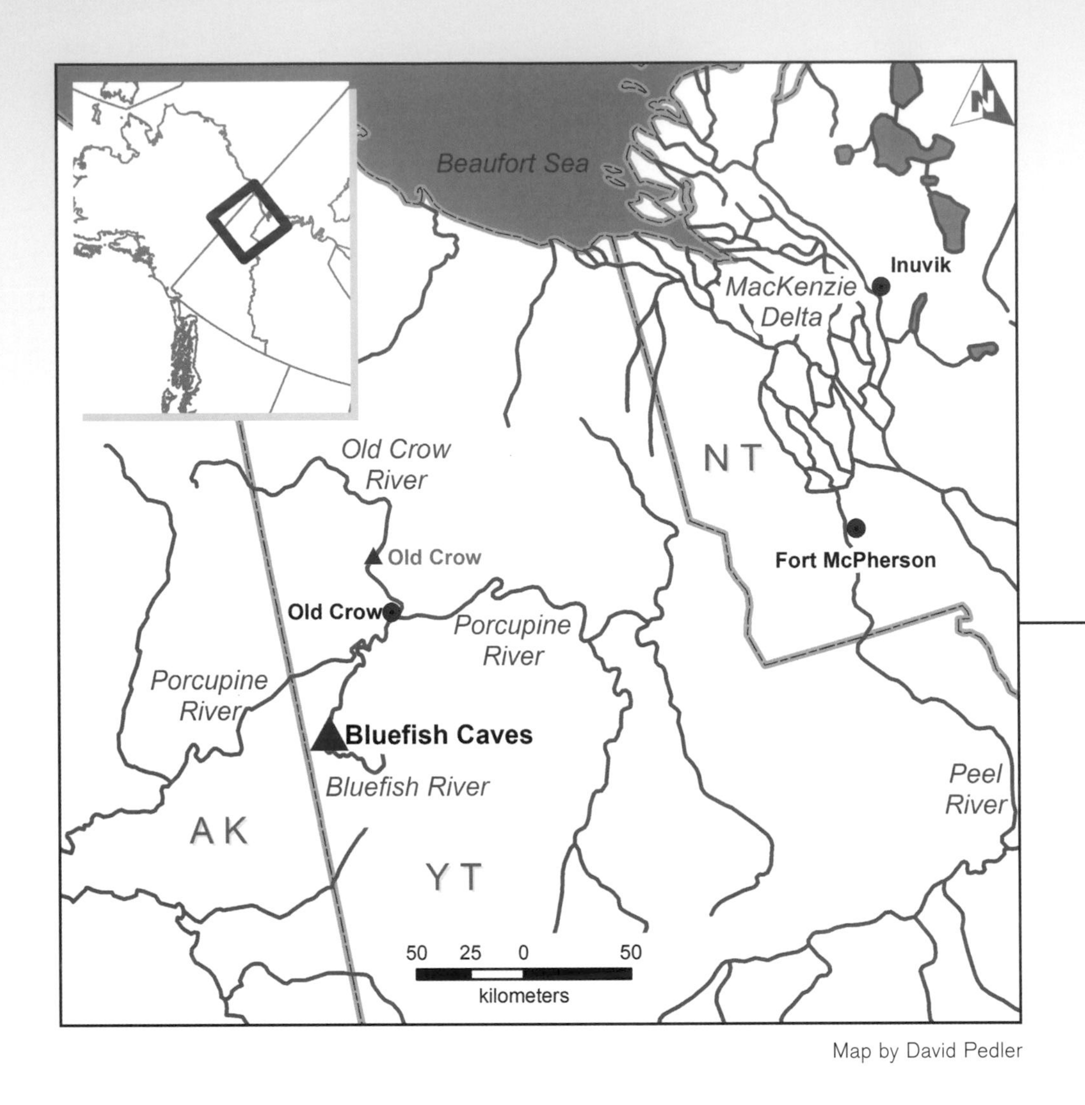

Map by David Pedler

The Bluefish Caves are located in the Canadian Yukon 70 kilometers north of the Arctic Circle. An allegedly human-worked mammoth bone spear point recovered at this site has been dated to 28,000 yr BP. This dating, like the even more ancient dates from Topper (see pages 274-285), remain controversial and unaccepted by most archaeologists.

24 BLUEFISH CAVES

LOCATION YUKON TERRITORY, CANADA

Coordinates 67°10'23.87"N, 140°34'6.09"W.

Elevation 580 meters above mean sea level.

Discovery Jacques Cinq-Mars in 1975.

The Bluefish Caves are among the first apparently intact, Late Pleistocene archaeological localities to be discovered in the North American Arctic Circle, and their investigation in the late 1970s occurred at a time when modern Beringian archaeology was in its infancy. Arguably, the discoveries at Bluefish Caves were instrumental in opening the way for the burgeoning research interest in the region over the past two decades, most notably in central and western Alaska. Despite the acknowledged validity of the artifacts and paleontological remains the sites produced, the Bluefish Caves are only infrequently mentioned in contemporary scholarly syntheses, perhaps due to claims that have been made for the sites' 25,000-year antiquity.

The Bluefish Caves are three proximal archaeological sites located in Canada's Yukon Territory about 770 kilometers northwest of Whitehorse, 800 kilometers northeast of Anchorage, Alaska, and 270 kilometers south of the Yukon's shoreline on the Arctic Ocean's Beaufort Sea. The caves occupy a limestone ridge in the Keele Range, a zone of flat-topped mountains and plateaus on the northern reaches of the North Ogilvie Mountains ecological zone. [Figure 24.1] The north-flowing Bluefish River, which lies about 6 kilometers west of the sites, meanders to its confluence with the Porcupine River basin about 45 kilometers north of the caves. The Porcupine River in turn joins the Yukon River over 200 kilometers to the west.

Unlike the lowland setting of the Old Crow archaeological sites just 80 kilometers to the northeast (see pages 142-151), the Bluefish Caves are situated in an upland zone at 580 meters above sea level and 250 meters above the present-day Bluefish River valley floor. [Figure 24.2] In Late Pleistocene times, the Bluefish River basin region was covered by a 2,000 square kilometer body of water that essentially formed the southern lobe of 13,000 square kilometer Glacial Lake Old Crow, which began to drain west into the Porcupine River sometime around 15,000 yr BP. The southern shore of this ancient lake would have been about 30 kilometers north of the

FIGURE 22.1 (left, top) The Bluefish Caves are three small cavities situated on the western edge of Devonian age (408–362 million yr BP) limestone ridges overlooking the Bluefish River. The ridges are what is left of an ancient landscape marked by sinkholes and caves, following the Early Pleistocene down-cutting of the Bluefish River. The entrance to Cave I is seen here. Courtesy of Ruth Gotthardt.

FIGURE 22.2 (left, bottom) Perched 250 meters above the meandering Bluefish River, the caves would have provided ideal lookouts for Paleoindian hunters to track animals foraging on the valley floor below. The remains of four fish species, twenty-three bird species, and thirty-five mammal species, including medium to large-size herbivores and carnivores, have been recovered from Bluefish Caves excavations. Courtesy of Ruth Gotthardt.

FIGURE 24.3 (below) Beringia during the Last Glacial Maximum (which ended around 20,000–19,000 yr BP), showing the location of the Bluefish Caves when the Bering land bridge connected Asia and North America. At its greatest extent this ancient land bridge was up to 1,000 kilometers wide and 1,600,000 square kilometers in area. The inset shows the position of the Bluefish Cave on a lobe of the Glacial Lake Old Crow and the approximate limits of the Laurentide glacial ice. Except in alpine zones, Beringia, like most of Siberia and all of northeastern China, was spared the glacial advances of the Pleistocene. Courtesy of Richard Harington.

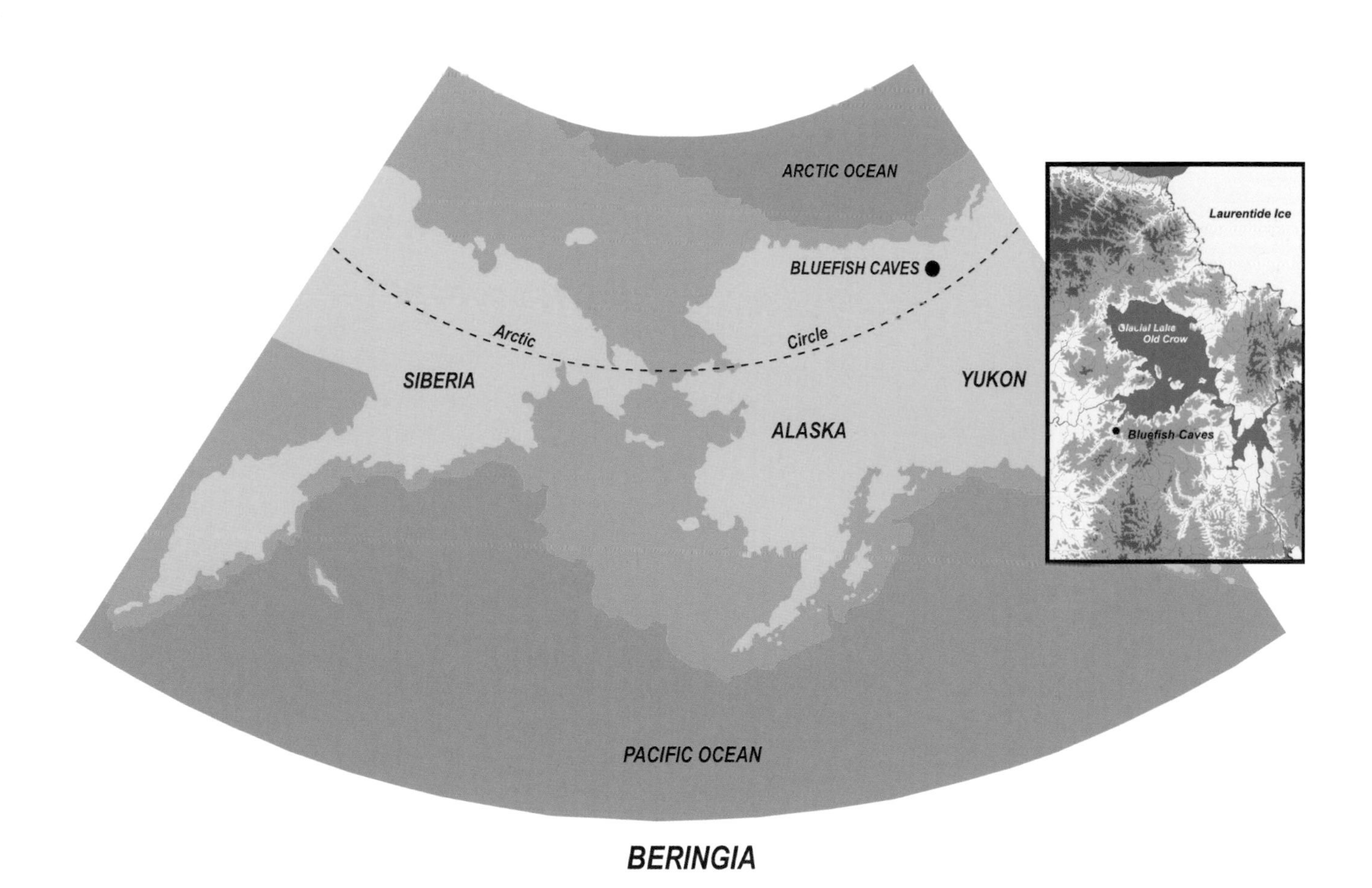

caves. [Figure 24.3] Today, the basin is dotted with hundreds of small lakes interspersed with patches of peat, shrubs, grasses, and moss, while the vegetation around the caves proper is predominantly alpine tundra that grades into the open spruce boreal forest of the adjacent Bluefish River valley. The caves are surrounded on the north, west, and south by steep, forested and rock-covered surfaces that can achieve local slopes of up to 30 percent.

The remote ridge containing the Bluefish Caves is composed of a series of barren, 14 meter tall limestone crags that rise above the upper limits of the present-day tree line. Situated at the base of that rock outcrop, the caves range in volume from about 30 cubic meters to 10 cubic meters and the depth of their sediments ranges from about 30 centimeters to 2 meters. The largest of the sites, Cave I, has a roof height that ranges 3.5–1.8 meters, a maximum width of 3 meters, and a depth of about 4 meters. The caves are so-called "solutional caves" that formed through the dissolution of the ridge's limestone bedrock via the erosion of flowing and freezing groundwater over great lengths of time.

The sites were discovered by Canadian Museum of History archaeologist Jacques Cinq-Mars during a helicopter reconnaissance of the Bluefish River area in 1975. [Figure 24.4] A brief ground survey of the ridge confirmed the presence of several small rockshelters and the caves—one of which, Cave I, produced a small number of bone fragments whose condition suggested human activity. Limited test excavations conducted in the late summer of 1977 recovered a number of very well preserved bones that ultimately proved to be those of extinct Late Pleistocene mammals. [Figure 24.5] As the cave deposits appeared to be a relatively thick and undisturbed primary deposit, a more comprehensive excavation project was conducted in 1978 under the auspices of the Canadian Museum of History's Northern Yukon Research Programme. The project concluded in the late 1980s and the last major publications on the sites appeared in the 1990s.

Cinq-Mars and his colleagues distinguished four discrete strata in their excavations inside, immediately outside, and downslope of all three caves. [Figure 24.6]

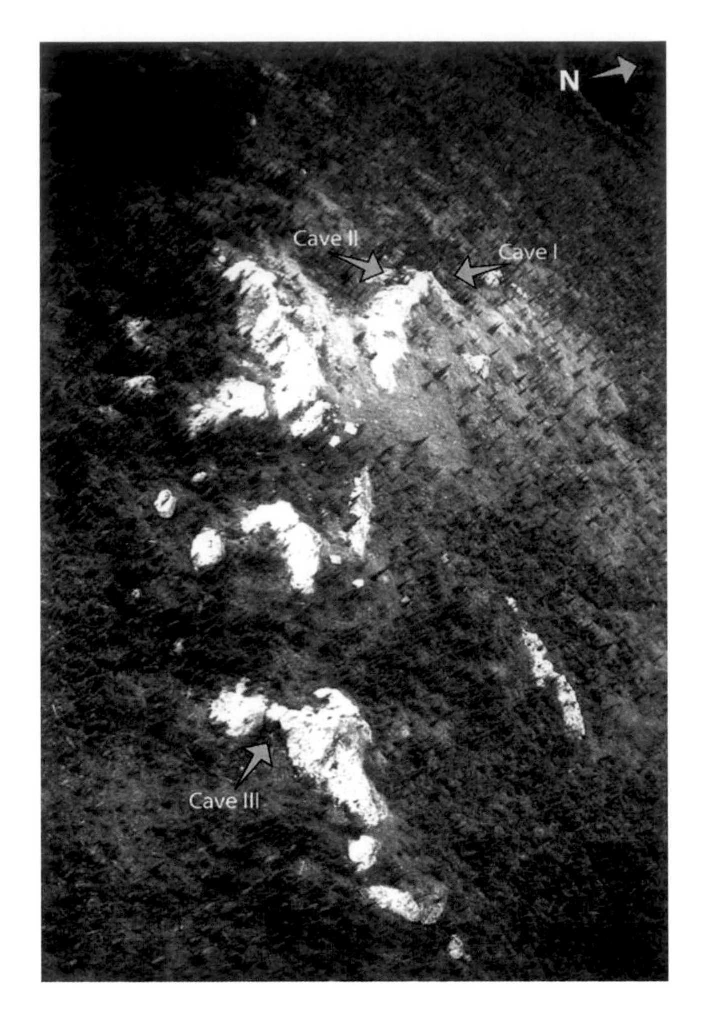

FIGURE 24.4 The Bluefish Caves were long known to the local Vuntut Gwichin community. They were brought to the attention of Canadian Museum of History archaeologist Jacques Cinq-Mars in 1975 during a helicopter reconnaissance of the Bluefish River basin. Cinq-Mars returned in 1977 with a small team under the auspices of the Northern Yukon Research Programme and began limited test excavations. Courtesy of Richard Harington.

The lowest stratum, called Unit A, represents the limestone bedrock of the cave formation and is thought to have been covered by Unit B sometime before 25,000 yr BP. Unit B is composed of wind-borne sediment (known as loess) that appears to have derived from erosion accompanying the receding shorelines of Glacial Lake Old Crow and, especially, its southern

FIGURE 24.5 (above) Jacques Cinq-Mars mapping the lower loess bone bed outside of Cave I. (see Figure 24.6) The wind-deposited loess contains a remarkable diversity of megafaunal remains, including horse (*Equus lambei*), caribou (*Rangifer tarandus*), sheep (*Ovis dalli*), bison (*Bison priscus*), moose (*cf. Alces alces*), elk (*Cervus elaphus*) and mammoth (*Mammuthus primigenius*). There are also saiga (*Saiga tatarica*), muskox (*Ovibos moschatus*), bear (*Ursus* sp.), wolf (*Canis lupus*), and lion (*Panther* sp.). Courtesy of Ruth Gotthardt.

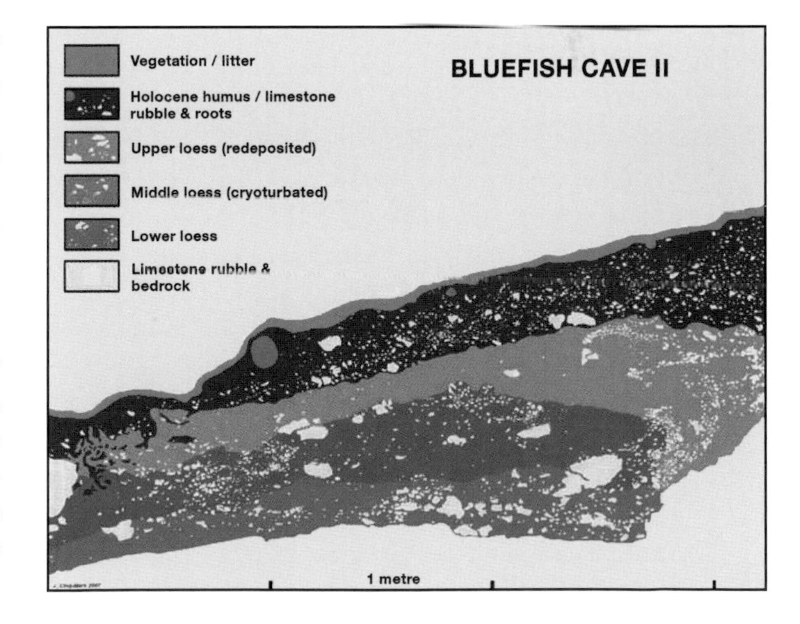

FIGURE 24.6 (right) The stratigraphic profile (sequence of sediments) of the downslope portion of Cave II. Limestone rubble & bedrock = Unit A. Lower loess = Unit B. Middle loess (cryoturbated) and Upper loess (redposited) = Unit C. Holocene human / limestone rubble and roots = Unit D. Before excavations began, the vegetation on the surface of the caves was ferns and lichen. Courtesy of Jacques Cinq-Mars.

lobe in the ancient Bluefish basin. Measuring up to 1 meter in thickness, this stratum yielded a wealth of Late Pleistocene faunal specimens including the remains of mammoth (*Mammuthus* sp.), Steppe Bison (*Bison priscus*), Yukon horse (*Equus lambei*), sheep (*Ovis dalli*), caribou (*Rangifer tarandus*), moose (probably *Cervalces scotti*), saiga antelope (*Saiga tatarica*), muskox (*Ovibus moschatus*), lion (*Felis concolor*), brown bear (*Ursus arctos*), wolf (*Canis lupus*), and numerous other smaller mammal, bird, and fish species. [Figure 24.7] This stratum is in turn overlain by Unit C, a thick layer of decomposed organic material, and the modern surface, Unit D, both of which appear to have formed over the last 10,000 years.

No cultural features (*e.g.*, fire pits, living floors, *etc.*) were identified at any of the caves, but a definitive human presence is indicated by a small assemblage of lithic artifacts, and perhaps less clearly, by putative bone tools, bone flakes, and butchered bones. The lithic artifacts have been attributed to two separate categories. The first group of lithic artifacts includes cores, microblades, burins, possible flake tools, and debitage that were recovered from the upper reaches of stratigraphic Unit B. The second lithic artifact category is composed of diminutive, 1–3 millimeter "microflakes" that occurred throughout the entire depth of Unit B. As no suitable toolstone is available in the limestone formations surrounding the sites, the raw materials employed in the manufacture of the artifacts appear to have been collected from distant sources, perhaps as far as 100 kilometers to the north. [Figure 24.8]

The precise dating of the lithic artifacts recovered from the caves, especially given the absence of cultural features (*e.g.*, fire pits, living floors, etc.) and datable charcoal, has proven somewhat problematic. Cinq-Mars and his Canadian Museum of History colleague Richard Morlan have broadly attributed the microblade and burin lithic tools to the "Paleoarctic-Diuktai complex," which is currently recognized as representing at least two discrete Beringian cultural horizons that date to 14,000–8500 yr BP. The recovery of the microflakes throughout the entire depth of the site deposits, on the other hand, suggests to Cinq-Mars and Morlan that they may date to at least 25,000 yr BP and the formation of Unit B.

FIGURE 24.7 A horse (*Equus lambei*) mandible found in the lower loess of Cave II. Courtesy of the Musée Canadien de l'Histoire.

The diminutive assemblage of bone tools from the caves appear to constitute, by turns, the most significant and controversial items recovered from the sites. Most of them are made from split long bones showing various traces of whittling, abrasion, or polish that have been attributed to human modification and use. One specimen made from a split caribou tibia has been interpreted as a fleshing tool that dates to 29,170–28,560 yr BP, which if valid would make it the oldest directly dated artifact recovered from an archaeological deposit in eastern Beringia and among the oldest from the New World. The two mammoth bone tools in the assemblage, both recovered from the deepest reaches of stratigraphic Unit B in Cave II, have been interpreted as a core from which at least three flakes have been detached and a specimen that appears to be one of those detached flakes. A series of AMS radiocarbon dates from these specimens indicate an average age of 28,080–27,350 yr BP. Collectively, the Bluefish Caves bone artifacts bear a striking similarity to items recovered from Old Crow.

FIGURE 24.8 (above) A chert burin recovered from Cave II Unit C deposits in 1979. (see Figure 24.6) The stratigraphic position of this artifact suggest in predates 10,000 yr BP. The source of the chert, which is not local, is unknown. Courtesy of Ruth Gotthardt.

FIGURE 24.9 (right, below) Université de Montréal archaeologist Lauriane Bourgeon undertook a re-examination of 5,600 caribou, horse, mammoth, and other animals bone fragments from Cave II. Most of them bore the marks of carnivores scavenging meals from gristle and marrow. At least two, however, showed evidence of having been modified by humans. The piece of caribou pelvic bone seen here has deep and parallel striations typical of a stone tool used to de-flesh a carcass. Although previously dated to about 25,000 yr BP, Bourgeon concluded that an age of 14,000 yr BP is more likely. Courtesy of Lauriane Bourgeon.

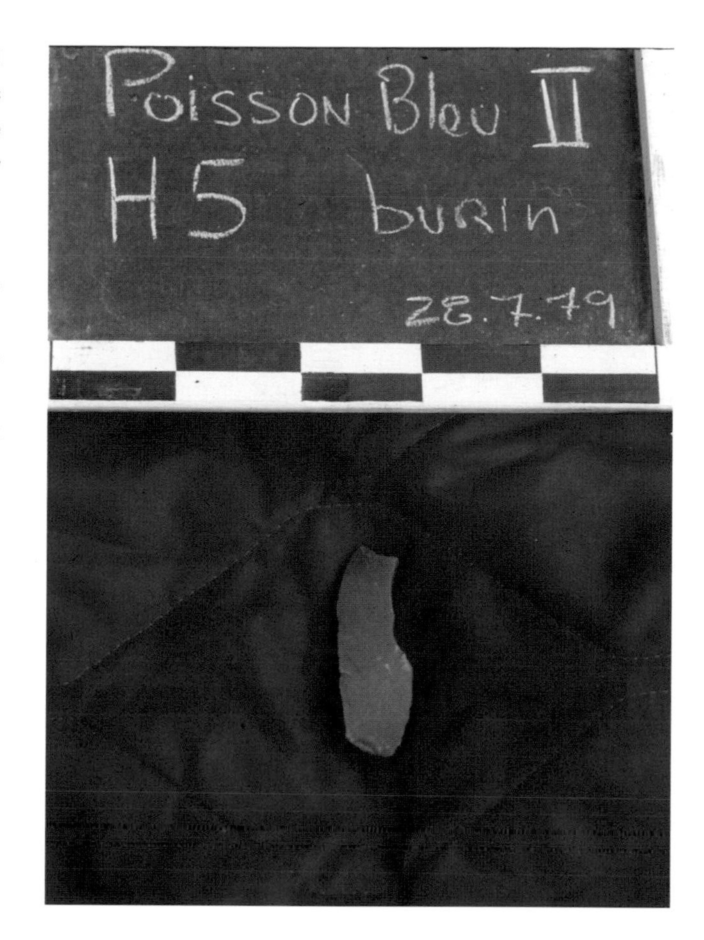

The best dated and most consistently ancient faunal remains from the sites were recovered from Cave II, which produced radiocarbon dates for five butchered bones (four mammoth and one caribou) with a median age of 24,070–23,170 yr BP. The validity of the butchered bone from Cave II, however, has been cast into serious doubt by a recent reanalysis which identified human-made cut marks on only two bones out of an analyzed sample of almost 6,000 specimens. [Figure 24.9]

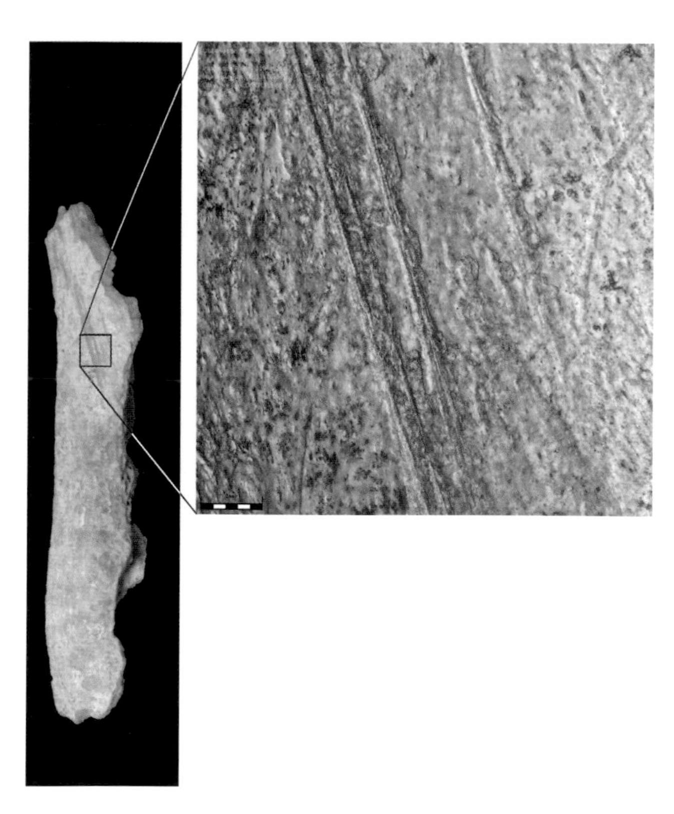

While it appears that Cinq-Mars and his colleagues have indeed identified an important archaeological site, complex persistent questions have remained concerning the relationships between the sites' cultural and paleontological materials. Research conducted over the decades since the findings at Bluefish Caves has firmly established an indisputable human presence in eastern Beringia—replete with well-documented and extensive lithic tool assemblages, cultural features, and verifiable human alterations to animal bone—that may date to as early as 14,000 yr BP (see pages 236-245). Without similarly solid evidence, a human presence at the Bluefish Caves around or even before 25,000 yr BP is widely considered to be highly problematic, as no contemporary sites are presently known elsewhere in eastern Beringia and only a handful are known in northeastern Asia, over 3,000 kilometers to the west.

Coda

With the collapse of Clovis First on both chronological and behavioral grounds, a number of questions have resurfaced which are nearly the same as those posed by Columbus in 1492. Additionally, new questions have arisen. While we know with confidence that the ancestors of Native Americans came from Northeast Asia, we still do not know precisely where their homeland was, or more likely, where their *homelands* were. Nor do we know how many times they crossed the Bering Strait. It now appears that many migration events occurred, some of which probably left no progeny and few material traces.

We may also confidently speculate that at least some of the colonists traveled by boat along the southern margins of Beringia and then proceeded down the coast of western North America. Some of these seafarers may have then traveled inland at various points, while others continued their journey south. Of course it is also clear that other populations may have journeyed exclusively, or nearly exclusively, by foot between the vast ice sheets before, after, or at the same time others were navigating the coasts.

These far-reaching considerations leave us with a fundamental question: what, then, is Clovis? Is it a unitary lifestyle, a monolithic archaeological culture, the paterfamilias of all New World cultures, or something else entirely? To address this question, we return to Housley and his colleagues. (see page 28) In their unfortunately seldom-cited study of the colonization of Late Pleistocene Northern Europe after the recession of the glaciers, these scholars identified an initial Pioneer phase with virtually no detectable signature and, hence, little archaeological visibility. After several thousand years, this phase was

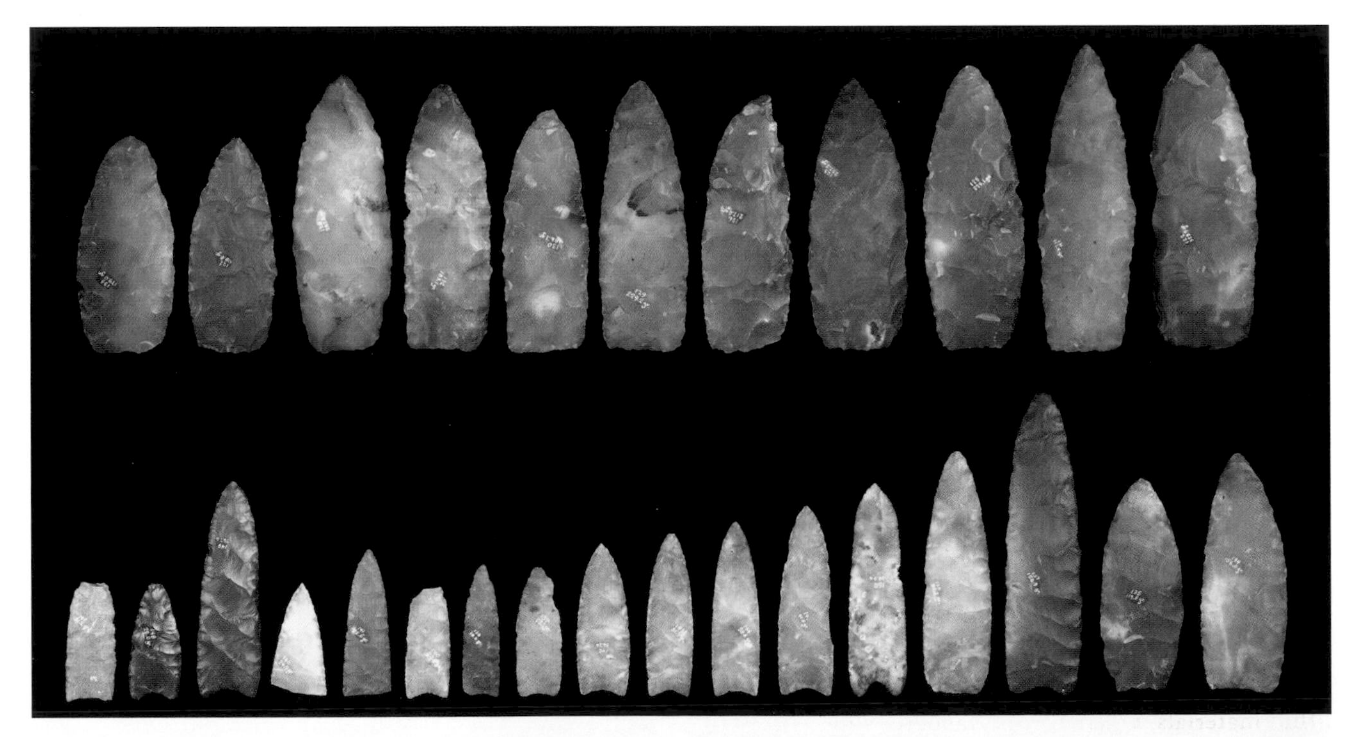

Clovis points from the Fenn Cache discovered sometime around 1902 in the three corners area of Utah, Idaho, and Wyoming. Courtesy of Heather Smith, Center for the Study of the First Americans, Department of Anthropology, Texas A&M University, College Station, Texas.

followed by a Residential phase in which sites were larger, and, consequently, more visible on the landscape.

This is what we believe Clovis represents, the cusp or threshold of archaeological visibility in North America—the establishment of a resident population in some portions of the hemisphere. At the same time, other contemporary populations were establishing themselves in other North and South American environments, also with attendant visibility.[1] In this perspective, the Clovis fluted projectile point is not the lithic banner of a rapidly moving group of migrants, but rather a stone icon passed quickly for still unknown reasons between and among some pre-existing populations. As there are no real functional advantages to fluting, either for hafting or killing, we presume its dispersal is somehow socio-ideological in origin.

In this regard, perhaps Clovis is analogous to the Hopewell tradition, which emerged from the various Woodland peoples of the American Midwest as a shared system of artifact styles and inter-regional exchange that operated throughout the Mississippi and Ohio River valleys and lower Great Lakes during a 600-year interval (2,150–1,550 yr BP) of the Middle Woodland period.

Whatever Clovis was, it was not alone. Other non-Clovis related groups existed who, for whatever reasons, did not participate in the Clovis horizon. Whatever their relationship to Clovis, if any, it is clear that Clovis point makers are not their ancestors. It is still unknown how many of these contemporaries existed, but they may well represent separate population pulses by genetically and linguistically different groups.

A related question, of course, is where did the Clovis phenomenon originate? It now appears that the fluting of projectile points did not begin in the American West, as has long been held by scholars, but may have evolved from the American Southeast as posited decades ago.[2] Unfortunately, while several pre-Clovis sites have yielded lithic materials which exhibit vague possible ancestral connections to Clovis, none demonstrate conclusive lineal affinities.

Even this small patch of the sky is populated by millions of galaxies. Even the smallest dots in this image are galaxies. Within each of these galaxies are millions of stars, some Sun-size, some of these with Earth-size planets and even Earth-like environments. The first humans journeying into the New World were perhaps drawn by herds of large game, or perhaps by what lay beyond the next horizon. Photograph courtesy of NASA.

And then there is Monte Verde I. Buried deeply below Monte Verde II and tentatively dated to greater than 38,000 yr BP, at a time when Neanderthals were contending with the arrival of modern humans in Europe and our species was establishing itself in Australia. This age is so far outside any other widely accepted pre-Clovis assays that Dillehay himself has expressed extreme caution about its validity. If in the future such dates are confirmed at Monte Verde or anywhere else, they will reshape not only the prehistory of the Western Hemisphere, but the prehistory of humans worldwide.

As with questions of who were the first humans and what they carried, we will probably never know for certain. What is certain is that the initial colonization of the New World was the last great continent-wide expansion of *Homo sapiens sapiens*. Indeed, it may well be the last great large-scale peopling episode of our kind, unless we ultimately choose or are forced to leave our planet to colonize other worlds and even stranger lands.

GLOSSARY

adze. Edged, and usually ground, stone tool used to work wood or dig in hard ground.

alluvial deposit. Sediment left in place by a stream or river. Also called *alluvium.*

AMS radiocarbon dating. One of two techniques (the other being conventional *radiocarbon dating*) for measuring radiocarbon in samples (archaeological, geological, *etc.*,) to determine their age. In some situations its results are more precise than conventional *radiocarbon dating*, and it can be successful with smaller samples. Both methods are used for dating most organic materials and some inorganic materials, but not metals. The acronym AMS, which stands for *accelerator mass spectrometry*, provides a means for counting the number of radioactive carbon 14 atoms present in a sample while conventional *radiocarbon dating* measures the radioactive decay of carbon 14 within a sample. See ***radiocarbon dating.***

anthropology. Scientific study of the origin and development of modern humans, their forebears, and their cultures.

anthropometry. Scientific study of the measurements and proportions of the human body.

archaeological context. Term used to describe an individual temporal event in an archaeological site's *stratigraphy*. An individual artifact's *archaeological context* would be its place in time and space. See ***stratigraphy.***

archaeological deposit. General collective term for buried material accumulations (*e.g., artifacts, cultural features*, human and butchered animal remains, *etc.*) left behind by human activity.

archaeometry. Field of archaeological science that applies techniques and analytical approaches based in physics, chemistry, biology, and earth sciences. The adjectival form is *archaeometric.*

Archaic period. See ***cultural period.***

argillite. Fine-grained sedimentary rock composed predominantly of *indurated* (firm or hardened) clay particles.

artifact. Object (*e.g.*, stone tool, ceramic pot, ornament, textile, *etc.*) made or used by humans that occurs in an *archaeological context.*

assay. Term (both noun and verb) indicating the *test* or *analysis* of a sample, as in *radiocarbon assay*. See ***AMS radiocarbon dating; radiocarbon dating.***

assemblage. Collection of interrelated artifacts recovered from an archaeological site.

association. Two or more objects become part of the archaeological record, more or less, at the same time as a result of the same process.

bedrock. Solid rock underlying unconsolidated surface materials such as soil and gravel.

Beringia. Presently submerged landform that was dry land during the last *Ice Age*, when lower ocean levels exposed the continental shelves of Alaska and Siberia in the Bering Sea and formed a "land bridge" connecting Asia and North America. At its greatest extent around 21,000 yr BP, Beringia is estimated to have been roughly 1,600 kilometers wide (north to south). Most archaeologists believe that humans from northeastern Asia first entered the New World though interior Beringia and/or along its coasts. Also known as the *Bering Platform*. See ***Ice age.***

biface. Flattish, edged stone tool that has been *flaked* or chipped on both of its faces. Bifaces can be fashioned as informal cutting and scraping tools or as more-formal, standardized implements such as *projectile points*. See ***projectile point; uniface.***

blade. Lithic artifact struck via percussion or pressure from a *core* using either a *hammerstone* or punch. Unlike *flakes*, blades are distinguished by being parallel-sided and at least twice as long as they are wide. Their shape and dimensions are believed to be intentionally uniform. Less formally, the term *blade* can also be used in reference to the main body (*i.e.*, minus the *stem*, if present) of a knife, biface, or projectile point. Small blades are sometimes called *blade flakes*. See ***core; flake; hammerstone.***

bone bed. A geological or archaeological deposit that contains a dense a concentration of the fossilized bones of extinct animals.

boreal forest. Belt of coniferous trees across North America and Eurasia that formed on formerly glaciated and permafrost areas. Also known as taiga.

BP. Abbreviation of years *before present,* rendered in this book as *yr BP*. In radiocarbon dating, the present is set at AD 1950.

burin. Tool often fashioned from a lithic *blade* with a pointed, chisel-like edge for engraving/carving wood or bone. The term is similar to, if not interchangeable with, the term *graver*. See ***flake; graver***.

Buttermilk Creek complex. Earlier than Clovis cultural horizon represented at the Debra L. Friedkin and Gault sites in central Texas.

calibrated radiocarbon age. Radiocarbon age is calculated by measuring the radiocarbon that has remained in an organism and determines its time of death by assuming a rate of decay (or *half-life*) of 5,730 years. Shortly after the discovery of *radiocarbon dating* in the late 1940s, however, physicists became aware that the level of atmospheric radiocarbon was not constant through time. The calibration of radiocarbon age uses three rings and other historic environmental data to compensate for variations in atmospheric radiocarbon. See Appendix I.

chalcedony. A crystalline sedimentary rock composed of quartz, valued as a material for the manufacture of stone tools. See ***chert; quartz***.

chert. Crystalline sedimentary rock composed of silicon dioxide that can occur as thick, layered deposits, concreted masses, nodules, or concretionary masses. Because the fracture of chert produces very sharp edges, it is an ideal material for the manufacture of stone tools. Chert is also informally called *flint*.

Chopper-Chopping Tool tradition. *Lower Paleolithic* period (about 3.3–0.2 million yr BP) lithic industry involving the creation and use of distinctive stone tools by early humans in Europe, Africa, the Middle East, and Asia.

Clovis complex. Paleoindian hunting and foraging groups that spread throughout most of the unglaciated reaches of North America and some areas further south around 13,300–12,800 yr BP. Readily identified on the basis of the distinctive *fluted Clovis projectile point*. See ***Folsom complex; fluted point***.

Clovis. Widely distributed North American prehistoric *Paleoindian* culture, named after the highly distinctive stone *tool kit* first found at sites near Clovis, New Mexico, and later identified at diverse sites throughout North America. Clovis appears to have existed during the relatively brief time frame of about 13,300–12,800 yr BP. See ***cultural period; tool kit***

colluvial deposit. Loose mass of soil and rock debris accumulated at the base of a slope as a result of various erosional processes (*e.g.*, rain wash, downslope creep, *etc.*) and gravity. Also called *colluvium*.

cordage. Collective term for string, cord, and rope made from plant and animal fibers.

Cordilleran ice sheet. Smaller of the two continental North American ice sheets. The *Cordilleran* covered far western North America during last glacial event of the *Quaternary period*. See ***Laurentide ice sheet; Quaternary period***.

core. Stone from which blades or flakes are detached via percussion or pressure to produce implements for expedient cutting tasks or subsequent refinement into more formalized tools such as *bifaces* and *projectile points*.

cultural complex. Term used to describe particular types of artifacts that consistently recur together in two or more sites within a well-defined region throughout the same time frame.

cultural feature. General term for a non-portable physical entity that indicates past human activity on an archaeological site. *Cultural features* include phenomena such as dwellings, living floors, walls, post molds, fire pits, storage pits, specialized activity areas, *etc.*

cultural period. Broadly defined term denoting the time span for shared networks of artifact types, archaeological site organization, apparent lifeway, *etc.*, across a common landscape. The specific names, spatial distributions, and durations of *cultural periods* vary considerably between regions in the Western Hemisphere, and indeed throughout the world. Generally speaking, however, for the western hemisphere the agreed-upon scheme begins with the Paleoindian period (earlier than 10,000 yr BP) followed by the Archaic period (about 10,000–1000 yr BP) and various post-Archaic temporal divisions known as the *Woodland, Formative, Classic*, and *Post-Classic periods*. These primary divisions also contain a number of subdivisions whose timeframes vary quite broadly between and within regions, but all end with the onset of the *Historic period*. The *Historic period* begins with contact between Europeans and Native Americans, the date of which—again—is highly variable and dependent upon the particular region and circumstance.

cultural zone. Geographic area within which one or more relatively homogeneous, related, and recognized human groups are/were present.

debitage. Small pieces of broken stone (*e.g.*, flakes, shatter, spalls, *etc.*) produced during stone tool manufacture, usually considered as waste or byproducts.

dripline. In caves and rockshelters, the point at which surface or rain water draining from the overlying rock formation makes contact with the ground surface

ecological zone. Area containing geographically defined and distinctive plant and animal communities.

endscraper. Lithic tool made from a *blade* or *flake* with a single working edge on one of its ends. See ***blade; flake.***

excavation. Systematic digging and recordation of archaeological materials, cultural features, and strata at an archaeological site.

fire pit. Pit dug in the ground or built on the ground surface at an archaeological site at the time of its occupation to contain and control fire. Charcoal from fire pits is used to assess the ages of sites via *radiocarbon dating*.

flake. Thin, often flat piece of stone intentionally removed by percussion or pressure during the manufacture of a lithic tool or the resharpening of an existing tool. Flakes may also be detached from a *core* for the production of expedient *flake tools* or more formalized artifacts such as *bifaces* or *projectile points*.

flaked stone. Prehistoric lithic technology focused on the production of cutting and scraping tools via the fracturing of lithic cores using percussion or pressure, which is known as *flaking*, chipping, or flintknapping. Because of stone's durability, it tends to be the best represented of *artifact* classes at an *archaeological site*. See ***artifact.***

fluted point. Stemless projectile point, with each face bearing a distinctive longitudinal groove (known as a *flute*) produced by the removal of a long channel flake from the point's base. See ***Clovis complex; Folsom complex; projectile point; unfluted point.***

Folsom complex. Collection of *Paleoindian* groups that spread across much of central North America around 12,400 yr BP, thought to have derived from the earlier *Clovis complex*. Readily identified on the basis of the distinctive *fluted Folsom projectile point*. See ***Clovis complex; fluted point.***

Formative period. See ***cultural period.***

geofact. Naturally occurring lithic item that has not been modified by humans. A *geofact* may sometimes be difficult to distinguish from a bona fide artifact. *See* ***artifact.***

genome. Organism's complete set of DNA, including all of its genes and containing all of the information needed to build and maintain that organism.

graver. See ***burin.***

grid. Three-dimensional coordinate system used to record the precise location of all objects and phenomena observed at an archaeological site. Modern archaeology in the United States and abroad uses metric units as the principal system of measure.

haft. To attach a shaft or handle to a tool. Used as a noun, the haft (sometimes called the *hafting element*) is the portion of the tool to which hafting was affixed.

hammerstone. Hard rock cobble or pebble used in the production of lithic tools and the processing of plant and animal resources via percussion.

handstone. Hand-held milling stone used to process plant materials. Also known by the Spanish term *mano*.

haplogroup. A particular configuration or cluster of genes inherited from one parent. See ***haplogroup.***

haplotype. Term used to classify a number of similar *haplotypes* which share a common ancestor. See ***haplotype.***

hearth. Remains of an ancient fire pit or campfire. Hearths may be lined with rock and/or clay, or they may simply be unlined.

Historic period. See ***cultural period.***

Holocene epoch. Geologic epoch that began at the end of the Pleistocene epoch (about 11,700 yr BP) and continues to the present. The epoch witnessed the beginning of written history and the development of the earth's major civilizations.

Ice age. A global cold period during which extensive *ice sheets* advance and retreat, such as the *Pleistocene epoch*. See ***Pleistocene epoch.***

ice sheet. Glacier of considerable thickness (sometimes over 3 kilometers) that covers an area of at least 50,000 square kilometers. Ice sheets form a continuous, moving cover of ice and snow over the landscape and are not confined by the underlying terrain.

in situ. Spatial condition of an archaeological object in its original position as deposited in an archaeological site. See ***archaeological context; archaeological deposit; secondary deposit.***

jasper. A variety of *chert*. See ***chert.***

knife. Flaked stone tool with a cutting and/or slicing function, sometimes (but not necessarily) *hafted* to a handle. See ***haft; projectile point.***

lanceolate. Term borrowed from botany to classify a flaked stone *projectile point* or *biface* whose shape is long, tapers at both ends, and is wider in the middle**.** See ***biface; projectile point.***

Last Glacial Maximum. The point during the most recent glaciation when the *ice sheets* reached their maximum area and thickness with a corresponding drop in global sea level. This event is roughly dated in the Northern Hemisphere to have occurred between about 26,500 yr BP and 20,000–19,000 yr BP. An abrupt rise in sea level accompanied the melting of the ice sheets at the end of this time period. See ***ice sheet; Cordilleran ice sheet; Laurentide ice sheet; Quaternary period.***

Laurentide ice sheet. Larger of the two continental North American *ice sheets*. The *Laurentide* covered central and eastern North America during the last glacial event of the *Quaternary period*. See ***Cordilleran ice sheet; Quaternary period.***

lithic. Relating to, or being produced from, stone. This term is informally used by archaeologists as a stand-in for lithic artifact. See ***artifact.***

loess. Aeolian *sediment* that has been eroded, transported, and/or deposited by wind. It is often associated with glaciers or arid landscapes.

megafauna. Collective noun referring to large animals. Archaeologists use the term to refer to large mammals that lived during the

Pleistocene epoch, but became extinct when the *ice age* ended. See ***Ice age; Pleistocene epoch.***

Mesolithic. Global term for the intermediate period between the *Paleolithic* and *Neolithic* intervals in Western Europe and Eurasia, spanning about 12,000–7000 yr BP. The term is also applied to sites in the Levant, beginning about 10,000 years earlier. The term is not used in New World archaeology. See ***Neolithic; Paleolithic.***

microlith. A very small lithic tool.

Miller complex. The earliest *cultural complex* at Meadowcroft Rockshelter and in the Cross Creek watershed of southwestern Pennsylvania. See ***cultural complex, cultural period.***

Montell complex. *Late Archaic cultural complex* noted for the Lower Pecos Canyonlands region of New Mexico. See ***cultural complex, cultural period.***

mitochondrial DNA. Mitochondrial deoxyribonucleic acid (mtDNA) is a small portion of a cell that transfers chemical energy from food. Mitochondrial DNA is inherited solely from the mother and can be used to determine the relatedness of populations. See ***Y-DNA.***

molecular biology. Branch of biology concerned with biological phenomena at the molecular level. See ***mitochondrial DNA; Y-DNA.***

Neolithic. Global term for the last stage (around 12,000–4000 yr BP) of the Stone Age in the Old World, including Europe, Africa, eastern Asia, and China. The term is not used in New World archaeology. See ***Mesolithic; Paleolithic.***

notch. Flaked U- or V-shaped indentation on the base of a lithic tool, usually to facilitate *hafting*. See ***haft.***

optically stimulated luminescence. Frequently referred to simply by the acronym *OSL*, optically-stimulated luminescence is a dating technique employed to determine the last time crystalline sediment (*e.g.*, quartz) was exposed to sunlight while being transported via wind, water, or ice prior to its burial.

osteometry. Study and measurement of the human or animal skeleton.

overkill hypothesis. Hypothesis proposed by University of Arizona geoscientist Paul Martin that argued humans were responsible for the Late Pleistocene extinction of *megafauna* while rapidly populating the interior of North America. See ***late Pleistocene; megafauna.***

outwash gravel. Gravel deposited by the flowing water of melting glacial ice.

Paleoindian. Informal term for Native Americans of the *Paleoindian period*, which generally precedes 10,000 yr BP. See ***cultural period; Paleoindian period.***

Paleoindian period. See ***cultural period.***

Paleolithic. Global term for the first stage of the Stone Age, beginning with the first use of stone tools by humans in Africa, Asia, and Europe, beginning about 3.3 million years ago and extending to about 12,000 yr BP. The *Upper Paleolithic* is broadly defined as the last 40,000 years of the *Paleolithic*. The term is not used in New World archaeology. See ***Mesolithic; Neolithic.***

paleosol. Otherwise known as a fossil soil, a paleosol is a soil that formed on the surface of a landscape in the past that is often found deeply buried.

Pebble Tool tradition. Crude stone tools produced by humans during the very earliest *Paleolithic*. See ***Paleolithic.***

Plainview complex. *Cultural complex* associated with the later *Paleoindian period* site located along a river known as Running Water Draw, near the town of Plainview, Texas. *Unfluted Plainview projectile points* have parallel or convex sides with a concave base. See ***cultural complex, cultural period.***

Pleistocene epoch. Geologic time epoch beginning around 2.6 million years ago and ending around 11,700 years ago, followed by the *Holocene epoch*. *Pleistocene* plant and animal communities were relatively similar to those of modern times, but often with different geographic distributions and combinations. The *Pleistocene* is distinguished from the *Holocene* by repeated glacial advances and the presence of distinctive large land mammals (such as *mammoths* and *mastodons*) and birds (such as *teratorns*) that became extinct around the end of the epoch. See ***Holocene epoch; Quaternary period.***

point. An alternate term for a *projectile point*. See ***projectile point.***

post mold. Soil stain in an *archaeological deposit* resulting from the decay of a wooden post associated with an architectural *cultural feature*. See ***cultural feature.***

pre-Clovis. Generic term used to designate all archaeological phenomena in the Western Hemisphere that pre-date the 13,300 yr BP beginning of the *Clovis period*. See ***Clovis; cultural period.***

preform. Artifact in an early stage of manufacture whose further development is presumed to result in a formal tool such as a *projectile point.*

projectile point. The stone tip *hafted* on arrows, darts, lances, spears, *etc.* They were used for hunting, fishing, and (probably much less frequently) in combat. Projectile points undoubtedly were also used as *knives*. Informally called "arrowheads" by avocational archaeologists and artifact collectors. See ***haft; knife; point.***

quartzite. Metamorphic rock that was originally pure quartz and is transformed into quartzite through heating and pressure resulting from tectonic activity.

Quaternary period. Current and most recent of the three periods of the *Cenozoic Era*, which began about 65 million years ago and is popularly known as the *Age of the Mammals*. Preceded by the *Tertiary period*, the *Quaternary* began about 2.6 million years ago. See ***Pleistocene epoch; Tertiary period.***

radiocarbon dating. One of two techniques (the other being *AMS radiocarbon dating*) for measuring radiocarbon in samples (archaeological, geological, *etc.*) to determine their age. In some situations, its results are less precise than those from *AMS radiocarbon dating* and larger samples are required to ensure accuracy. Both methods are used for dating most organic materials and some inorganic materials, but not metals. *Radiocarbon dating* measures the radioactive decay of carbon 14 within a sample while *AMS radiocarbon dating* counts the number of radioactive carbon 14 atoms present in a sample. See ***AMS radiocarbon dating; calibrated radiocarbon age.***

retouch. General term relating to the modification or refurbishment (such as sharpening) of a *flaked stone artifact* following its primary manufacture and during the course of its use.

rhyolite. An igneous rock that can be used to manufacture stone tools.

riparian. *Ecological zone* at the contact between stream courses and the adjacent ground surface. See ***ecological zone.***

scraper. *Flaked stone* tool taking a wide variety of shapes and sizes used for the preparation of animal hides and skins, woodworking, butchering, *etc.*

secondary deposit. An archaeological phenomenon or object whose original (*in situ*) position in an archaeological deposit has been altered or disturbed by natural or subsequent human agencies. See ***archaeological deposit; in situ.***

sediment. Unconsolidated fragmentary material from the weathering of rock that is transported via air, water, ice, or other natural process, and has accumulated on the landscape over time.

site. Specific place that contains material evidence of human activity.

soil horizon. A layer of sediment that can be distinguished from adjacent layers on the basis of chemistry, structure, color, texture, or composition.

stratum. Individual layer or body of sediment among other stacked layers of an *archaeological deposit.* Its plural form is *strata.* See ***soil horizon.***

stratigraphic sequence. Geologic term describing the chronological succession of sedimentary rocks or sediments, from older to younger without interruption. In *archaeology*, the term *stratigraphic sequence* refers to the chronologic succession of *strata* in the *archaeological deposit* at a given *site.* See ***archaeological deposit; stratum; stratigraphy.***

stratigraphy. Term borrowed from geology to describe the stacked arrangement of *strata* in an *archaeological deposit.* See ***archaeological deposit; strata.***

talus. Accumulation of rock fragments at the base of a cliff or a very steep, rocky slope.

taxon. Named and formally recognized group of organisms. The plural form of the term is *taxa.*

terrace. Long, often relatively narrow land surface bounded on either side by a steeper descending and ascending slope. A terrace commonly occurs along a body of water and marks a former water level.

Tertiary period. Former term for the geologic period ranging about 66–2.6 million yr BP, still widely in use. Geologists have more recently divided this interval into the Paleogene and Neogene periods. The earliest documented members of the genus *Homo* emerged toward the end of the *Tertiary period.* See ***Quaternary period.***

tool kit. Suite of stone tools that characterize an archaeological *site*, *cultural complex*, or *cultural period.* See ***cultural period; cultural complex; site.***

type site. The site where objects or materials regarded as defining the characteristics of a particular culture, period, *etc.*, were found.

type. Specific classification of archaeological objects, such as *projectile points*, based upon their physical characteristics and occurrence in time. Such classification uses a recognized and agreed-upon system known as *typology.* The site at which a particular type of artifact and its associated array of archaeological remains (*e.g.*, *cultural features*, domestic structures such as *longhouses*, etc.) is referred to as the *type site.*

unfluted point. *Projectile point* lacking the distinctive longitudinal groove (or *flute*) produced by the removal of a large, wide channel flake from the point's base. See ***fluted projectile point; projectile point.***

uni-beveled. Edge of a tool, typically made of bone, that gradually slopes away from one of the tool's surfaces.

uniface. Flattish, edged flaked stone tool that has been *flaked* or chipped on only one of its faces. See ***projectile point; biface.***

use wear. Evidence of minute modifications to a tool resulting from their use for a specific task.

watershed. The region or area of land that is drained by or contributes water to a body of water such as a stream or lake.

Western Stemmed tradition. Apparently independent contemporary of the *Clovis complex* identified in the North American region known as the Intermountain West, which is composed of the Columbia Plateau, the Snake River Plain, and the Great Basin.

Wisconsinan glacial stage. The final, most-recent glacial stage of the *Pleistocene epoch* in North America, occurring about 85,000–11,000 yr BP.

Woodland period. See *cultural period.*

worked. Portion of an artifact that has been shaped or altered by humans, such as through the removal of *flakes* from a *biface.*

Y-DNA. Deoxyribonucleic acid composition of the Y chromosome, inherited solely from the father, which can be used to determine the relatedness of populations. See ***mitochondrial DNA.***

RADIOCARBON DATING

The radiocarbon dating method was developed in the late 1940s by University of Chicago physical chemist Willard Libby, who received the 1960 Nobel Prize in Chemistry for his work. Radiocarbon dating became a standard analysis in archaeological and paleontological investigations soon thereafter. As the practice became more widespread through the ensuing decades, it became apparent that variability in age assessments would require some measure of correction or calibration to make results comparable, particularly those from the late Pleistocene. The radiocarbon ages of all sites discussed in this book are reported in calibrated radiocarbon years.

Libby's discovery was based on the observation of natural processes associated with the interaction between cosmic rays and Earth's atmosphere. Cosmic rays are high-speed particles that originate mainly from outside our solar system in the Milky Way Galaxy, and to a lesser degree from more distant galaxies and as near as our Sun. When cosmic rays make contact with the upper atmosphere, nitrogen in the atmosphere is broken down into several carbon isotopes, one of which is an unstable isotope called radiocarbon or carbon-14 (abbreviated as ^{14}C or C-14). This exchange leads to the creation of carbon dioxide, which thereafter diffuses throughout the atmosphere. Carbon dioxide, and thus the carbon-14 incorporated within it, is then absorbed by virtually all organisms in Earth's food chain, either directly via photosynthesis—as in the case of plants—or indirectly, as in the case of animals that eat those plants or other animals that have done so.

Carbon-14 is said to be unstable because once formed, it begins to go through radioactive decay—that is, it emits radiation (in the form of beta particles) as it reverts back to nitrogen. Continued consumption on the part of living organisms replenishes their "supply" of carbon-14, but that replenishment ends when an organism dies. The carbon-14 in that dead organism then begins to decay at a rate governed by its radioactive *half-life*, which is set at 5,730 years. Hence, after 5,730 years only half (50 percent) of the organism's original carbon-14 is present, and after 5,730 more years the amount of original carbon-14 stands at 25 percent. The amount of remaining original carbon-14 becomes more difficult to measure as that amount approaches zero, which means that the practical limit for radiocarbon dating is about 50,000 years, or almost nine half-lives.

The decay of carbon 14 in a sample of organic material, such as a piece charcoal from a fire pit or a plant-fiber textile, can be measured via two basic methods. The older, traditional method involves rendering the sample out of its solid state into a gas or liquid and counting the beta particles that are released by the continued decay of the carbon-14 that has remained in the sample. The accelerator mass spectrometry (AMS) method, on the other hand, employs an atomic particle accelerator to isolate the remaining carbon-14 atoms from other atoms in the sample. Those atoms are then directly counted relative to the stable carbon isotopes in the sample, namely carbon-12 and carbon-13.

To derive an age estimate, the values for carbon-14 content obtained via these methods are compared to a modern, constant value for atmospheric radiocarbon content that was established by Libby in 1950. All radiocarbon analyses use this constant value to maintain comparability among radiocarbon age determinations that have been calculated since 1950, as the amount cosmic radiation bombarding the atmosphere—and, hence, the amount of original carbon-14 present in dead organisms—is now known to have varied through time.

The statistical comparison of a sample's radiocarbon content to the constant value, therefore, produces a *probable* age for the sample that is accompanied by a numerical margin of error (accompanied by the symbol ±) and a confidence interval (measured as a percentage of probability that the age of a sample falls within the indicated age range). In other words, rather than producing a *specific* age, radiocarbon dating provides a probable age *range* for a given specimen.

When Libby developed the radiocarbon dating method he was working under the assumption that that rate of cosmic ray bombardment, the rate of carbon-14 production, and the amount of carbon-14 in the atmosphere were constant. It has since become apparent that these values fluctuate, and that radiocarbon years and calendar years are not equivalent. In order to adjust the variable length of radiocarbon years, international scholars have collaborated on developing calculation methods and building global data bases to permit the calibration of radiocarbon ages. Calibration data sets rely on the interrelationships between organisms of great and known age, such as living trees (whose ages are calculated from annular growth rings), archaeological wood samples of known age, coral reefs, and varves (*i.e.*, annual layers of sediment or sedimentary rock). The calibrated radiocarbon ages in this book were calculated using CALIB (version 7.0.2) radiocarbon calibration program (employing the IntCal13 calibration curve) developed by geochemist Minze Stuiver and archaeologist and palaeoecologist Paula J. Reimer. Approximate equivalencies between millennial radiocarbon ages, calibrated ages, and calendar ages are shown in Table 1. The radiocarbon and calibrated ages for the sites described in this book are listed in Table 2. The calibrated ages that appear elsewhere in the book have been rounded to the nearest decade.

Table 1. Approximate equivalencies between millennial radiocarbon ages, calibrated ages, and calendar ages.

Radiocarbon Age (radiocarbon years BP)	Mean Calibrated Age (calibrated years BP)	Calendar Age (BCE)
24,000	28,010	26,060
23,000	27,330	25,380
22,000	26,200	24,250
21,000	25,350	23,400
20,000	24,060	22,110
19,000	22,880	20,930
18,000	21,790	19,840
17,000	20,500	18,550
16,000	19,330	17,380
15,000	18,220	16,270
14,000	16,990	15,040
13,000	15,540	13,590
12,000	13,810	11,860
11,500	13,340	11,390
11,000	12,840	10,890
10,000	11,420	9470
9000	10,200	8250
8000	8810	6860
7000	7830	5880
6000	6840	4890
5000	5730	3780
4000	4490	2540

Table 2. Radiocarbon dating provides a probable age for ancient organic materials by calculating the amount of radiocarbon (or carbon-14) that has remained in them since they were exposed to atmospheric carbon-14 as living things. Because the amount of carbon-14 in the atmosphere fluctuates (or is variable) through time rather than being constant, a year as measured by an object's carbon-14 content is not necessarily equivalent to the length of our 365-day year. To account for this variation, scientists have developed complex methods to calibrate radiocarbon ages so that they are more or less constant and roughly equivalent to calendar years over great spans of time. This table lists the calibrated equivalents of the radiocarbon ages by site reported in this book.

Site	Context	Radiocarbon Age (^{14}C years BP)	Calibrated Age (cal years BP [2s])	Material Dated	Lab Number
Blackwater Draw	Unit C_0	11,170 ± 360	13,771–12,367	carbonized plant remains	A-481
Blackwater Draw	Unit C (mean)	11,290 ± 290	13,733–12,684	carbonized plant remains	A-481 and A-491
Blackwater Draw	UnitC_1	11,630 ± 400	14,708–12,706	carbonized plant remains	A-491
Bluefish Caves	Cave II	19,640 ± 170	24,069–23,168	butchered mammoth bone	RIDDL-330
Bluefish Caves	Cave II	23,555 ± 225	28,081–27,348	mammoth bone core and flake	Average of RIDDL-224 and RID-DL-225
Bluefish Caves	Cave II	24,820 ± 115	29,169–28,558	caribou bone artifact (flesher)	RIDDL-226
Bonfire Shelter	Bone Bed 3	2780 ± 110	3210–2725	hearth charcoal	TX-106
Bonfire Shelter	Bone Bed 2	10,230 ± 160	12,430–11,330	hearth charcoal	TX-153
Bonfire Shelter	Bone Bed 1	12,430 ± 490	16,060–13,370	charcoal	AA-344
Broken Mammoth	CZ3, east hearth (most secure)	10,290 ± 70	12,394–11,809	charcoal	CAM-5357 (AMS)
Broken Mammoth	CZ4, east hearth	11,420 ± 70	13,416–13,115	charcoal	CAMS-5358 (AMS)
Broken Mammoth	CZ4, central hearth	11,510 ± 120	13,563–13,115	charcoal	WSU-4262
Cactus Hill	Clovis hearth, Area B	10,920 ± 250	13,301–12,368	charcoal	Beta-81589
Cactus Hill	pre-Clovis Hearth, Area B	15,070 ± 70	18,526–18,079	charcoal	Beta-81590
Cactus Hill	pre-Clovis charcoal concentration, Area B	16,670 ± 730	21,933–18,486	charcoal	Beta-97708
Calico	Lake Manix shoreline	19,300 ± 400	24,157–22,398	tephra	UCLA-121
Diuktai Cave	Unit 7b	13,070 ± 90	15,959–15,320	charcoal from eroded hearth	LE-784

Site	Context	Radiocarbon Age (^{14}C years BP)	Calibrated Age (cal years BP [2s])	Material Dated	Lab Number
Diuktai Cave	Unit 7b	14,000 ± 100	17,355–16,623	charcoal from eroded hearth	GIN-404
Dry Creek	Unit 7b	10,060 ± 75	11,840–11,283	charcoal	AA-11727 (AMS)
Dry Creek	Dry Creek site, Component II (Paleosol 1/Loess 3, Denali complex)	10,615 ± 100	12,731–12,376	charcoal	AA-11728 (AMS)
Dry Creek	Dry Creek site, Component I (Loess 2, Nenana complex)	11,120 ± 85	13,129–12,773	charcoal	SI-2880
El Fin del Mundo	upper Stratum 4, Locality 1	8870 ± 60	10,184–9738	shell	AA-88885
El Fin del Mundo	lower Stratum 4, Locality 1	9715 ± 64	11,245–11,065	charcoal	AA-80084A
El Fin del Mundo	upper Stratum 3B, Locality 1, bone bed	11,550 ± 60	13,489–13,265	charcoal	AA-100181A
Folsom	McJunkin Formation m2 (youngest)	4470 ± 90	5318–4859	charcoal	TX-1272
Folsom	McJunkin Formation m1 (oldest)	6910 ± 110	7947–7579	charcoal	TX-1271
Folsom	Folsom Formation f3 (youngest), unrelat-ed to the occupation	9220 ± 50	10,513–10,249	bison bone	CAMS-74654
Folsom	Folsom Formation f2 (youngest)	10,010 ± 50	11,717–11,273	charcoal	CAMS-74645
Folsom	f2 bison kill (mean)	10,490 ± 20	12,540–12,404	bison bone	CAMS-74655 through -74659, CAMS-96034
Folsom	Folsom Formation f2 (oldest)	11,500 ± 40	13,439–13,267	charcoal	Average of CAMS-57513 and CAMS-57514
Folsom	Folsom Formation f1 (oldest)	12,355 ± 210	15,150–13,786	charcoal	AA-7090
Hebior	—	12,480 ± 60	15,040–14,267	purified bone collagen	CAMS-28303
Hebior	bone bed	12,590 ± 50	15,174–14,668	purified bone collagen	CAMS 61137
Lehner	Hearth 2	7022 ± 450	8787–6940	charcoal	A-32
Lehner	Hearth 2	8330 ± 450	10,406–8299	charcoal	A-30

Site	Context	Radiocarbon Age (^{14}C years BP)	Calibrated Age (cal years BP [2s])	Material Dated	Lab Number
Lehner	Hearth excavated by Haynes	10,940 ± 40	12,905–12,707	charcoal	Average of 12 samples
Lehner	Hearth 2	11,850 ± 50	13,766–13,559	charcoal	Average of A-40a and A-40b
Lehner	Hearth 2	12,000 ± 450	15,385–12,989	charcoal	A-40b
Meadowcroft	Stratum IIa	11,300 ± 700	15,160–11,250	charcoal	SI-2491
Meadowcroft	Stratum IIa	12,800 ± 870	17,580–13,060	charcoal	SI-2489
Meadowcroft	Stratum IIa	13,240 ± 1010	18,360–13,220	charcoal	SI-2065
Meadowcroft	Stratum IIa hearth/ fire feature	16,175 ± 975	21,975–17,247	charcoal	SI-2354
Meadowcroft	lower Stratum IIa	19,100 ± 810	25,036–21,088	charcoal	SI-2062
Meadowcroft	lower Stratum IIa	19,600 ± 2400	28,510–17,960	bark	SI-2060
Meadowcroft	lower Stratum IIa	21,070 ± 475	26,210–24,140	charcoal	DIC-2187
Monte Verde	Monte Verde II	11,990 ± 200	14,587–13,404	bone (mast-odon)	TX-3760
Monte Verde	Monte Verde II	12,450 ± 150	15,121–14,031	Wood (artifact)	OxA-381
Monte Verde	Monte Verde I	12,980 ± 40	15,259–15,680	bone collagen	Beta-37,5838
Monte Verde	Monte Verde I	13,200 ± 60	15,850–16,520	burned plant stem	Beta-36,9125
Monte Verde	Monte Verde I	15,210 ± 30	18,304–18,578	burned plant stem	Beta-37,2893
Monte Verde	Monte Verde I	16,000 ± 60	19,150–19485	bone collagen	Beta-37,2889
Monte Verde	Monte Verde I	33,370 ± 530	38,797–36,262	burned wood	Beta-6745
Murray Springs	Graveyard Member, Murray Springs Formation	10,900 ± 50	12,868–12,694	charcoal	Mean of Tx-1462, A-805A, Tx-1413, SMU-42, SMU-18, SMU-41, Tx-1459, SMU-27
Old Crow	Stream channel	1350 ± 150	1550–957	bone (cari-bou, artifact [flesher])	RIDDL-145 (AMS)
Old Crow	—	11,350 ± 110	13,432–13,030	bone (caribou antler artifact [punch])	Beta-27512

Site	Context	Radiocarbon Age (^{14}C years BP)	Calibrated Age (cal years BP [2s])	Material Dated	Lab Number
Old Crow	—	24,800 ± 650	30,388–27,698	bone (caribou antler artifact [pestle])	CRNL-1233
Old Crow	Stream channel	27,000 ± 400	31,624–30,356	bone (caribou, artifact [flesher])	RIDDL-232 (original conventional radiocarbon date)
Old Crow	—	30,810 ± 975	36,991–32,701	bone (steppe bison tibia [modified])	Beta-33192
Old Crow	—	39,700 ± 1400	45,837–41,417	bone (steppe bison tibia [modified])	UCIAMS-71652
Paisley Five Mile Point Caves	Cave 2	2295 ± 15	2348–2315	human coprolite	UCIAMS-79714
Paisley Five Mile Point Caves	Hearth, Cave 2	11,625 ± 35	13,563–13,384	human coprolite	UCIAMS-77104
Paisley Five Mile Point Caves	Cave 2	11,740 ± 25	13,601–13,460	bone (horse maxilla)	UCIAMS-86251
Paisley Five Mile Point Caves	Cave 2	11,930 ± 25	13,822–13,704	bone (artiodactyl rib)	UCIAMS-90593
Paisley Five Mile Point Caves	Cave 5 (associated with Western Stemmed point)	11,070 ±30	13,046–12,816	plant material	UCIAMS-80378
Paisley Five Mile Point Caves	Bone Pit feature (youngest), Cave 5	11,130 ± 40	13,094–12,867	bone (horse phalanx)	Beta-185942
Paisley Five Mile Point Caves	Cave 5 (associated with Western Stemmed point)	11,500 ± 30	13,425–13,274	plant material	UCIAMS-80381
Paisley Five Mile Point Caves	Cave 5	11,795 ± 30	13,741–13,544	bone (camel)	UCIAMS-79657
Paisley Five Mile Point Caves	Cave 5	12,195 ± 30	14,198–13,978	bear bone artifact	UCIAMS-68017
Paisley Five Mile Point Caves	Cave 5	12,380 ± 70	14,851–14,112	butchered mountain sheep bone	Beta-239087
Paisley Five Mile Point Caves	Cave 5	12,385 ± 30	14,711–14,170	horse tooth	UCIAMS-90592
Paisley Five Mile Point Caves	Bone Pit feature (oldest), Cave 5	12,400 ± 60	14,857–14,144	human coprolite	Beta-213424
Pedra Furada	Unit 3, Serra Talhada phase (youngest)	6150 ± 60	7179–6887	charcoal	GIF-8108
Pedra Furada	Unit 3, Serra Talhada phase (oldest)	10,400 ± 180	12,703–11,611	charcoal	GIF-5862

Site	Context	Radiocarbon Age (^{14}C years BP)	Calibrated Age (cal years BP [2s])	Material Dated	Lab Number
Pedra Furada	PF3, youngest	14,300 ± 210	17,944–16,795	charcoal	GIF-6159
Pedra Furada	PF3, oldest	21,400 ± 400	26,487–24,626	charcoal	GIF-6160
Pedra Furada	PF2, youngest	25,200 ± 320	30,173–28,594	charcoal	GIF-6147
Pedra Furada	PF2, oldest	32,160 ± 1000	37,840–33,420	charcoal	GIF-6653
Pedra Furada	PF1, youngest	42,400 ± 2600	50,000–42,340	charcoal	GIF-TAN89097
Pedra Furada	PF1, oldest	55,570 +1590/-1320	(>59,000)	charcoal	ANUA-16325
Pendejo Cave	Zone C	1780 ± 50	1821–1569	sandal (*Yucca* sp.)	UCR-2642
Pendejo Cave	Zone C	2580 ± 40	2770–2690	textile (cordage)	UCR-3611/ CAMS-44669
Pendejo Cave	Zone c	2740 ± 40	2925–2760	textile (cordage)	UCR-3612/ CAMS-
Pendejo Cave	Zone C	5840 ± 60	6677–6636	textile (sandal)	UCR-2643
Pendejo Cave	Zone C1	11,300 ± 110	13,440–13,065	wood	UCR-2602
Pendejo Cave	Zone C1	11,900 ± 150	14,100–13,440	wood	UCR-2641
Pendejo Cave	Zone C2	12,370 ± 80	14,875–14,090	human hair	UCR-3276A/ CAMS-12366
Pendejo Cave	Zone D	16,440 ± 650	21,492–18,391	charcoal	UCR-2504
Pendejo Cave	Zone F	28,430 ± 960	34,274–30,833	charcoal	UCR-2501
Pendejo Cave	Zone I, human imprints	32,000 ± 1200	38,820–33,780	charcoal	UCR-2645
Pendejo Cave	Zone N	>55,000	(>55,000)	wood	QL-4625
Saltville	middle horizon	13,950 ± 70	17,175–16,614	woody twigs *in situ*	Beta-65209
Saltville	oldest horizon	14,510 ± 80	17,921–17,464	bone ("modified" [Specimen VMNH 721])	Beta-117541
Schaefer	Mandible	12,290 ± 60	14,625–14,029	purified bone collagen	CAMS 72140
Schaefer	left humerus	12,570 ± 45	15,152–14,627	purified bone collagen	CAMS 95521
Shawnee-Minisink	hearth in Kline's Paleoindian levels	10,590 ± 300	13,061–11,595	charcoal	W-2994 (USGS)

Site	Context	Radiocarbon Age (^{14}C years BP)	Calibrated Age (cal years BP [2s])	Material Dated	Lab Number
Shawnee-Minisink	hearth in Kline's Paleoindian levels, direct assoc w/ artifacts	10,900 ± 40	12,826–12,701	charred hawthorn seed	Beta-127162 (AMS)
Shawnee-Minisink	Hearth 2 in Gingerich's 2006 excavation, direct association with Paleoindian artifacts and Clovis point	10,970 ± 50	12,979–12,721	charred hawthorn seed	OxA-1731
Shawnee-Minisink	hearth in Kline's Paleoindian levels, direct assoc w/ artifacts	11,020 ± 30	13,005–12,772	charred plum seed	UCIAMS-24866 (AMS)
Swan Point	Swan Point CZ3, north hearth 1(youngest)	10,050 ± 60	11,824–11,288	charcoal	Beta-170458
Swan Point	CZ3, south hearth 4 (oldest)	10,570 ± 40	12,659–12,514	charcoal	Beta-209885
Swan Point	CZ4, Hearth 1b	12,040 ± 40	14,027–13,760	charcoal	Bet-QA-619 (AMS)
Swan Point	CZ4, Hearth 2	12,290 ± 40	14,485–14,051	charcoal	Beta-209882 (AMS)
Taima-taima	Upper Unit III	9650 ± 90	11,219–10,737	soil	IVIV-658
Taima-taima	Upper Unit III	10,290 ± 80	12,402–11,767	soil	IVIC-667
Taima-taima	Unit I	12,580 ± 150	15,309–14,178	soil	IVIC-627
Taima-taima	Unit I mastodon (*Haplomastodon waringi*), youngest	12,980 ± 85	15,809–15,239	wood	SI-3316
Taima-taima	Unit I	13,390 ± 130	16,506–15,735	soil	IVIC-668
Tequendama	Occupation Zone 1	10,920 ± 260	13,307–12,142	charcoal	GrN-6539
Tibitó	—	11,740 ± 110	13,775–13,340	bone	GrN-9375
Topper	Stratum 1a	>54,700	(>54,700)	nutshell (hickory [*Carya* sp.])	CAMS-79022
Walker Road	Walker Road site, Component I, Hearth 1	11,010 ± 230	13,356–12,523	charcoal	AA-1683 (AMS)
Walker Road	Same as above	11,300 ± 120	13,416–12,899	charcoal	AA-2264 (AMS)
Yana RHS	Northern Point locality cultural layer	28,500 ± 200	33,126–31,715	mammoth ivory (artifact)	Beta-191326

ENDNOTES FOR PART ONE

1 Oxford English Dictionary 2015.
2 Alexseev 1979.
3 Chatters 2001; Doran 2002; Jantz and Owsley 1998; Neves and Blum 2001; Neves et al. 1999; Powell and Neves 1999.
4 Owsley and Jantz 2014.
5 Neumann 1952.
6 Turner 1986, 1989.
7 Greenberg 1987.
8 Szathmáry 1994:17.
9 Spuhler 1979.
10 Szathmáry 1979, 1981, 1984; Szathmáry et al. 1978.
11 Gallatin 1848.
12 Brinton 1891.
13 Powell 1891.
14 Boas 1911.
15 Sapir 1921.
16 Voeglin and Voeglin 1965.
17 Goddard and Campbell 1994.
18 Nichols 1995.
19 Vajda 2011.
20 Schurr 2002, 2004.
21 Adovasio and Pedler 2008:39-40.
22 Klein 2009.
23 Mochanov 1988.
24 Slobodin 2011.
25 cf. Pitulko et al. 2013.
26 Derevianko 1998; Waters et al. 1997, 1999
27 Pitulko et al. 2013.
28 Mochanov 1977.
29 Pitulko et al. 2013.
30 Hakenbeck 2009.
31 Fladmark 1979.
32 Bryan 1980, 1988.
33 Gruhn .1994.
34 Dixon 1997.
35 Erlandson 2013; Erlandson et al. 2007.
36 Anderson 2013; Anderson et al. 2013.
37 Adovasio et al. 2007.
38 Hamilton and Goebel 1999; Potter et al. 2013; Smith et al. 2013.
39 Fedge 2002; Mackie et al. 2013.
40 Erlandson 2013, Erlandson and Braje 2011.
41 Erlandson 2013.
42 Geoarchaeological Explorations on the Inner-Continental Shelf of the Florida Gulf of Mexico, J. M. Adovasio and C. A. Hemmings, research ongoing.
43 Stanford and Bradley 2012.
44 Mason 1962.
45 See Meltzer 2006 for a thorough rendering of this story.
46 Martin 1967.
47 Fiedel 1999.
48 Meltzer 2009:192.
49 Housley et al. 1997.
50 Damas 1984; Helm 1981
51 Collins 1937; Croes 1977; Doran 2002.
52 Kvavadze et al. 2009; Soffer 2004; Soffer et al. 2000; Stone 2011.
53 Hyland et al. 2002.
54 Adovasio et al. 2009; Frison et al. 1986; Sandweiss et al. 1998.
55 Sondaar et al. 1994.
56 Leppard 2014; Phoca-Cosmeateau and Rabett 2014; Runnels 2014.
57 Adovasio and Pedler 2008.

ENDNOTES FOR CODA

1 Housley et al. 1997.
2 Mason 1962.

SOURCES FOR PART TWO SITE ENTRIES

Site(s)	Sources
Clovis and Folsom Age Sites	
Folsom, New Mexico	Anderson and Haynes 1979; Figgins, 1927; Haynes 1969; Meltzer 2006; Meltzer, Todd, and Holliday 2002.
Blackwater Draw, New Mexico	Boldurian 1990; Boldurian and Cotter 1999; Cotter 1937; Haynes 1995; Howard 1936; Stock and Bode 1937.
Naco, Lehner, and Murray Springs, Arizona	Haury 1986; Haury, Sayles, and Wasley 1986; Haynes and Huckell 2007; Mehringer and Haynes 1965.
Shoop, Pennsylvania	Cox, 1986; Carr, Adovasio, and Vento 2013; Carr and Adovasio 2002; Meltzer 1988.
Shawnee-Minisink, Pennsylvania	Carr and Adovasio 2002; Dent 2002; Gingerich 2013; McNett 1985; Wisner 2007.
Kimmswick Bone Bed, Missouri	Graham, Haynes, Johnson, and Kay 1981; Grayson ad Meltzer 2002; Graham and Kay1988.
Bonfire Shelter, Texas	Bryant 1969; Byerly, Cooper, Meltzer, Hill, and LaBelle 2007; Byerly, Meltzer, Cooper, and Theler 2007; Dibble 1970; Dibble and Lorrain 1968; Robinson 1977.
Central Alaska (Broken Mammoth, Dry Creek, Swan Point, and Walker Road)	Byer 2006; Dixon 1985; Goebel and Buvit 2011; Graf and Bigelow 2011; Graf et al. 2015; Holmes, VanderHoek, and Dilley 1996; Pearson 1999; Potter, Holmes, and Yesner 2013; Powers and Hoffecker 1989; West 1996.
El Fin del Mundo, Mexico	Hill 2015; Meltzer 2014; Sanchez and Carpenter 2012; Sanchez, Holliday, Gaines, Arroyo-Cabrales, Martínez-Tagüeña, Kowler, Lange, Hodgins, Mentzer, and Sanchez-Morales 2014.

Site(s)	Sources
Disputed Pre-Clovis Sites	
Old Crow, Canadian Yukon	Bower 1987; Irving 1985; Irving and Harrington 1973; Jopling, Irving, and Beebe 1981; Morlan, Nelson, Brown, Vogel, and Southon 1990; Nelson, Morlan, Vogel, Southon, and Harington 1988.
Calico Mountain, California	Bamforth and Dorn 1988; Haynes 1973; Leakey, Simpson, and Clements 1968, 1970; Simpson 1980.
Pendejo Cave, New Mexico	Betancourt, Rylander, Peñalba, and McVickar 2001; Chrisman, MacNeish, Mavalwala, and Savage 1996; Dincauze 1997; MacNeish and Libbey 2004; Shaffer and Baker 1997.
Tule Springs, Nevada	Bryan 1964; Harrington and Simpson 1961; Haynes, Doberenz, and Allan 1966; Shutler 1965; Taylor 2009; Wormington and Ellis 1967.
Pedra Furada, Brazil	Aimola, Andrade, Mota, and Parenti 2014; Gruhn 2007; Guidon, Pessis, Parenti, Guérin, Peyre, and dos Santos 2002; Lourdeau 2015; Meltzer, Adovasio, and Dillehay 1994; Parenti 2002; United Nations Educational, Scientific and Cultural Organization (UNESCO) 2015.
Pre-Clovis Sites	
Meadowcroft Rockshelter, Pennsylvania	Adovasio, Gunn, Donahue and Stuckenrath 1977; Adovasio, Donahue, Carlisle, Cushman, Stuckenrath, and Wiegman 1984; Adovasio, Carlisle, Cushman, Donahue, Guilday, Johnson, Lord, Parmalee, Stuckenrath and Wiegman 1985; Adovasio, Boldurian, and Carlisle 1988; Adovasio, Donahue, and Stuckenrath 1990; Adovasio, Gunn, Donahue, and Stuckenrath 1978; Carlisle and Adovasio 1984; Goldberg and Arpin 1999; Haynes 1991.
Monte Verde, Chile	Adovasio and Pedler 1997; Collins 1999; Dillehay 1989a, 1989b, 1997; Dillehay et al. 2015; Fiedel 1999; Meltzer 1997; Meltzer, Grayson, Ardila, Barker, Dincauze, Haynes, Mena, Nunez, and Stanford 1997.
Cactus Hill, Virginia	Boyd 1998; Johnson 2013; McAvoy 1992; McAvoy and McAvoy 1997; McAvoy, Baker, Feathers, Hodges, McWeeney, and Whyte 2000.

Site(s)	Sources
Paisley Five Mile Point Caves, Oregon	Jenkins, Davis, Stafford, Campos, Hockett, Jones, Cummings, Yost, Connolly, Yohe, Gibbons, Raghavan, Rasmussen, Paijmans, Hofreiter, Kemp, Barta, Monroe, Gilbert, and Willerslev 2012; Jenkins, Davis, Stafford, Campos, Connolly, Cummings, Hofreiter, Hockett, McDonough, Luthe, O'Grady, Reinhard, Swisher, White, Yates, Yohe, Yost, and Willerslev 2013; Licciardi 2001; Saban and Jenkins 2015.
Schaefer and Hebior Mammoth, Wisconsin	Fredlund, Johnson, Porter, Revane, Schmidt, Overstreet, and Kolb 1996; Overpeck, Webb, and Webb 1992; Overstreet 1993, 1996, 1998, 2005; Overstreet and Stafford 1997.
Buttermilk Creek Complex (Debra L. Friedkin and Gault), Texas	Collins 2002; Collins and Bradley 2008; Pinson 2012a, 2012b; Waters, Pevney, and Carlson 2011; Waters and Stafford 2007; Waters, Forman, Jennings, Nordt, Driese, Feinberg, Keene, Halligan, Lindquist, Pierson, Hallmark, Collins, and Wiederhold 2011.
Controversial Pre-Clovis Sites	
Topper, South Carolina	Goodyear 2000, 2001a, 2001b, 2005, Jennings 2012; Sain 2015; Smallwood 2010; Smallwood, Miller, and Sain 2013; Waters, Forman, Stafford, and Fosse 2009.
Saltville, Virginia	Anderson 2005; Goodyear 2005; McDonald 2000; McDonald and Bartlett 1983.
Taima-taima, Venezuela, and Tibitó, Columbia	Bryan, Casamiquela, Cruxent, Gruhn, and Ochsenius 1978; Correal 1990; Dillehay 2000; Gruhn and Bryan 1984, 1998; Lavallée 2000; Lynch 1990; Ochsenius and Gruhn 1979.
Bluefish Caves, Canadian Yukon	Bourgeon 2015; Cinq-Mars 1979, 1990; Cinq-Mars and Morlan 1999; Harington and Cinq-Mars 2008; Jopling, Irving, and Beebe 1981.

BIBLIOGRAPHY

Acosta, José de. 2002 [1590]. *Natural and Moral History of the Indies*. Edited by J. E. Mangan and translated by F. M. López-Morillas. Duke University Press, Durham, North Carolina.

Adovasio, J. M., R. L. Andrews, and J. S. Illingworth. 2009. Netting, Net Hunting, and Human Adaptation in the Eastern Great Basin. In *Past Present and Future Issues in Great Basin Archaeology: Papers in Honor of Don D. Fowler*, edited by B. Hockett, pp. 84–102. Cultural Resource Series 20. Bureau of Land Management, Elko, Nevada.

Adovasio, J. M., A. T. Boldurian, and R. C. Carlisle. 1988. Who Are Those Guys? Some Biased Thoughts on the Initial Peopling of the New World. In *Americans before Columbus: Ice-age Origins*, edited by R. C. Carlisle, pp. 45–61. Ethnology Monographs 12. Department of Anthropology, University of Pittsburgh, Pittsburgh.

Adovasio, J. M., R. C. Carlisle, K. A. Cushman, J. Donahue, J. E. Guilday, W. C. Johnson, K. Lord, P. W. Parmalee, R. Stuckenrath, and P. W. Wiegman. 1985. Paleoenvironmental Reconstruction at Meadowcroft Rockshelter, Washington County, Pennsylvania. In *Environments and Extinctions: Man in Late Glacial North America*, edited by J. I. Mead and D. J. Meltzer, pp. 73–110. Peopling of the Americas Series. Center for the Study of Early Man, University of Maine, Orono.

Adovasio, J. M., J. Donahue, R. C. Carlisle, K. Cushman, R. Stuckenrath, and P. Wiegman. 1984. Meadowcroft Rockshelter and the Pleistocene/Holocene Transition in Southwestern Pennsylvania. In *Contributions in Quaternary Vertebrate Paleontology: A Volume in Memorial to John E. Guilday*, edited by Hugh H. Genoways and Mary R. Dawson, pp. 347–369. Special Publication No. 8. Carnegie Museum of Natural History, Pittsburgh.

Adovasio, J. M., J. Donahue, and R. Stuckenrath. 1990. The Meadowcroft Rockshelter Radiocarbon Chronology 1975–1990. *American Antiquity* 55:348–354.

Adovasio, J. M., J. D. Gunn, J. Donahue, and R. Stuckenrath. 1977. Meadowcroft Rockshelter: A 16,000 Year Chronicle. In *Amerinds and their Paleoenvironments in Northeastern North America*, edited by W. S. Newman and B. Salwen, pp. 137–159. Annals of the New York Academy of Sciences, Vol. 288. New York Academy of Sciences, New York.

———. 1978. Meadowcroft Rockshelter, 1977: An Overview. *American Antiquity* 43:632–651.

Adovasio, J. M., and D. R. Pedler. 1997. Monte Verde and the Antiquity of Humankind in the Americas. *Antiquity* 71(273):573–580.

———. 2008. The Peopling of North America. In *North American Archaeology*, edited by T. R. Pauketat and D. DiPaolo Loren, pp. 30–55. Blackwell, Malden, Massachusetts.

Adovasio, J. M., O. Soffer, and J. Page. 2007. *The Invisible Sex*. Harper Collins, New York.

Aimola, G., C. Andrade, L. Mota, and F. Parenti. 2014. Final Pleistocene and Early Holocene at Sitio do Meio, Piauí, Brazil: Stratigraphy and Comparison with Pedra Furada. *Journal of Lithic Studies* 1(2):5–24.

Alexseev, V. P. 1979. Anthropometry of Siberian People. In *The First Americans: Origins, Affinities, Adaptations*, edited by W. S. Laughlin and A. B. Harper, pp. 57–90. G. Fisher, New York.

Anderson, A. B., and C. V. Haynes, Jr. 1979. How old is Capulin Mountain?: Correlation between Capulin Mountain Volcanic Flows and the Folsom type site, Northeastern New Mexico. In *Proceedings of the First Conference on Scientific Research in the National Parks*, Volume 2, edited by R. M. Linn, pp. 893–899. Transactions and Proceedings Series No. 5. National Park Service, Washington.

Anderson, D. G. 2005. Pleistocene Human Occupation of the Southeastern United States: Research Directions for the Early 21st Century. In *Paleoamerican Origins: Beyond Clovis*, Edited by R. Bonnichsen, B. T. Lepper, D. Stanford, and M. R. Waters, pp. 29–42. Center for the Study of the First Americans, Texas A&M University, College Station.

Anderson, D. G. 2013. Paleoindian Archaeology in Eastern North America: Current Approaches and Future Directions. In *The Eastern Fluted Point Tradition*, edited by J. A. M. Gingerich, pp. 371–403. University of Utah Press, Salt Lake City.

Anderson, D. G., T. G. Bissett, and S. J. Yerka. 2013. The Late-Pleistocene Human Settlement of Interior North America: The Role of Physiography and Sea-Level Change. In *Paleoamerican Odyssey*, edited by K. E. Graf, C. V. Ketron, and M. R. Waters, pp. 183–203. Center for the Study of the First Americans, Texas A&M University, College Station.

Bamforth, D. B., and R. I. Dorn. 1988. On the Nature and Antiquity of the Manix Lake Industry. *Journal of California and Great Basin Anthropology* 10(2):209–226.

Bednarik, R. G. 2014. The Beginnings of Maritime Travel. *Advances in Anthropology* (4):209–221.

Betancourt, J. L., K. A. Rylander, C. Peñalba, and J. L. McVickar. 2001. Late Quaternary Vegetation History of Rough Canyon, South-Central New Mexico, USA. *Palaeogeography, Palaeoclimatology, Palaeoecology* 165:71–95.

Bever, M. R. 2006. Too Little, Too Late? The Radiocarbon Chronology of Alaska and the Peopling of the New World. *American Antiquity* 71(4):595–620.

Boas, F. 1911. *Handbook of American Indian Languages*. Vol. 1. Bureau of American Ethnology, Bulletin 40, Smithsonian Institution. United States Government Printing Office, Washington.

Boldurian, A. T. 1990. Lithic Technology at the Mitchell Locality of Blackwater Draw: A Stratified Folsom Site in Eastern New Mexico. *Plains Anthropologist Memoir* 24:1-115.

Boldurian, A. T., and J. L. Cotter. 1999. *Clovis Revisited: New Perspectives on Paleoindian Adaptations from Blackwater Draw, New Mexico.* Monograph 103. University of Pennsylvania Museum of Archaeology and Anthropology, Philadelphia.

Bourgeon, L. 2015. Bluefish Cave II (Yukon Territory, Canada): Taphonomic Study of a Bone Assemblage. *PaleoAmerica* 1(1):105–108

Bower, B. 1987. Flakes, Breaks and the First Americans. *Science News* 131(11):172–173.

Boyd, C. C. 1998. Review of *Archaeological Investigations of Site 44SX202, Cactus Hill, Sussex County, Virginia,* by J. M. McAvoy and L. D. McAvoy. *Southeastern Archaeology* 17(1):106–107.

Bradley, B., and D. Stanford. 2004. The North Atlantic Ice-Edge Corridor: a Possible Palaeolithic Route to the New World. *World Archaeology* 36(4):459–478.

Brinton, D. G. 1891. *The American Race.* N. D. C. Hodges, New York.

Bryan, A. L. 1964. New Evidence Concerning Early Man in North America. *Man* 64:152–153

———, 1980. The Stemmed Point Tradition: An Early Technological Tradition in Western North America. In *Anthropological Papers in Memory of Earl H. Swanson, Jr.*, edited by L. B. Harten, C. N. Warren, and D. R. Tuohy, pp. 77–107. Idaho State Museum of Natural History, Pocatello.

———. 1988. The Relationship of the Stemmed Point and Fluted Point Traditions in the Great Basin. In *Early Human Occupation in Far Western North America: The Clovis-Archaic Interface*, edited by J. A. Willig, C. M. Aikens, and J. L. Fagan, pp. 53–74. Anthropological Papers No. 21. Nevada State Museum, Carson City.

Bryan, A. L., R. M. Casamiquela, J. M. Cruxent, R. Gruhn, and C. Ochsenius 1978 An El Jobo Mastodon Kill at Taima-taima, Venezuela. *Science* 200:1275–1277

Bryant, V. M. 1969. *Late Full-Glacial and Postglacial Pollen Analysis of Texas Sediments.* Ph.D. dissertation. University of Texas, Austin.

Byerly, R. M., J. R. Cooper, D. J. Meltzer, M. E. Hill, and J. M. LaBelle. 2007. A Further Assessment of Paleoindian Site-Use at Bonfire Shelter. *American Antiquity* 72(2):373–381.

Byerly, R. M., D. J. Meltzer, J. R. Cooper, and J. Theler. 2007. Exploring Paleoindian Site-Use at Bonfire Shelter (41VV218). *Bulletin of the Texas Archeological Society* 78:125–147.

Carlisle, R. C., and J. M. Adovasio (editors). 1984. *Meadowcroft: Collected Papers on the Archaeology of Meadowcroft Rockshelter and the Cross Creek Drainage.* Department of Anthropology, University of Pittsburgh, Pittsburgh.

Carr, K. W., and J. M. Adovasio. 2002. Paleoindians in Pennsylvania. In *Ice Age Peoples of Pennsylvania*, edited by K. W. Carr and J. M. Adovasio, pp. 1–50. Pennsylvania Historical and Museum Commission, Harrisburg.

Carr, K. W., J. M. Adovasio, and F. J. Vento. 2013. A Report on the 2008 Field Investigations at the Shoop Site (36DA20). In *The Eastern Fluted Point Tradition*, edited by J. A. M. Gingerich, pp. 75–103. University of Utah Press, Salt Lake City.

Chatters, J. C. 2001. *Ancient Encounters: Kennewick Man and the First Americans.* Simon and Schuster, New York.

Chrisman, D., R. S. MacNeish, J. Mavalwala, and H. Savage. 1996. Late Pleistocene Human Friction Skin Prints from Pendejo Cave, New Mexico. *American Antiquity* 61(2):357–376.

Cinq-Mars, J. 1979. Bluefish Cave 1: A Late Pleistocene Eastern Beringian Cave Deposit in the Northern Yukon. *Canadian Journal of Archaeology* 3:1–32.

———. 1990. La Place des Grottes du Poisson-Bleu dans la Préhistoire Béringienne. *Révista de Arqueologia Americana* 1:9–32.

Cinq-Mars, J., and R. E. Morlan. 1999. Bluefish Caves and the Old Crow Basin. In *Ice Age People of North America: Environments, Origins, and Adaptations*, edited by R. Bonnichsen and K. L. Turnmire, pp. 200–212. Center for the Study of the Fist Americans, Oregon State University Press, Corvallis.

Clark, P. U., A. S. Dyke, J. D. Shakun, A. E. Carlson, J. Clark, B Wohlfarth, J. X. Mitrovica, S. W. Hostetler, and A. M. McCabe. 2009. The Last Glacial Maximum. *Science* 325:710–714.

Collins, H. B. Jr. 1937. Culture Migrations and Contacts in the Bering Sea Region. *American Anthropologist* 39(3):375–384.

Collins, M. B. 1999. Monte Verde Revisited: Reply to Fiedel, Part II. *Discovering Archaeology* (November/December):14–15.

———. 2002. The Gault Site, Texas, and Clovis Research. *Athena Review* 3(2):24–36.

Collins, M. B., and B. A. Bradley. 2008. Evidence for Pre-Clovis Occupation at the Gault Site (41BL323), Central Texas. *Current Research in the Pleistocene* 25:70–72.

Correal, G. 1990. Evidencias Culturales Durante El Pleistoceno y Holoceno de Colombia. *Revista de Arqueología Americana* 1:69–71, 73–89.

Cotter, J. L. 1937. The Occurrence of Flints and Extinct Animals in Pluvial Deposits Near Clovis, New Mexico, Part IV—Report on the Excavations at the Gravel Pit in 1936. *Proceedings of The Academy of Natural Sciences of Philadelphia* 89:1-16.

———. 1938. The Occurrence of Flints and Extinct Animals in Pluvial Deposits near Clovis, New Mexico, Part VI: Report on the Field Season of 1937. *Proceedings Philadelphia Academy of Natural Sciences* 90:113–117.

Cox, S. L. 1986. A Re-Analysis of the Shoop Site. *Archaeology of Eastern North America* 14:101–170.

Croes, D. 1977. *Basketry from the Ozette Village Archaeological Site: a Technological, Functional and Comparative Study.* Ph.D. dissertation, Washington State University. University Microfilms, Ann Arbor, Michigan.

Damas, D. (editor). 1984. *Arctic.* Handbook of North American Indians, Vol. 5, William C. Sturtevant, general editor. Smithsonian Institution, Washington, D.C.

Darwin, C. 2009 [1859]. *The Origin of Species.* Cambridge University Press, New York.

Dent, R. J. 2002. Paleoindian Occupation of the Upper Delaware Valley: Revisiting Shawnee-Minisink and Nearby Sites. In *Ice Age peoples of Pennsylvania*, edited by K. W. Carr and J. M. Adovasio, pp. 51–78. Pennsylvania Historical and Museum Commission, Harrisburg.

Derevianko, A. P. 1998. *The Paleolithic of Siberia: New Discoveries and Interpretations.* University of Illinois Press, Urbana.

Dibble, D. S. 1970. On the Significance of Additional Radiocarbon Dates from Bonfire Shelter, Texas. *Plains Anthropologist* 15:251–254.

Dibble, D. S., and D. Lorrain. 1968. *Bonfire Shelter: A Stratified Bison Kill Site, Val Verde County, Texas.* Texas Memorial Museum Miscellaneous Papers 1. University of Texas, Austin.

Dillehay, T. D. 1989a. *Monte Verde: A Late Pleistocene Settlement in Chile, Volume 1: Paleoenvironmental and Site Context.* Smithsonian Institution, Washington, D.C.

———. 1989b. Monte Verde. *Science* 245:1436.

———. 1997. *Monte Verde: A Late Pleistocene Settlement in Chile, Volume 2: The Archaeological Context and Interpretation.* Smithsonian Institution, Washington, D.C.

———. 2000. *The Settlement of the Americas: A New Prehistory.* Basic Books, New York.

Dillehay, T. D., C. Ocampo, J. Saavedra, A. O. Sawakuchi, R. M. Vega, M. Pino, M. B. Collins, L. Scott Cummings, I. Arregui, X. S. Villagran, G. A. Hartmann, M. Mella, A. González, and G. Dix. 2015. New Archaeological Evidence for an Early Human Presence at Monte Verde, Chile. *PLOS ONE* November 18, 2015. Electronic document available at http://journals.plos.org/plosone/article?id=10.1371/journal.pone.0141923.

Dillehay, T. D., C. Ramírez, M. Pino, M. B. Collins, Chatters, James C. 2001. Ancient Encounters: Kennewick Man and the First Americans. Simon and Schuster, New York.

Dillehay, T. D., C. Ramírez, M. Pino, M. B. Collins, J. Rossen, and J. D. Pino-Navarro. 2008. Monte Verde: Seaweed, Food, Medicine, and the Peopling of South America. *Science* 320:784–786.

Dincauze, D. F. 1997. Regarding Pendejo Cave: Response to Chrisman et al. *American Antiquity* 62(3):554–555.

Dixon, E. J. 1985. Cultural Chronology of Central Interior Alaska. *Arctic Anthropology* 22(1):47–66.

———. 1999. *Bones, Boats, and Bison: Archeology and the First Colonization of Western North America.* University of New Mexico Press, Albuquerque

Doran, G. H. (editor). 2002. *Windover: Multidisciplinary Investigations of an Early Archaic Florida Cemetery.* University Press of Florida, Gainesville.

Erlandson, J. M. 2013. After Clovis-First Collapsed: Reimagining the Peopling of the Americas. In *Paleoamerican Odyssey*, edited by K. E. Graf, C. V. Ketron, and M. R. Waters, pp. 127–132. Center for the Study of the First Americans, Texas A&M University, College Station.

Erlandson, J. M., M. H. Graham, B. J. Bourque, D. Corbett, J. A. Estes, and R. S. Steneck. 2007. The Kelp Highway Hypothesis: Marine Ecology, the Coastal Migration Theory, and the Peopling of the Americas. *The Journal of Island and Coastal Archaeology* 2(2):161–174.

Fedje, D., Q. Mackie, T. Lacourse, and D. McLaren. 2011. Younger Dryas Environments and Archaeology on the Northwest Coast of North America. *Quaternary International* 242:452–462.

Fiedel, S. J. 1999. Artifact Provenience at Monte Verde: Confusion and Contradictions. *Discovering Archaeology* 1(6):1–12.

Figgins, J. D. 1927. The Antiquity of Man in America. *Natural History* 27:229–239.

Fladmark, K. R. 1979. Routes: Alternative Migration Corridors for Early Man in North America. *American Antiquity* 44:55–69.

Fredlund, G. G., R. B. Johnson, G. S. Porter, T. A. Revane, H. K. Schmidt, D. F. Overstreet, and M. Kolb. 1996. Late Pleistocene Vegetation of Hebior Mammoth Site, Southeastern Wisconsin. *Current Research in the Pleistocene* 13:87–89.

Frison, G. C., R. L. Andrews, J. M. Adovasio, R. C. Carlisle, and R. Edgar. 1986. A Late Paleoindian Animal Trapping Net from Northern Wyoming. *American Antiquity* 51(2):352–361.

Gallatin, A. 1848. Hale's Indians of North-West North America, and Vocabularies of North America. *Transactions of the American Ethnographic Society* 2:1–130. New York.

Gingerich, J. A. M. 2013 Revisiting Shawnee-Minisink. In *The Eastern Fluted Point Tradition*, edited by J. A. M. Gingerich, pp. 218–256. University of Utah Press, Salt Lake City.

Goldberg, P., and T. L. Arpin, 1999. Micromorphological Analysis of Sediments from Meadowcroft Rockshelter, Pennsylvania: Implications for Radiocarbon Dating. *Journal of Field Archaeology* 26:325–342.

Goddard, I., and L. Campbell, 1994. The History and Classification of American Indian Languages: What are the Implications for the Peopling of the Americas? In *Method and Theory for Investigating the Peopling of the Americas*, edited by R. Bonnichsen and D. G. Steele, pp. 189–207. Center for the Study of the First Americans, Oregon State University, Corvallis.

Goebel, T., and I. Buvit (editors). 2011. *From the Yenisei to the Yukon: Interpreting Lithic Assemblage Variability in Late Pleistocene/Early Holocene Beringia.* Texas A&M University Press, College Station.

Goodyear, A. C. 1999a. Results of the 1999 Allendale Paleoindian Expedition. *Legacy* 4(1–3):8–13. Newsletter of the South Carolina Institute of Archaeology and Anthropology, University of South Carolina.

———. 1999b. The Early Holocene Occupation of the Southeastern United States: A Geoarchaeology Summary. In *Ice Age Peoples of North America*, edited by R. Bonnichsen and K. L. Turnmire, pp. 432–81. Oregon State University Press, Corvallis.

———. 2000. The Topper Site 2000: Results of the 2000 Allendale Paleoindian Expedition. *Legacy* 5 (2):18–25. Newsletter of the South Carolina Institute of Archaeology and Anthropology, University of South Carolina, Columbia.

———. 2001a. The 2001 Allendale Paleoindian expedition and beyond. *Legacy* 6(2):18–21. Newsletter of the South Carolina Institute of Archaeology and Anthropology, University of South Carolina, Columbia.

———. 2001b. The Topper Site: Beyond Clovis at Allendale. *Mammoth Trumpet* 16 (4): 10–15.

———. 2005. Evidence for Pre–Clovis Sites in the Eastern North United States. In *Paleoamerican Origins: Beyond Clovis*, edited by R. Bonnichsen, B. T. Lepper, D. Stanford, and M. R. Waters, pp. 103–112. Center for the Study of First Americans, Texas A&M University, College Station.

Goodyear, A. C., and T. Charles. 1984. *An Archaeological Survey of Chert Quarries in Western Allendale County, South Carolina.* Research Manuscript Series 195. Institute of Archaeology and Anthropology, University of South Carolina, Columbia.

Goodyear, A. C., and K. Steffy. 2003. Evidence for a Clovis Occupation at the Topper Site, 38AL23, Allendale County, South Carolina. *Current Research in the Pleistocene* 20:23–25.

Graf, K. E., L. M. DiPietro, K. E. Krasinski, A. K. Gore, H. L. Smith, B. J. Culleton, D. J. Kennett, and D. Rhode. 2015. Dry Creek Revisited: New Excavations, Radiocarbon Dates and Site Formation Inform on the Peopling of Eastern Beringia. *American Antiquity* 80(4):671–694.

Graf, K. E., and N. H. Bigelow. 2011. Human Response to Climate During the Younger Dryas Chronozone in Central Alaska. *Quaternary International* 242:434–45.

Graham, R. W., C. V. Haynes, D. L. Johnson, and M. Kay. 1981. Kimmswick: A Clovis-Mastodon Association in Eastern Missouri. *Science* 213(4512):1115–1117.

Graham, R. W., and M. Kay. 1988. Taphonomic Comparisons of Cultural and Noncultural Faunal Deposits at the Kimmswick and Barnhart Sites, Jefferson County Missouri. In *Late Pleistocene and Early Holocene Paleoecology and Archaeology of the Eastern Great Lakes*, edited by R. S. Laub, N. G. Miller, and D. W. Steadman, pp. 227–240. Bulletin of the Buffalo Society of Natural Sciences 33. Buffalo Museum of Science, Buffalo.

Grayson, D. K., and D. J. Meltzer. 2002. Clovis Hunting and Large Mammal Extinction: A Critical Review of the Evidence. *Journal of World Prehistory* 16(4):313–359.

Greenberg, J. H. 1987. *Language in the Americas*. Stanford University Press, Stanford.

Greenman, E. F. 1960. The North Atlantic and Early Man in the New World. *Michigan Archaeologist* 6:19–39.

Gruhn, R. 1994. The Pacific Coast Route of Initial Entry: An Overview. In *Method and Theory for Investigating the Peopling of the Americas*, edited by R. Bonnichsen and D. G. Steele, pp. 249–256. Center for the Study of the First Americans, Oregon State University, Corvallis.

———. 2007. The Earliest Reported Archaeological Sites in South America. *Mammoth Trumpet* 22(1):14–18.

Gruhn, R., and A. L. Bryan. 1984. The Record of Pleistocene Megafaunal Extinctions at Taima-taima, Northern Venezuela. In *Quaternary Extinctions*, edited by P. S. Martin and R. G. Klein, pp 128–137. University of Arizona Press, Tucson.

Gruhn, R., and A. L. Bryan. 1998. A Reappraisal of the Edge-Trimmed Tool Tradition. In *Explorations in American Archaeology: Essays in Honor of Wesley R. Hurt*, edited by M. G. Plew. University Press of America, Lanham, Maryland.

Guidon, N., A-M. Pessis, F. Parenti, Cl. Guérin, E. Peyre, and G. M. dos Santos. 2002. Pedra Furada, Brazil: Paleoindians, Paintings and Paradoxes, an Interview. *Athena Review* 3(2):42–52.

Hakenbeck, S. 2009. 'Hunnic' Modified Skulls: Physical Appearance, Identity and Transformative Nature of Migrations. In *Mortuary Practices and Social Identities in the Middle Ages, Essays in Burial Archaeology in Honour of Heinrich Härke*, edited by D. Sayer and H. M. R. Williams, pp. 64–80. University of Exeter Press, Exeter.

Hamilton, T. D., and T. Goebel. 1999. Late Pleistocene Peopling of Alaska. In *Ice Age Peoples of North America: Environments, Origins, and Adaptations*, edited by R. Bonnichsen and K. L. Turnmire, pp. 156–99. Oregon State University Press, Oregon.

Harington, C. R., and J. Cinq-Mars. 2008. Bluefish Caves—Fauna and Context. *Beringian Research Notes* 19:1–8.

Harrington, M. R., and R. D. Simpson. 1961. *Tule Springs, Nevada, with Other Evidences of Pleistocene Man in North America*. Southwest Museum Papers No. 18. Southwest Museum, Los Angeles.

Haury, E. W. 1986. Artifacts with Mammoth Remains: Discovery of the Naco Mammoth and the Associated Projectile Points. In *Emil W. Haury's Prehistory of the American Southwest*, edited by J. J. Reid and D. E. Doyel, pp. 78–98. University of Arizona Press, Tucson.

Haury, E. W., E. B. Sayles, and W. W. Wasley. 1986. The Lehner Mammoth Site, Southeastern Arizona. In *Emil W. Haury's Prehistory of the American Southwest*, edited by J. J. Reid and D. E. Doyel, pp. 99–145. University of Arizona Press, Tucson.

Haynes, C. V., Jr., 1969. The Earliest Americans. *Science* 166:709–715.

———. 1973. The Calico Site: Artifacts or Geofacts? *Science* 181(4097):305–310.

———. 1991. More on Meadowcroft Radiocarbon Chronology (Review of Adovasio, Donahue, and Stuckenrath). *Review of Archaeology* 12(1):8–14.

———. 1995. Geochronology of Paleoenvironmental Change, Clovis Type Site, Blackwater Draw, New Mexico. *Geoarchaeology* 10: 317-388.

Haynes, C. V., Jr., A. R. Doberenz, and J. A. Allan. 1966. Geological and Geochemical Evidence Concerning the Antiquity of Bone Tools from Tule Springs, Site 2, Clark County, Nevada. *American Antiquity* 31(4):517–521.

Haynes, C. V., Jr., and B. B. Huckell,(editors). 2007. *Murray Springs: A Clovis Site with Multiple Activity Areas in the San Pedro Valley, Arizona*. University of Arizona Press, Tucson.

Helm, J. (editor). 1981. *Subarctic*. Handbook of North American Indians, Vol. 6, William C. Sturtevant, general editor. Smithsonian Institution, Washington, D.C.

Hill, K. 2015. News from the End of the World—As We Know It? *Mammoth Trumpet* 30(3):4–7.

Holliday, V.T., M. Bever, and D. J. Meltzer. 2009. Paleoindians in the American Southwest and Northern Mexico. *Archaeology Southwest* 23(3):1–3.

Holmes, C. E., R. VanderHoek, and T. E. Dilley. 1996. Swan Point. In *American Beginnings: The Prehistory and Paleoecology of Beringia*, edited by F. H. West, pp. 319–323, University of Chicago Press, Chicago.

Housley, R. A. C. S. Gamble, M. Street, and P. B. Pettitt. 1997. Radiocarbon Evidence for the Lateglacial Human Recolonisation of Northern Europe. *Proceedings of the Prehistoric Society* 63:25–54.

Howard, E. B. 1936. The Occurrence of Flints and Extinct Animals in Pluvial Deposits near Clovis, New Mexico, Part I, Introduction. *Proceedings of the Academy of Natural Sciences* 87:299–303.

Hyland, D. C., I. S. Zhushchikhovskaya, V. E. Medvedev, A. P. Derevianko, and A. V. Tabarev. 2002. Pleistocene Textiles in the Russian Far East: Impressions from Some of the World's Oldest Pottery. *Anthropologie* 40(1):1–10.

Irving, W. N. 1985. Context and Chronology of Early Man in the Americas. *Annual Review of Anthropology* 14:529–555.

Irving, W. N., and C. R. Harington. 1973. Upper Pleistocene Radiocarbon-Dated Artifacts from the Northern Yukon. *Science* 179(4071):335–340.

Jantz, R. L., and D. W. Owsley. 1998. How Many Populations of Early North America Were There? *American Journal of Physical Anthropology,* Supplement 26:128

Jenkins, D. L., L. G. Davis, T. W. Stafford, Jr., P. F. Campos, T. J. Connolly, L. S. Cummings, M. Hofreiter, B. Hockett, K. McDonough, I. Luthe, P. W. O'Grady, K. J. Reinhard, M. E. Swisher, F. White, B. Yates, R. M. Yohe II, C. Yost, and E. Willerslev. 2013. Geochronology, Archaeological Context, and DNA at the Paisley Caves. In In *Paleoamerican Odyssey*, edited by K. E. Graf, C. V. Ketron, and M. R. Waters, pp. 485–510. Texas A&M University Press, College Station.

Jenkins, D. L., L. G. Davis, T. W. Stafford, Jr., P. F. Campos, B. Hockett, G. T. Jones, L. Scott Cummings, C. Yost, T. J. Connolly, R. M. Yohe II, S. C. Gibbons, M. Raghavan, M. Rasmussen, J. L. A. Paijmans, M. Hofreiter, B. M. Kemp, J. L. Barta, C. Monroe, M. T. P. Gilbert, and E. Willerslev. 2012. Clovis Age Western Stemmed Projectile Points and Human Coprolites at the Paisley Caves. *Science* 337(5877):223–238.

Jennings, T. A. 2012. Clovis, Folsom, and Midland Components at the Debra L. Friedkin Site, Texas: Context, Chronology, and Assemblages. *Journal of Archaeological Science* 39(10):3239–3247

Johnson, M. F. 2013. *Cactus Hill, Rubis-Pearsall and Blueberry Hill: One is an Accident; Two is a Coincidence; Three is a Pattern—Predicting "Old Dirt" in The Nottoway River Valley of Southeastern Virginia, U.S.A.* Unpublished PhD dissertation, Archaeology Department, University of Exeter, Exeter, United Kingdom.

Jopling, A. V., W. N. Irving and B. F. Beebe. 1981. Stratigraphic, Sedimentological and Faunal Evidence for the Occurrence of Pre-Sangamonian Artefacts in Northern Yukon. *Arctic* 34(1):3–33.

Klein, R. G. 2009. *The Human Career: Human Biological and Cultural Origins.* University of Chicago Press, Chicago.

Koch, A. C. 1841. *Description of Missourium, or Missouri Leviathan.* Prentice and Weissinger, Louisville.

Kvavadze, E., O. Bar-Yosef, A. Belfer-Cohen, E. Boaretto, N. Jakeli, Z. Matskevich, and T. Meshveliani. 2009. 30,000-Year-Old Wild Flax Fibers. *Science* 325(5946):1359.

Lavallée, D. 2000. *The First South Americans: The Peopling of a Continent from the Earliest Evidence to High Culture.* Translated by Paul G. Bahn. University of Utah Press, Salt Lake City.

Leakey, L. S. B., R. D. Simpson, and T. Clements. 1968. Archaeological Excavations in the Calico Mountains, California: Preliminary Report. *Science* 160(3831):1022–1023.

———. 1970. Early Man in America: The Calico Mountains Excavation. *1970 Encyclopedia Britannica Yearbook of Science and the Future*:64–79.

Leppard, T. P. 2014. Modeling the Impacts of Mediterranean Island Colonization by Archaic Hominins: The Likelihood of an Insular Lower Palaeolithic. *Journal of Mediterranean Archaeology* 27(2):231–254.

Licciardi, J. M. 2001. Chronology of Latest Pleistocene Lake-level Fluctuations in the Pluvial Lake Chewaucan Basin, Oregon, USA. *Journal of Quaternary Science* 16(6): 545–553.

Lourdeau, A. 2015. Lithic Technology and Prehistoric Settlement in Central and Northeast Brazil: Definition and Spatial Distribution of the Itaparica Technocomplex. *PaleoAmerica* 1(1):52–67.

Lyell, C. 1997[1830–1833]. *Principles of Geology.* Penguin, New York.

Lynch, T. F. 1990. Glacial-Age Man in South America? A Critical Review. *American Antiquity* 55(1):12–36.

Mackie, Q., L. Davis, D. Fedje, D. McLaren, and A. Gusick. 2013. Locating Pleistocene-age Sites on the Northwest Coast: Current Status of Research and Future Directions. In *Paleoamerican Odyssey*, edited by K. E. Graf, C. V. Ketron, and M. R. Waters, pp. 133–147. Center for the Study of the First Americans, Texas A&M University, College Station.

MacNeish, R. S., and J. G. Libbey (editors). 2004. *Pendejo Cave.* University of New Mexico Press, Albuquerque.

Martin, P. S. 1967. Pleistocene Overkill. *Natural History* 76(10):32–38.

Mason, R. J. 1962. The Paleo-Indian Tradition in Eastern North America. *Current Anthropology* 3(3): 227–278.

McAvoy, J. M. 1992. *Nottoway River Survey Part I: Clovis Settlement Patterns: The 30 Year Study of a Late Ice Age Hunting Culture on the Southern Interior Coastal Plain of Virginia.* Special Publication Number 28. Archeological Society of Virginia, Richmond.

McAvoy, J. M., J. C. Baker, J. K. Feathers, R. L. Hodges, L. J. McWeeney, and T. R. Whyte. 2000. *Summary of Research at the Cactus Hill Archaeological Site, 40SX202, Sussex County, Virginia.* Report to the National Geographic Society in Compliance with Stipulations of Grant #6345-98. Nottoway River Archaeological Research, Sandston, Virginia.

McAvoy, J. M, and L. D. McAvoy. 1997. *Archaeological Investigations of Site 44SX202, Cactus Hill, Sussex County, Virginia.* Department of Historic Resources, Commonwealth of Virginia, Richmond. Nottoway River Archaeological Research, Sandston, Virginia.

McDonald, J. N. 2000. An Outline of the Pre-Clovis Archaeology of SV-2, Saltville, Virginia, with Special Attention to a Bone Tool Dated 14,510 yr B.P. *Jeffersoniana* 9:1–59.

McDonald, J. N., and C. S. Bartlett, Jr. 1983. An Associated Musk Ox Skeleton from Saltville, Virginia. *Journal of Vertebrate Paleontology* 2(4):453–470.

McNett, C. W. 1985. *A Stratified Paleoindian–Archaic Site in the Upper Delaware Valley of Pennsylvania.* Academic Press, Orlando.

Mehringer, P. J., Jr., and C. V. Haynes, Jr. 1965. The Pollen Evidence for the Environment of Early Man and Extinct Mammals at the Lehner Mammoth Site, Southeastern Arizona. *American Antiquity* 31(1):17–23.

Meltzer, D. J. 1988. Late Pleistocene Adaptations in Eastern North America. *Journal of World Prehistory* 2:1–52.

———. 1997. Monte Verde and the Pleistocene Peopling of the Americas. *Science* 276:754–755.

———. 2006. *Folsom: New Archaeological Investigations of a Classic Paleoindian Bison Kill.* University of California Press, Berkeley.

———. 2009. *First Peoples in a New World: Colonizing Ice Age America.* University of California Press, Berkeley.

———. 2014. Clovis at the End of the World. *Science* 111(34):12276–12277.

Meltzer, D. J., J. M. Adovasio, and T. D. Dillehay. 1994. On a Pleistocene Human Occupation at Pedra Furada, Brazil. *Antiquity* 68:695–714.

Meltzer, D. J., D. K. Grayson, G. Ardila, A. W. Barker, D. F. Dincauze, C. Vance Haynes, F. Mena, L. Nunez, and D. J. Stanford. 1997. On the Pleistocene Antiquity of Monte Verde, Southern Chile. *American Antiquity* 62(4): 659–663.

Meltzer, D. J., L. C. Todd, and V. T. Holliday. 2002. The Folsom (Paleoindian) Type Site: Past Investigations, Current Studies. *American Antiquity* 67(1):5–36.

Mochanov, Y. A. 1977. *Drevneishie Etapy Zaseleniya Chelovekom Severo-Vostochnoi Azii*. [The Earliest Stages of Settlement by Man of Northeast Asia]. Nauka, Novosibirsk.

———. 1993. The Most Ancient Paleolithic of the Diring and the Problem of Nontropical Origin for Humanity. *Arctic Anthropology* 30:22–53.

Morlan, R. E., D. E. Nelson, T. A. Brown, J. S. Vogel, and J. R. Southon. 1990. Accelerator Mass Spectrometry Dates on Bones from Old Crow Basin, Northern Yukon Territory. *Canadian Journal of Archaeology/Journal Canadien d'Archéologie* 14:75–92.

Nelson, D. E., Richard E. Morlan, J. S. Vogel, J. R. Southon, and C. R. Harington. 1988. New Dates on Northern Yukon Artifacts: Holocene not Upper Pleistocene. *Science* 232(4751):749–751.

Neumann, G. K. 1952. Archeology and Race in the American Indian. In *Archeology of the Eastern United States*, edited by J. B. Griffin, pp. 13–34. University of Chicago Press, Chicago.

Neves, W. A., and M. Blum. 2001. "Luzia" is Not Alone: Further Evidence of a Non-mongoloid Settlement of the New World. *Current Research in the Pleistocene* 18:73–78.

Neves, W. A., J. F. Powell, A. Prous, Erik G. Ozolins, and M. Blum. 1999. Lapa Vermelha IV Hominid 1: Morphological Affinities of the Earliest Known American. *Genetics and Molecular Biology* 22(4):461–469.

Nichols, J. 1990. Linguistic Diversity and the First Settlement of the New World. *Language* 66(3):475–521.

———. 1995. Diachronically Stable Structural Features. In *Historical Linguistics 1993: Papers from the 11th International Conference on Historical Linguistics*, edited by H. Anderson, pp. 337–356. John Benjamins, Amsterdam.

Ochsenius, C., and R. Gruhn (editors). 1979. *Taima-taima: A Late Pleistocene Paleo-Indian Kill Site in Northernmost South America—Final Reports of 1976 Excavations*. South American Quaternary Documentation Program, Baden-Württemberg, Germany. Reissued by The Center for the Study of First Americans, Texas A&M University Press, College Station.

Overpeck, J., R. S. Webb, and T. Webb. 1992. Mapping Eastern North American Vegetation Change of the Past 18 ka: No-Analogs and the Future. *Geology* 20:1071–1074.

Overstreet, D. F. 1993. *Chesrow: A Paleoindian Complex in the Southern Lake Michigan Basin*. Great Lakes Archaeological Press, Milwaukee, Wisconsin.

———. 1996. Still More on Cultural Contexts of Mammoth and Mastodon in the Southwestern Lake Michigan Basin. *Current Research in the Pleistocene* 13:36–38.

———. 1998. Late Pleistocene Geochronology and the Paleoindian Penetration of the Southwestern Lake Michigan Basin. *Wisconsin Archaeologist* 79:28–52.

———. 2005. Late-Glacial Ice-Marginal Adaptation in Southeastern Wisconsin. In *Paleoamerican Origins: Beyond Clovis*, edited by R. Bonnischsen, B. T. Lepper, D. Stanford, and M. R. Waters, pp. 183–195. Center for the Study of First Americans, Department of Anthropology, Texas A&M University, College Station

Overstreet, D. F., and T. W. Stafford. 1997. Additions to a Revised Chronology for Cultural and Non-Cultural Mammoth and Mastodon Fossils in the Southwestern Lake Michigan Basin. *Current Research in the Pleistocene* 14:70–71.

Owsley, D. W., and R. L. Jantz (editors). 2014. *Kennewick Man: The Scientific Investigation of an Ancient American Skeleton*. Peopling of the Americas Publications. Texas A&M University Press, College Station.

Oxford English Dictionary Online. 2015. Anthropometry. Electronic document available at http://dictionary.oed.com.

Parenti, F. 2002. *Le Gisement Quarternaire de Pedra Furada: Stratigraphie, Chronologie, Evolution Culturelle*. Editions Recherche sur les Civilisations. Eisenbrauns, Winona Lake, Indiana.

Pearson, G. A. 1999. Early Occupations and Cultural Sequence at Moose Creek: A Late Pleistocene Site in Central Alaska. *Arctic* 52(4):332–345.

Phoca-Cosmetatou, N., and R. J. Rabett. 2014. Reflections on Pleistocene Island Occupation. *Journal of Mediterranean Archaeology* 27(2):255–278.

Pinson, A. O. 2012a. Buttermilk Creek Part I: A Pre-Clovis Occupation along the Margin of the Southern High Plain. *Mammoth Trumpet* 27(2):1–6.

———. 2012b. Buttermilk Creek. *Mammoth Trumpet* 27(3):5–11.

Pitulko, V., P. Nikolskiy, A. Basilyan, and E. Pavolova. 2013. Human Habitation in Arctic Western Beringia Prior to the LGM. In *Paleoamerican Odyssey*, edited by K. E. Graf, C. V. Ketron, and M. R. Waters, pp. 13–44. Center for the Study of the First Americans, Texas A&M University, College Station.

Potter, B. A., C. E. Holmes, and D. R. Yesner. 2013. Technology and Economy among the Earliest Prehistoric Foragers in Interior Eastern Beringia. In *Paleoamerican Odyssey*, edited by K. E. Graf, C. V. Ketron, and M. R. Waters, pp. 81–103. Center for the Study of the First Americans, Texas A&M University, College Station.

Powell, J. F., and W. A. Neves. 1999. Craniofacial Morphology of the First Americans: Pattern and Process in the Peopling of the New World. *Yearbook of Physical Anthropology* 42:153–188.

Powell, J. W. 1891. Indian Linguistic Families of America North of Mexico. In *7th Annual Report of the Bureau of [American] Ethnology for 1885-1886*, pp. 1–142. Washington, D.C.

Powers, W. R., and J. F. Hoffecker. 1989. Late Pleistocene Settlement in the Nenana Valley, Central Alaska. *American Antiquity* 54(2):263–287.

Robinson, D. G. 1997. Stratigraphic Analysis of Bonfire Shelter, Southwest Texas: Pilot Studies of Depositional Processes and Paleoclimate. *Plains Anthropologist* 42(159): 33–43.

Runnels, C. 2014. Early Palaeolithic on the Greek Islands? *Journal of Mediterranean Archaeology* 27(2):211–230.

Saban, C. V., and D. L. Jenkins. 2015. Late Pleistocene to Early Holocene Paleoenvironmental Conditions at Paisley Caves (35LK3400): Pollen Evidence from Cave 2. Electronic document, available at https://www.academia.edu/5933672/Late_Pleistocene_to_Early_Holocene_Paleoenvironmental_Conditions_at_Paisley_Caves_35LK3400_Pollen_Evidence_from_Cave_2, accessed November 2014.

Sain, D. 2015. *Pre Clovis at Topper (38AL23): Evaluating the Role of Human versus Natural Agency in the Formation of Lithic Deposits from a Pleistocene Terrace in the American Southeast*. Ph.D. dissertation, University of Tennessee, Knoxville.

Sanchez, G., and J. Carpenter. 2012. Paleoindian and Archaic Traditions in Sonora, Mexico. In *From the Pleistocene to the Holocene: Human Organization and Cultural Transformations in Prehistoric North America*, edited by C. B. Bousman and B. J. Vierra, pp. 125–147. Texas A&M University Press, College Station.

Sanchez, G., V. T. Holliday, E. P. Gaines, J. Arroyo-Cabrales, N. Martínez-Tagüeña, A. Kowler, T. Lange, G. W. L. Hodgins, S. M. Mentzer, and I. Sanchez-Morales. 2014. Human (Clovis)–Gomphothere (Cuvieronius sp.) Association ~13,390 Calibrated yBP in Sonora, Mexico. *Science* 111(30):10972–10977.

Sandweiss, D. H., H. McInnis, R. L. Burger, A. Cano, B. Ojeda, R. Paredes, M. C. Sandweiss, and M. D. Glascock. 1998. Quebrada Jaguay: Early South American Maritime Adaptations. *Science* 281:1830–1832.

Sapir, E. 1911. A Bird's-Eye View of American Languages North of Mexico. *Science* 54:408.

Schurr, T. G. 2002. A Molecular Anthropological View of the Peopling of the Americas. *Athena Review* 3(2):59–77.

———. 2004. Genetic Diversity in Siberians and Native Americans Suggests an Early Migration to the New World. In *Entering America: Northeast Asia and Beringia Before the Last Glacial Maximum*, edited by D. B. Madsen, pp. 187–238. University of Utah Press, Salt Lake City.

Shaffer, B. S. and B. W. Baker. 1997. How Many Epidermal Ridges per Linear Centimeter? Comments on Possible pre-Clovis Human Friction Skin Prints from Pendejo Cave. *American Antiquity* 62:559-560.

Shutler, R., Jr. 1965. Tule Springs Expedition. *Current Anthropology* 6(1):110–111. 186.

Simpson, R. D. 1980. *The Personal History of the Early Years of the Calico Mountains Archaeological Site*. Electronic document, accessed 2 June 2015. Electronic document available at http://calicoarchaeology.com/pdf/deesimpson.pdf.

Slobodin, S. B. 2011. Late Pleistocene and Early Holocene Cultures of Beringia: The General and the Specific. In *From the Yenisei to the Yukon, Interpreting Lithic Assemblage Variability in Late Pleistocene/Early Holocene Beringia*, edited by T. Goebel and I. Buvit, pp. 91–118. Texas A&M University Press, College Station

Smallwood, A. M. 2010. Clovis Biface Technology at the Topper Site, South Carolina: Evidence for Variation and Technological Flexibility. *Journal of Archaeological Science* 37:2413–2425.

Smallwood, A. M., D. S. Miller, and D. Sain. 2013. Topper Site, South Carolina: An Overview of the Clovis Lithic Assemblage. In *The Eastern Fluted Point Tradition*, edited by J. A. M. Gingerich, pp. 280–298. University of Utah Press, Salt Lake City.

Smith, H. L., J. T. Rasic, and T. Goebel. 2013. Biface Traditions of Northern Alaska and Their Role in the Peopling of the Americas. In *Paleoamerican Odyssey*, edited by K. E. Graf, C. V. Ketron, and M. R. Waters, pp. 105–123. Center for the Study of the First Americans, Texas A&M University, College Station.

Soffer, O. 2004. Recovering Perishable Technologies through Use Wear on Tools: Preliminary Evidence for Upper Paleolithic Weaving and Net Making. *Current Anthropology* 45(34):407–413.

Soffer, O., J. M. Adovasio, and D. C. Hyland. 2000. The "Venus" Figurines: Textiles, Basketry, Gender, and Status in the Upper Paleolithic. *Current Anthropology* 41(4):511–537.

Sondaar, P. Y., G. D. Van den Bergh, B. Mubroto, F. Aziz, J. De Vos, U. L. Batu. 1994. Middle Pleistocene Faunal Turnover and Colonization of Flores (Indonesia) by Homo-Erectus. *Comptes Rendus de l'Académie des Sciences Paris* 319:1255–1262.

Sphuler, J. N. 1979. Genetic Distances, Trees and Maps of North American Indians. In *The First Americans: Origins, Affinities, Adaptations*, edited by W. S. Laughlin and A. B. Harper, pp. 135–183. G. Fisher, New York.

Stanford, D. J., and B. A. Bradley. 2012. *Across Atlantic Ice: The Origin of America's Clovis Culture*. University of California Press, Berkeley.

Stock, C., and F. D. Bode. 1937. The Occurrence of Flints and Extinct Animals in Pluvial Deposits near Clovis, New Mexico, Part III: Geology and Vertebrate Paleontology of the Late Quaternary. *Proceedings Philadelphia Academy of Natural Sciences* 88:219-241.

Stone, E. A. 2011. *Through the Eye of the Needle: Investigations of Ethnographic, Experimental, and Archaeological Bone Tool Use Wear from Perishable Technologies*. Ph.D. dissertation, University of New Mexico, Albuquerque.

Szathmáry, E. J. E. 1979. Blood groups of Siberians, Eskimos, and Subarctic and Northwest Coast Indians: the Problem of First Origins and Genetic Relations. In *The First Americans: Origins, Affinities, and Adaptations*, edited by W. S. Laughlin and A. B. Harper, pp. 185–209. Gustav Fischer, New York.

———. 1981. Genetic Markers in Siberian and Northern North American Populations. *Yearbook of Physical Anthropology* 24: 37–73.

———. 1984. Human Biology of the Arctic. In *Arctic*, edited by D. Damas, pp. 64–71. Handbook of North American Indians, Vol. 5, William C. Sturtevant, general editor. Smithsonian Institution, Washington, D.C.

Szathmáry, E. J. E., N. S. Ossenberg, M. S. Clabeaux, D. C. Cook, M. H. Crawford, D. E. Dumond, R. L. Hall, A. B. Harper, M. G. Hurlich, P. L. Jamison, B. Jørgensen, K. A. Korey, M. Kowta, R. McGhee, C. Meiklejohn, T. A. Murad, C. B. Pereira, S. Pfeiffer, F. M. Salzano, D. S. Weaver, S. L. Zegura and M. L. Fleischman. 1978. Are the Biological Differences Between North American Indian and Eskimos Truly Profound? *Current Anthropology* 19(4):673–701.

Taylor, R. E. 2009. Six Decades of Radiocarbon Dating in New World Archaeology. *Radiocarbon* 51(1):173–212.

Torben, C. R., J. M. Erlandson, R. L. Vellanoweth, and T. J. Braje. 2005. From Pleistocene Mariners to Complex Hunter-Gatherers: The Archaeology of the California Channel Islands. *Journal of World Prehistory* 19:169–228.

Turner, C. G., II. 1985. The Dental Search for Native American Origins. In *Out of Asia: Peopling the Americas and the Pacific*, edited by R. Kirk and E. J. E. Szathmáry, pp. 31–78. Journal of Pacific History, Australian National University, Canberra.

———. 1987. Late Pleistocene and Holocene Population History of East Asia Based on Dental Variation. *American Journal of Physical Anthropology* 73(3): 5–32.

United Nations Educational, Scientific and Cultural Organization (UNESCO). 2015. Serra da Capivara National Park. Electronic document available at http://whc.unesco.org/en/list/606, last accessed August 2015.

Vajda, E. 2011. A Siberian Link to Na-Dene languages. In *The Dene-Yeniseian Connection*, edited by J. Kari and B. Potter, pp. 33–99. Anthropological Papers of the University of Alaska, Vol. 5. University of Alaska Fairbanks Department of Anthropology, Fairbanks.

Voegelin, C. F., and F. M. Voegelin. 1965. Classification of American Indian Languages. *Anthropological Linguistics* 7 (7):121–150.

Waters, M. R., S. L. Forman, T. A. Jennings, L. C. Nordt, S. G. Driese, J. M. Feinberg, J. L. Keene, J. Halligan, A. Lindquist, J. Pierson, C. T. Hallmark, M. B. Collins, and J. E. Wiederhold. 2011. The Buttermilk Creek Complex and the Origins of Clovis at the Debra L. Friedkin Site, Texas. *Science* 331:1599–1603.

Waters, M. R., S. L. Forman, J. M. Pierson. 1997. Diring Yuriakh: A Lower Paleolithic Site in Central Siberia. *Science* 275:1281–1284.

Waters, M. R., S. L. Forman, J. M. Pierson. 1999. Late Quaternary Geology and Geochronology of Diring Yuriakh, an Early Paleolithic Site in Central Siberia. *Quaternary Research* 51:195–211.

Waters, M. R., S. L. Forman, T. W. Stafford, Jr., and J. Fosse. 2009. Geoarchaeological Investigations at the Topper and Big Pine Tree Sites, Allendale County, South Carolina. *Journal of Archaeological Science* 36:1300–1311.

Waters, M. R., C. D. Pevny, and D. Carlson. 2011. *Clovis Lithic Technology: Investigation of a Stratified Workshop at the Gault Site, Texas.* Texas A&M University Press, College Station.

Waters, M. R., and T. W. Stafford, Jr. 2007. Redefining the Age of Clovis: Implications for the Peopling of the Americas. *Science* 315:1122–1126.

West, F. H. (editor). 1996. *American Beginnings: The Prehistory and Palaeoecology of Beringia.* University of Chicago Press, Chicago.

Wisener, G. 2007. The Shawnee-Minisink Site. *Mammoth Trumpet* 22(2): 4–7.

Wormington, H. M., and D. Ellis. 1967. *Pleistocene Studies in Southern Nevada.* Anthropological Papers No. 13. Nevada State Museum, Carson City.

ACKNOWLEDGEMENTS

A book like this one cannot be written or published without the assistance, co-operation, and collaboration of a large number of individuals and institutions, for whose efforts we are extremely grateful. We are also grateful for the level of scholarship and dedication evident in the body of archaeological research of our colleagues which, when viewed from the perspective of writing a book like this, has again proven itself to be truly remarkable.

Perhaps more fundamentally, this book would not have existed at all without the encouragement, interest, and perseverance of Peter Névraumont. Peter originally proposed the idea for this book to author Adovasio several years ago, with David Pedler becoming involved in the in the project more recently. We also especially appreciate Peter's dogged pursuit of the hundreds of images that appear in this book, which would have fallen short of the mark without his tireless efforts.

By any normal standards, David Pedler would be the primary author of this book. He was responsible for assembling and synthesizing the sometimes arcane information presented in Part 2, which represents the far larger portion of this book. He also managed to craft the site entries in a way that avoided "ax-grinding" in any form, instead attempting to provide objective, balanced presentations rather than sophistic screed. That David was not designated primary author reflects both his fundamental humility and the fact that he firmly believed that the current author order was somehow "more appropriate," while also slyly adding "why settle for Ringo when McCartney is still drawing breath?" Both authors, however, assume full and equal responsibility for any errors, omissions, oversights, or unintended slights that might appear in this book. While offering no apologies for our opinions, we sincerely apologize for any misrepresentations.

In any case, the authors wish to acknowledge the many scholars who provided illustrations and/or other relevant data for this book as well the reviewers who kindly provided insightful observations. We specifically wish to thank the following contributors. **Part 1**: Kathryn Barca, Pete Bostrom, Yan Axel Gómez Coutouly, Ariane de Pree-Kajfez, Michael H. Graham, Michael Kunz, Duncan McLaren, Douglas Owsley, Vladimir V. Pitulko, Pamela Ronsaville, Richard Thornton, Michael Waters, Collin Woodward. **Blackwater Draw**: George Crawford. **Bluefish Caves**: Lauriana Bourgeon, Jacques Cinq-Mars, Ruth Gotthardt, C.R. Harington, Chelsea Jeffery, Vincent Lafond. **Bonfire Shelter**: Steve Black, Susan Dial, David J. Meltzer, Elton Prewitt, Jack Skiles. **Cactus Hill**: Michael Johnson, Larry Kimball, Carol Nash. **Calico Early Man**: Sandy Karhu, Robin Laska, Adella Schroth, James Shearer, Claude Short. **Central Alaska**: Yan Axel Gómez Coutouly, James Dixon, Ted Goebel, Kelly Graf, Liz Haberkorn, Charles Holmes, Robert King, Kathryn Krasinski, Mike Kunz, KJ Muschovic, Kristen Quarles, Jeff Rasic, Joshua Reuther, Dennis Stanford, Richard VandeHoek. **El Fin del Mundo**: Vance T. Holliday, Guadalupe Sánchez, Kate Sarther, Henry Wallace. **Folsom:** Kimberley Evans, Michele Koons, Barbara Mathé, David J. Meltzer, René O'Connell, René Payne, Gregory August Raml. **Debra L. Friedkin**: Joshua Keene, Michael Waters. **Gault**: Michael Collins, Clark Wernecke. **Kimmswick**: Jessica Boldt, Jaime Bourassa, James Chandler, Ryan Ellsworth, Jonathan Haas, Lauren Hancock, Marvin Kay, Brooke Mahar. **Lehner, Murray Springs, and Naco**: Julia Balestracci, Bob Bryson,

C. Vance Haynes, Bruce Huckell, Chris Kaplan, Jonathan Mabry, Nora Evans-Reitz, Jannelle Weakly. **Meadowcroft Rockshelter**: Don Hitchcock. **Monte Verde**: Kristin Benson, Tom D. Dillehay. **Old Crow**: Dominique Dufour, C. R. Harington, Vincent Lafond, Karen McCullough, Sarah Prower, Alberto Reyes, Natalia Rybczynski, Robert Thorson, Grant Zazula. **Paisley Five Mile Point Caves**: Dennis Jenkins. **Pedra Furada**: Niède Guidon, Fabio Parenti, Rosa Trakalo. **Pendejo Cave**: Barry Rolett, Bonnie K. Sousa, Ryan Wheeler. **Saltville**: Jerry McDonald. **Schaefer and Hebior Mammoth**: Claudia Jacobson, Dan Joyce, Sara Podejko. **Shawnee-Minisink**: Joseph Gingerich, Keith Heinrich. **Shoop**: Kurt Carr, Elizabeth Wagner. **Taima-Taima and Tibitó**: Humberto Arias, Richard Cooke, Tom D. Dillehay, Ruth Gruhn, José Oliver. **Topper**: Albert Goodyear, Shane Miller, Sean Taylor. **Tule Springs**: Eugene Hattori, Rachel Kaleilehua Malloy, Jay Matternes, Rebecca Palmer, Liza Posas, Sali Underwood, Marilyn Van Winkle. **Coda**: Heather Smith, NASA.

We also thank Lionel Koffler, Michael Woek, Jacqueline Raynor, Diane Vanderkooy and Parisa Michailidis at Firefly Books for their patient and ongoing publishing support. We are grateful to Cathleen Elliott for her gorgeous book design.

We very much appreciate David Madsen for his careful review of *Strangers in a New Land*.

Finally the authors wish to acknowledge the Senator John Heinz History Center with which both authors have been informally (and now formally) connected for a long time.

INDEX